Generation IV Nuclear Reactors

Design, operation and prospects for future energy production

Online at: https://doi.org/10.1088/978-0-7503-6069-2

Generation IV Nuclear Reactors

Design, operation and prospects for future energy production

Richard A Dunlap
Department of Physics and Atmospheric Science, Dalhousie University, Halifax, Nova Scotia, Canada

IOP Publishing, Bristol, UK

ISBN 978-0-7503-6069-2 (ebook)
ISBN 978-0-7503-6067-8 (print)
ISBN 978-0-7503-6070-8 (myPrint)
ISBN 978-0-7503-6068-5 (mobi)

DOI 10.1088/978-0-7503-6069-2

Version: 20241001

IOP ebooks

British Library Cataloguing-in-Publication Data: A catalogue record for this book is available from the British Library.

Published by IOP Publishing, wholly owned by The Institute of Physics, London

IOP Publishing, No.2 The Distillery, Glassfields, Avon Street, Bristol, BS2 0GR, UK

US Office: IOP Publishing, Inc., 190 North Independence Mall West, Suite 601, Philadelphia, PA 19106, USA

Contents

Preface

The use of nuclear fission as a source of power began in the 1950s with the construction of the first commercial nuclear power reactors. At that time, it was imagined that nuclear energy could provide an inexpensive and virtually limitless source of electrical power. The realization in the early-1970s that fossil fuel resources would not last indefinitely motivated a significant increase in nuclear power plant construction. However, major nuclear accidents beginning with Three Mile Island in 1979 and, more significantly, with Chernobyl in 1986 raised concerns for nuclear safety. Additionally, the problem of nuclear waste disposal and the question of nuclear security had adverse effects on the further development of commercial nuclear power. Since the early-1990s, growth of the nuclear generating capacity has been slow and at times new capacity has barely compensated for the decommissioning of aging reactors constructed in the 1970s.

In recent years, the idea that nuclear fission power could provide a source of low-carbon electricity and contribute to decreasing greenhouse gas emissions has gained momentum. However, reactor safety, nuclear waste disposal, nuclear proliferation, and high reactor construction cost, as well as the longevity of uranium resources were still concerns that needed to be addressed. Proposed advanced reactors, designated Generation IV reactors, were designed to deal these concerns.

International collaboration on the future of nuclear power began in 2000 when the U.S. Department of Energy convened a meeting of representatives from nine countries with active nuclear power research programs to establish the Generation IV International Forum (GIF). At present, the Forum has 14 member states,

- Argentina
- Australia
- Brazil
- Canada
- China
- Euratom (European Atomic Energy Community)
- France
- Japan
- Republic of Korea
- Russia
- South Africa
- Switzerland
- United Kingdom
- United States

The goals of the GIF fall into four broad areas, as follows,

Sustainability: To provide sustainable energy generation that makes effective use of fuel resources and to provide effective management of nuclear waste.

Economics: To provide energy at an advantageous life-cycle-cost and to minimize financial risk.

Safety and reliability: To operate in a manner that excels in safety and reliability with a very low likelihood of core damage and to eliminate the need for off-site emergency response.

Proliferation resistance and physical protection: To minimize the risk of theft of weapons-grade radioactive material and provide physical protection against acts of terrorism.

The GIF has identified six basic fission reactor designs that can, in principle, satisfy the goals as stated above (Generation IV International Forum 2002 *A Technology Roadmap for Generation IV Nuclear Energy Systems* GIF-002-00). These are summarized below and each reactor design is designated with its official GIF three or four letter acronym. The design that can be realized as a thermal neutron reactor is,

- Very high-temperature reactor (VHTR), sometimes called high-temperature gas-cooled reactor (HTGR) or just high-temperature reactor (HTR).

The designs that can be realized as either thermal neutron reactors or fast neutron reactors are,

- Molten salt reactors (MSRs), sometimes called molten salt fast reactors (MSFRs) when fast neutrons are utilized.
- Supercritical water-cooled reactors (SCWRs).

The designs that can be implemented as fast neutron reactors are,

- Gas-cooled fast reactor (GFR).
- Sodium-cooled fast reactor (SFR).
- Lead-cooled fast reactor (LFR).

The development of new reactor technologies that meet the criteria specified by the GIF is a significant challenge. The philosophy of the Forum is that this challenge can only be effective met by extensive international collaboration. It is by defining six promising reactor technologies that can best meet the Generation IV goals that efforts can be focused and the possibility of success can be optimized. The Forum has established international steering committees for each of the six reactor technologies and each committee has prepared a roadmap that provides guidance to the approach that is appropriate for the development of reactors in each area.

This book begins with an assessment of the future energy options that are available and which can help to mitigate climate change that results from greenhouse gas emissions. It also provides arguments for the importance of including nuclear fission energy as a component of our future energy mix. Chapters 2 through 6 provide an overview of nuclear power. This includes chapters that review basic nuclear physics, cover the history of nuclear fission reactors, discuss the details of the design and operation of nuclear reactors currently in use, and describe the risks associated with nuclear fission energy. Chapters 6 through 12 describe the six GIF reactor designs and chapter 13 provides a brief overview of how these reactors could be integrated into our future low-carbon energy mix.

Author biography

Richard A Dunlap

Richard A Dunlap received a BS in Physics from Worcester Polytechnic Institute in 1974, an AM in Physics from Dartmouth College in 1976, and a PhD in Physics from Clark University in 1981. Since receiving his PhD, he has been on the Faculty at Dalhousie University. He was appointed Faculty of Science Killam Research Professor in Physics from 2001 to 2006, and served as Director of the Dalhousie University Institute for Research in Materials from 2009 to 2015. He currently holds an appointment as Research Professor in the Department of Physics and Atmospheric Science. Prof. Dunlap has published more than 300 refereed research papers and his research interests have included magnetic materials, amorphous alloys, critical phenomena, hydrogen storage, quasicrystals, superconductivity, and materials for advanced batteries. Prof. Dunlap's previous books include *Experimental Physics: Modern Methods* (Oxford 1988), *The Golden Ratio and Fibonacci Numbers* (World Scientific 1997), *Sustainable Energy* (Cengage, 1st edn 2015, 2nd edn 2019), *Novel Microstructures for Solids* (IOP/Morgan & Claypool 2018), *Particle Physics* (IOP/Morgan & Claypool 2018), *The Electrons in Solids: Contemporary Topics* (IOP/Morgan & Claypool 2019), *Energy from Nuclear Fusion* (IOP Publishing 2021), *Transportation Technologies for a Sustainable Future* (IOP Publishing 2023), *Lasers and their Application in the Cooling and Trapping of Atoms (Second Edition)* (IOP Publishing 2023), *An Introduction to the Physics of Nuclei and Particles (Second Edition)* (IOP Publishing 2023), and *The Mössbauer Effect (Second Edition)* (IOP Publishing 2023).

IOP Publishing

Generation IV Nuclear Reactors
Design, operation and prospects for future energy production
Richard A Dunlap

Chapter 1

Our energy needs and the case for nuclear fission

Chapter 1 begins with an overview of our past and present energy use and provides a breakdown of our primary energy use by source. The trends in energy use and energy use per capita in different countries is considered, along with a prediction for future energy use trends. The environmental consequences of our energy use, in particular global climate change resulting from greenhouse gas emissions, are considered. The results of greenhouse gas emissions include, global warming, loss of Arctic and Antarctic ice coverage, increasing ocean heat content, raising sea level, and ocean deoxygenation and acidification. This chapter reviews possible low-carbon energy options for the future and provides the basis for the use of nuclear energy to provide a major component of our base load electricity requirements.

1.1 Introduction

The development and use of nuclear energy during the latter part of the twentieth century has provided a means of reducing our dependence on fossil fuels. In order to understand the potential for the utilization of nuclear energy, it is important to consider how it fits into our current energy needs and how those needs are likely to change in the future. This chapter provides an overview of our energy use, with particular emphasis on the use of nuclear energy.

1.2 Our past and present energy use

Since our ancestors first made use of fire as a source of heat, humans have been dependent on the harvesting of energy from the environment. Over the years, our energy needs have increased and the sources of energy that we utilize have changed. Our first source of energy, the combustion of wood for heat and cooking, remained our principal source of energy up until the beginning of the twentieth century, as illustrated in figure 1.1, when it was replaced by coal. The higher energy density of coal, compared with wood, and its greater convenience in utilization, made it preferable for heating and industrial applications and for the increasing need for

doi:10.1088/978-0-7503-6069-2ch1

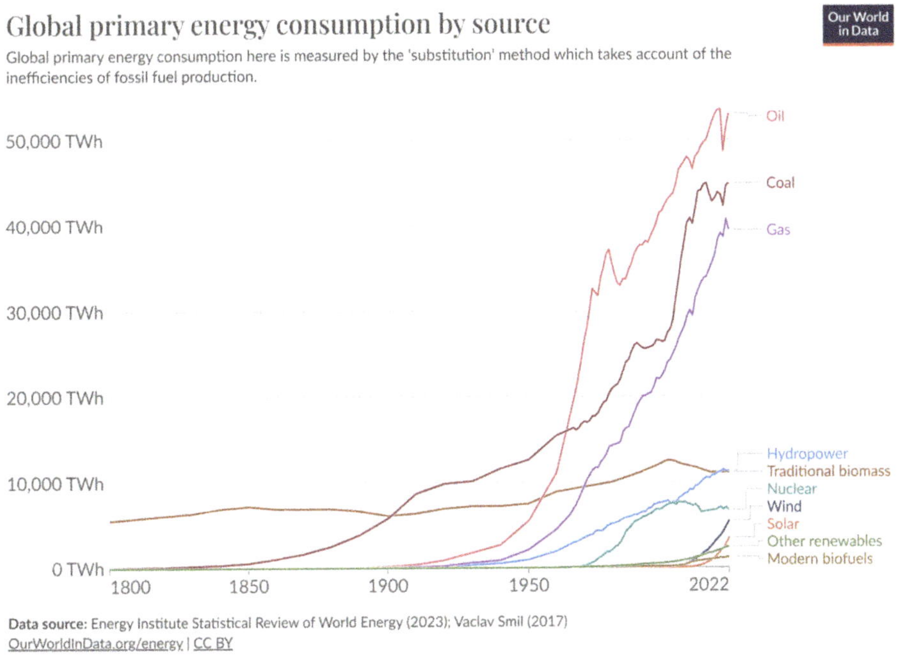

Figure 1.1. Global annual primary energy sources from 1800 to the present. 'Traditional biomass' consists primarily of wood. Data are from Energy Institute (2023) and Smil (2017). Reproduced from Our World in Data (2023). CC BY 4.0.

electricity generation. By around the middle of the past century, petroleum products, specifically oil, became our dominant source of energy because of its convenience for heating and transportation. As figure 1.1 shows, oil has remained the most important source of energy to the present day, followed by coal and natural gas. Overall, these fossil fuels account for more than 80% of the current annual global primary energy supply of about 168 000 TWh.

It is interesting to note that the energy that has been plotted in figure 1.1 is referred to as primary energy. This energy is the energy that is extracted from nature, such as the chemical energy available from coal or oil. It is distinguished from end user (or final) energy because of the inefficiency of the energy conversions involved in changing one type of energy to another. For example, if the primary chemical energy content of coal is used to generate electricity that is used in our homes, then the end user energy is only about 40% of the primary energy. It is essential to consider this relationship between primary energy and end user energy in any analysis of our energy use. Figure 1.2 shows the various stages of energy conversions. It is typically the final energy that consumers associate with their energy use because it is energy in this form that the user purchases.

In recent years, our total global primary energy use has increased fairly consistently, as shown in figure 1.3. There are three factors that affect changes in the total primary energy consumption of society. These are

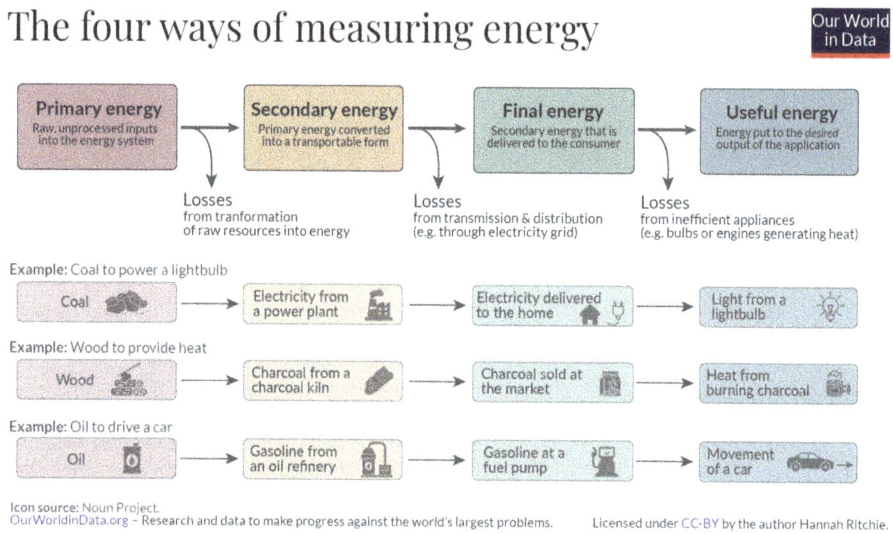

Figure 1.2. Four ways of measuring energy, i.e., primary, secondary, final, and useful. Reproduced by Ritchie (2022). CC BY 4.0.

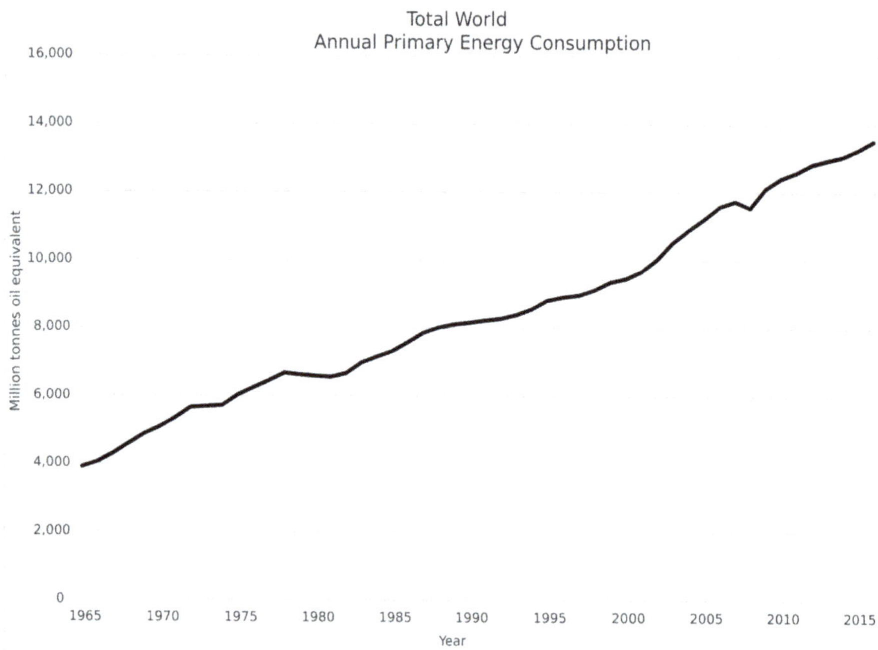

Figure 1.3. Total annual global primary energy consumption from 1965 to 2015. Note the vertical scale is in million tonnes oil equivalent (Mtoe) where 1 Mtoe = 11.63 TWh. This Total World Annual Primary Energy Consumption graph drawn from the data in the 2011 report image has been obtained by the author from the Wikimedia website where it was made available by Lbeaumont (2011) under a CC BY-SA 3.0 licence. It is included within this chapter on that basis. It is attributed to Lbeaumont.

- The average per capita energy consumption
- Population
- The relative proportion of energy from different sources and the distribution of end uses

The effects of the first two factors are fairly obvious. The total energy use will be the product of the average per capita energy use and the population. The average per capita energy use includes not only personal energy use that can be accounted for on an individual basis, such as home electricity use and gasoline for a family vehicle, but also the per capita contribution to societal energy use, such as energy used by industry and businesses.

The third point itemized above is related to total overall primary energy use in a less direct way. Different sources of primary energy, such as wind energy compared with fossil fuel energy, have different efficiencies for conversion into the form of energy required by the end user. In addition, different approaches to using a specific form of energy have different efficiencies. For example, if oil (in the form of fuel oil) is used for space heating, the efficiency of converting the chemical energy stored in the oil into heat is fairly high (perhaps 85%). On the other hand, if oil (in the form of diesel fuel) is used to propel a vehicle using an internal combustion engine, then the efficiency of converting the chemical energy of the fuel into mechanical energy to drive the wheels is fairly low (perhaps 20%).

Combining the primary energy provided by the various sources shown in figure 1.1, provides the total global primary energy use as shown in figure 1.3. While the total energy shows a roughly linear increase in time over the past half a century, this linear increase is the result of several competing factors, as outlined above. We can see one aspect of this in figure 1.4, where the per capita energy use in the United States is shown from 1650 to present. There was a substantial increase in per capita energy use during some periods of time shown in the graph, particularly during the early part of the twentieth century, where there was an increase in industrialization, as well as an increase in personal energy use for home utilization and transportation. However, for most of the period shown in figure 1.3, there is no clear trend in the per capita energy use. In fact, it has remained relatively constant from around 1970 to the present. This is, in part, due to a trade-off between increased personal needs related to a more modern lifestyle and improved efficiency of energy utilizing devices, such as home electronics, heating, and vehicles, as well as active energy conservation efforts.

Overall, the effects discussed above with regard to energy use in the United States, combined with a relatively slow population growth, has resulted in a relatively small increase in energy use over the past 25 or 30 years. The same features apply to energy use in Europe.

The linear energy growth shown in figure 1.3 can be understood by looking at the energy use in developing countries. Figure 1.5 shows a breakdown of energy use in different parts of the world. It is clear from this graph that the net increase in global energy is largely dominated by a rapid increase in energy use in China. This is the result of the combined increase in population and an increase in per capita energy

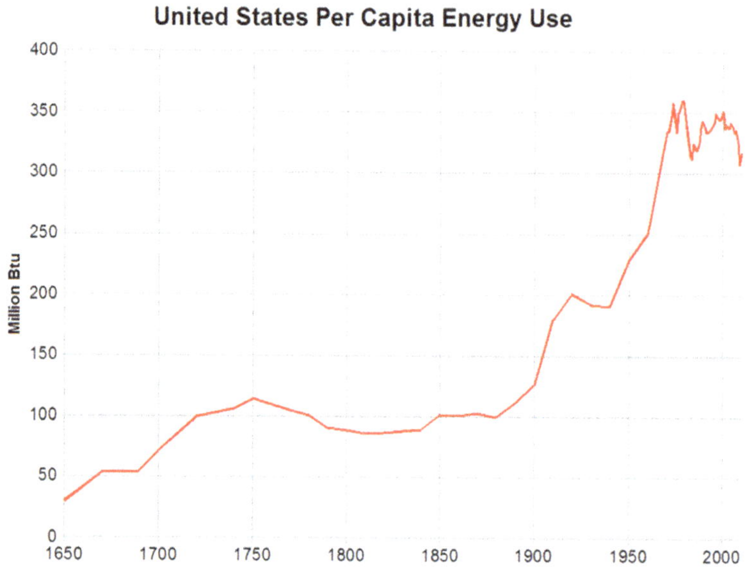

Figure 1.4. Per capita annual energy consumption in the United States from 1650 to the present. Note the vertical scale is in million Btu (British thermal units) where 1 MBtu = 293.1 kWh. Reproduced from (Delphi234 2012). CC0.

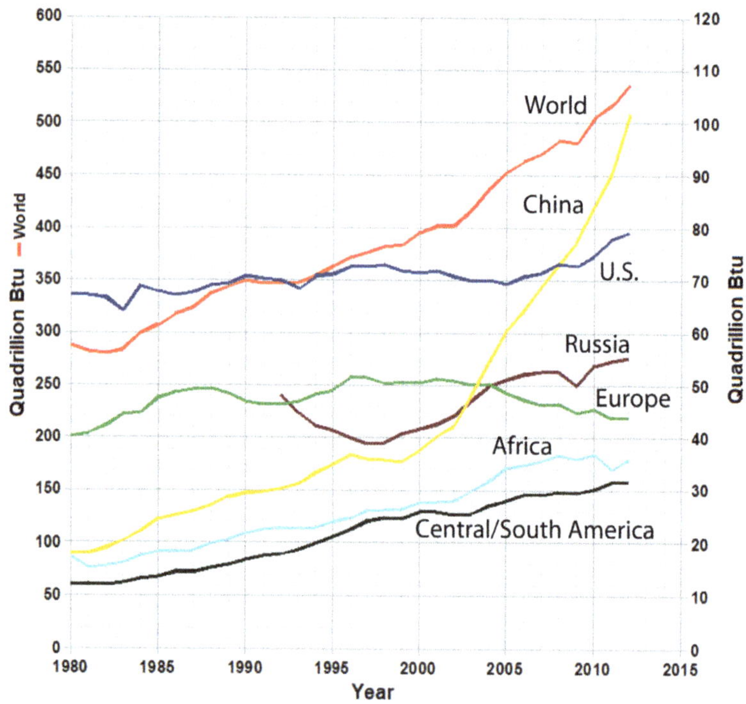

Figure 1.5. Breakdown of global annual primary energy use by region. 1 QBtu = 2.93×10^{11} kWh. Reproduced from (Delphi234 2014). Image stated to be in the public domain.

use as a result of increased industrialization, analogous to the situation in (for example) the United States during the early and middle parts of the twentieth century.

Figure 1.1 shows that there have been increases in low-carbon energy sources in recent years. For example, hydroelectric energy has shown a consistent increase since the mid-twentieth century. The use of other forms of renewable energy, such as wind and solar, has increased since around 2000. However, it is quite obvious that the increase in energy use shown in figure 1.1 has been dominated by the increase in energy derived from fossil fuel sources. The single largest portion of this fossil fuel component has been from coal use for electricity generation in China.

Of particular note in figure 1.1 is the utilization of nuclear energy. Since its emergence around 1960, the use of nuclear energy increased consistently until the mid-1980s. Since that time, it has been a relatively constant contribution of about 4% of our primary energy use. The details of the development of nuclear power and its use are presented in chapter 5.

1.3 Environmental consequences of our energy use

Since the 1950s, it has become increasingly obvious that fossil fuel resources are not unlimited. The longevity of fossil fuel resources has been the subject of considerable work dating back more than half a century. The first detailed description of resource utilization was proposed by M King Hubbert (1903–89; Hubbert 1956). For an overview of the Hubbert model, see Dunlap (2020a). This model is based on the assumption that resource use is limited by resource availability. As illustrated in figure 1.6, the model shows a resource utilization rate (i.e., annual production) that follows a symmetric curve which peaks when half of the resource has been utilized. This approach, as shown in the figure, has been successful in modeling the production of specific resources, in this case Norwegian oil.

More recently, it has been suggested that resource utilization may be limited more by demand than availability (Sorrell *et al* 2009). Figure 1.7 shows an example of future oil utilization in terms of demand. Future use based on resource availability shows a curve characteristic of the predictions of the Hubbert model. However, if demand decreases for reasons that are not directly related to availability, then utilization curves such as 3 and 4 in figure 1.7 can occur. These curves correspond to reduced demand for oil as a result of the development of low-carbon energy sources to replace oil and other fossil fuels. The motivation for this approach and the meaning of the curves in figure 1.7 that represent demand driven utilization is discussed in the remainder of the present section.

The use of fossil fuels has a variety of adverse environmental effects. Pollution from the burning of fossil fuels is a major factor for illness and death associated with our use of energy, as discussed in section 6.9. However, the most serious consequence of the use of fossil fuels is global climate change that results from greenhouse gas emissions. As figure 1.8 shows, fossil fuel use in the energy sector is responsible for about two-thirds of greenhouse gas emissions.

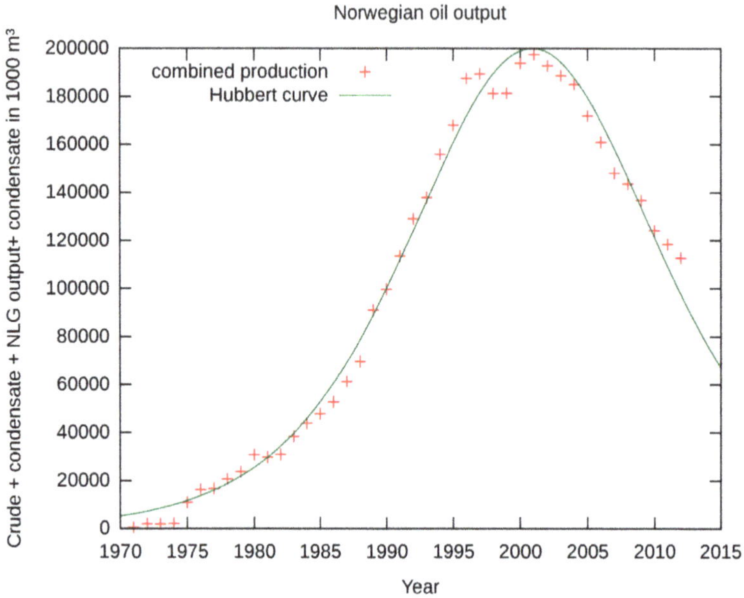

Figure 1.6. Annual Norwegian oil production and fitted Hubbert model curve. This Norwegian oil output and a fitted bell curve image has been obtained by the author from the Wikimedia website where it was made available by Kockmeyer (2013) under a CC BY 3.0 licence. It is included within this chapter on that basis. It is attributed to Kockmeyer.

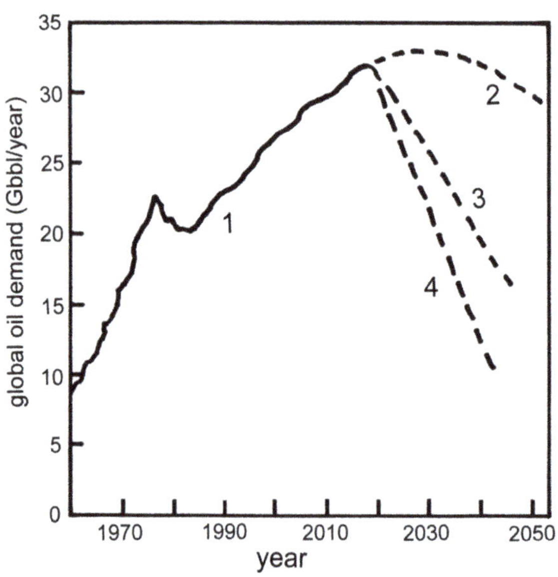

Figure 1.7. Global oil demand as a function of time: (1) historical production driven by demand, (2) future demand limited by resources, (3) sustainable future demand, and (4) net zero future demand. Reproduced from Dunlap (2023).

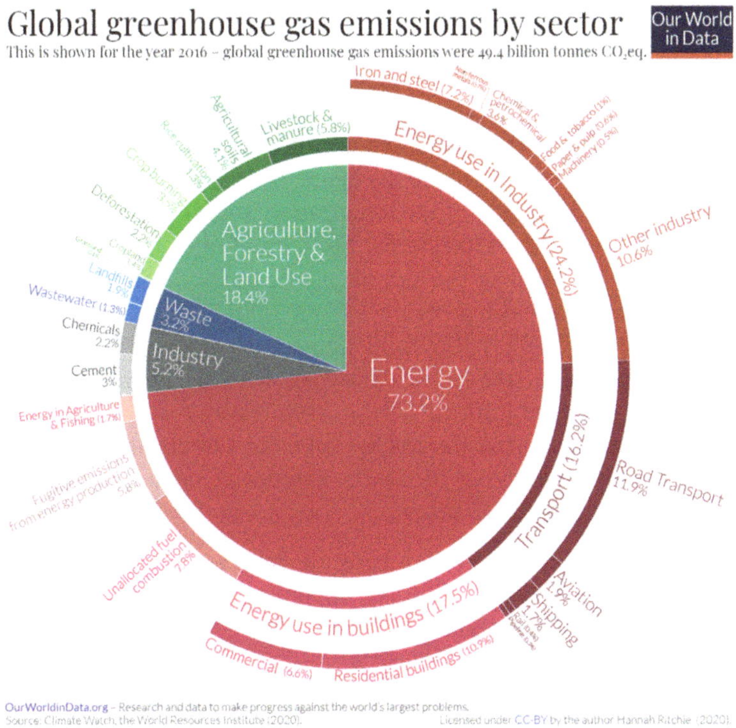

Figure 1.8. Breakdown of world greenhouse gas emissions by sector for 2016. Reproduced from Ritchie (2020a). CC BY 4.0.

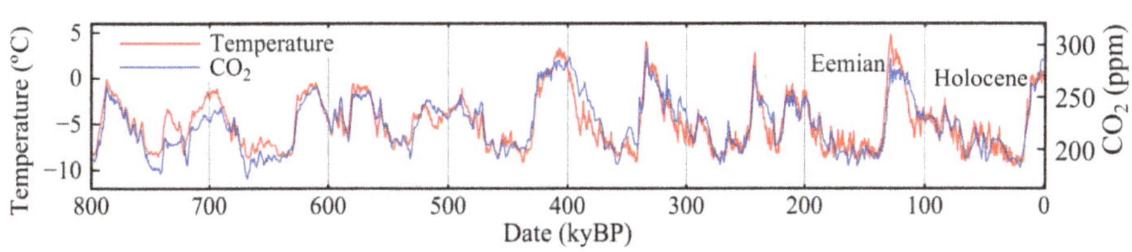

Figure 1.9. Antarctic Dome C temperature for past 800 ky from Jouzel *et al* (2007) relative to the mean of the last 10 ky and Dome C CO_2 amount from Luthi *et al* (2008) (kyBP = kiloyears before present). Reproduced from Hansen *et al* (2023). CC BY 4.0.

The relationship between atmospheric CO_2 concentration and average global temperature has been well established and is illustrated in figure 1.9. This graph shows the relationship between CO_2 and global temperature over the past 800 000 years. While the correlation between these two quantities is quite obvious, the reasons for the fluctuations that occur on a time scale of tens of thousands of years is not well known, although these variations are clearly not anthropogenic in nature. Variations in the cosmic ray flux that results from galactic events has been considered as one possibility (see Ney 1959). This figure shows that over the past

800 000 years the atmospheric CO_2 content has ranged from a low of about 170 ppm to a high of about 300 ppm.

Figure 1.10 shows direct measurements of atmospheric CO_2 at Mauna Loa, Hawaii from 1958 to the present. The atmospheric CO_2 concentration shows a consistent increase from around 315 ppm to the present (2023) level of 420 ppm. The annual variations, as illustrated by the insert in the figure, are the result of seasonal local plant growth. The current atmospheric CO_2 concentration is significantly higher than at any previous time during the past 800 000 years. The current rate of increase is about 2 ppm per year, and this is greater than measurable increases shown in figure 1.9.

The annual average global temperature from 1880 to 2022, as referenced to the average from 1900 to 2000, is shown in figure 1.11. From the late-1950s to the present, these temperature trends are consistent with the increase in atmospheric CO_2 that is shown in figure 1.10.

The recent increase in atmospheric CO_2 and the associated increase in average global temperature has a number of clearly observable consequences. Some of the most important are summarized below.

Ice sheet loss: Increasing global temperature has caused the loss of ice mass in both the Arctic and the Antarctic. This is most noticeable in the Greenland ice sheet, where the current loss rate is about 286 Gt y^{-1}.

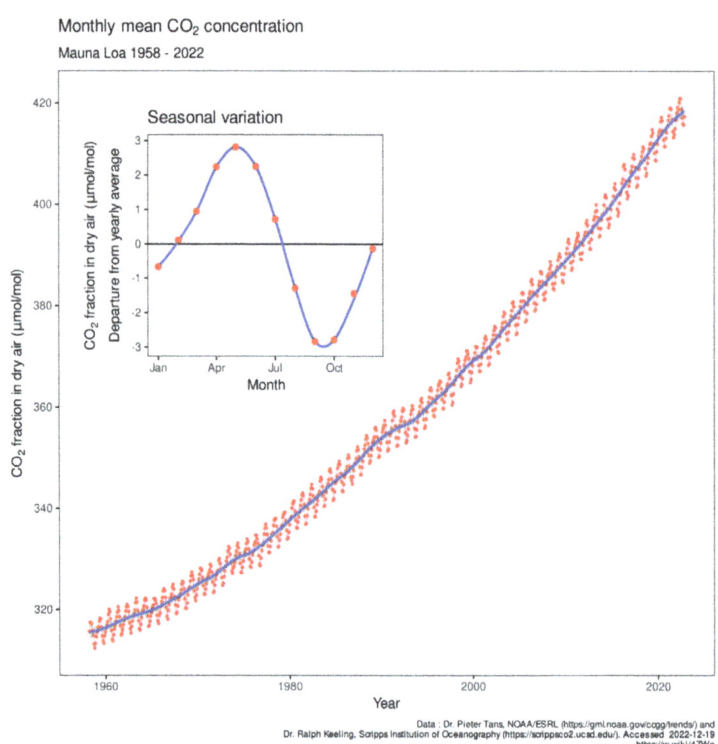

Figure 1.10. Atmospheric CO_2 from 1958 to the present as measured at Mauna Loa, Hawaii. The seasonal variations result from seasonal changes in local vegetation. Reproduced from Delorme (2019). CC BY-SA 4.0.

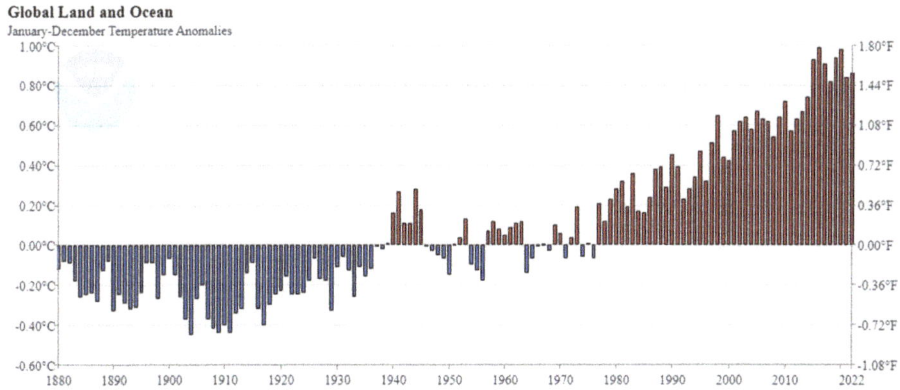

Figure 1.11. Global average temperature anomaly since 1880. Data are referenced to the average for 1900–2000. Reproduced from NOAA National Centers for Environmental Information (2023). Image stated to be in the public domain.

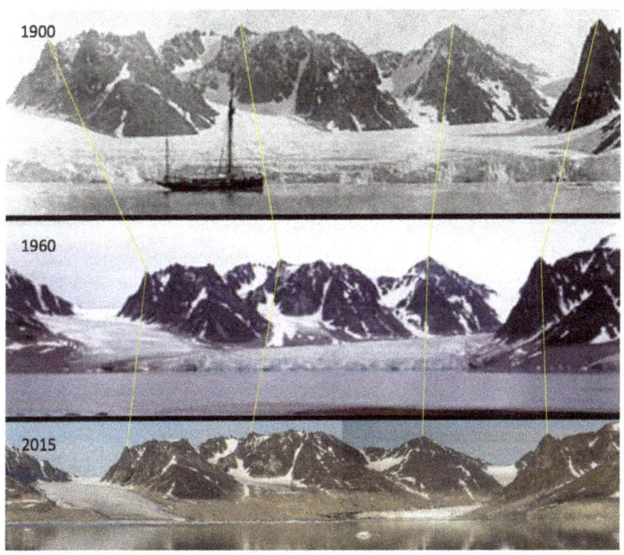

Figure 1.12. The glacier Waggonwaybreen on Magdalenenfjord in Spitzbergen, Svalbard over the years from 1900 to 2015. Reproduced from Andreas Weith (2016). CC BY-SA 4.0.

Increased ocean heat content: Atmospheric temperature increases will result in increased heat absorption by the upper layers of the ocean, and this will be reflected in the total heat content of the oceans.

Arctic sea ice decline: Melting of sea ice, particularly in the Arctic, has resulted from increasing ocean temperatures.

Decreased number and extent of glaciers: Increasing global temperatures have caused the melting of glaciers. The decrease in glacial extent is illustrated by the glacier Waggonwaybreen as observed from 1900 to 2015 (see figure 1.12).

Thawing of permafrost: Permanently frozen ground occurs in the Arctic regions of Alaska, Canada, Greenland, and Siberia. Greenhouse gases such as carbon dioxide and methane that are trapped in the gas are released when permafrost thaws. This leads to further global warming.

Increased sea level: Increased sea level results from the melting of sea ice and ice sheets, and also from increased water volume due to thermal expansion.

Ocean deoxygenation: Ocean stratification results from increased surface temperatures, as noted above, and decreased surface salinity from melting ice sheets. This stratification reduces ocean mixing and reduces oxygen concentration in lower layers of the ocean. This is detrimental to the survival of deep-water marine organisms.

Ocean acidification: Increasing atmospheric carbon dioxide will lead to increased carbon dioxide dissolved in the oceans. Some dissolved carbon dioxide will produce carbonic acid by the process

$$CO_2 + H_2O \rightarrow H_2CO_3.$$

This can be followed by dissociation of the carbonic acid,

$$H_2CO_3 \rightarrow 2H^+ + CO_3^{2-}.$$

The H^+ ions in the ocean will lead to increased acidity and can have adverse effects on marine organisms.

Altered ocean circulation patterns: As noted above, salinity gradients can lead to reduced ocean mixing. It can also lead to altered circulation patterns such as a weakening of the Gulf Steam.

1.4 Approaches to climate change mitigation

There are several approaches to mitigating climate change. These have recently been reviewed by Fawzy *et al* (2020) and the Royal Society (2018). These approaches fall into three categories:

- Reduction of greenhouse gas emissions
- Reduction of atmospheric greenhouse gases
- Radiative forcing geoengineering

Reduction of greenhouse gas emissions includes transitioning from fossil fuel use to the use of low-carbon energy sources. These sources include renewable energy such as wind and solar, as well as nuclear. Reduction of greenhouse gas emissions also includes carbon capture from sources that emit greenhouse gases, such as fossil fuel burning generating stations. Finally, this approach can include less severe measures such as switching from high-carbon fossil fuels (i.e., coal) to lower-carbon fossil fuels (i.e., natural gas) and improving the efficiency of fossil fuel combustion. The latter would include not only the improvement in the efficiency of the combustion process and the conversion of heat to electrical energy but also measures such as combined cycle generation to make effective use of waste heat.

Reduction of atmospheric greenhouse gases involves the removal of existing greenhouse gases, primarily carbon dioxide, from the atmosphere. This can be accomplished by forestation, i.e., either afforestation (establishing new forests) or reforestation (restoring deforested areas). Other methods of removing carbon dioxide from the atmosphere include ocean fertilization and ocean alkalinity enhancement. The former approach involves the spreading of fertilizers in the ocean surface layer to increase the growth of marine microorganisms (i.e., phytoplankton) that utilize carbon dioxide. The latter approach involves spreading of alkalines in the ocean surface layer to counteract the effects of ocean acidification, as noted above.

Radiative forcing geoengineering involves techniques such as stratospheric aerosol injection, marine sky brightening, and space-based mirrors to increase the Earth's albedo, and thereby decrease the absorption of solar radiation. Stratospheric aerosol injection utilizes aircraft to disperse reflective aerosol particles in the stratosphere to reduce the amount of sunlight reaching the Earth's surface. This method mimics the natural effects of particles that are distributed in the atmosphere after a volcanic eruption. The most suitable material for dispersion is still under investigation. Marine sky brightening uses a similar philosophy. Seawater is sprayed into the atmosphere. The water evaporates leaving very fine salt particles suspended in the air and these particles increase the reflectivity of the clouds over the ocean. Space-based mirrors are, perhaps, the most extreme approach. Mirrors deployed in space by satellites reflect sunlight before it enters the Earth's atmosphere. All of these methods are in the very early stages of development, and their economic and technical viability are questionable at present.

Various government policies aimed at mitigating climate change exist in different countries. The international treaty known as the Paris Agreement is the most extensive approach to reducing atmospheric greenhouse gases and is aimed at minimizing the overall increase in global temperature above preindustrial levels. Principal among the proposed policies to reduce atmospheric greenhouse gases is the reduction of fossil fuel use by the implementation of low-carbon energy sources such as renewables and nuclear.

Figure 1.13 shows the total estimated additional CO_2 that can be emitted and still limit global warming to 1.5 °C or 2.0 °C. For comparison, the 2022 CO_2 emission, 4.1×10^{10} t, is shown in the figure. In order to have a 2/3 chance of maintaining temperatures to within 1.5 °C of preindustrial levels, the estimate in figure 1.13 indicates that we can continue emitting CO_2 at the present rate for a maximum of another three years or so (from 2022). However, it has recently been reported by the EU's Copernicus Climate Change Service (Poynting 2024) that the average global temperature for the 12-month period from February 2023 to January 2024 was 1.52 °C above preindustrial levels. Thus, we are already at this level of global warming. The figure also shows that to maintain the global temperature increase below 2.0 °C, we can continue emitting CO_2 at the present level for another 20–25 years (with 67% probability). It is, therefore, essential that on this time scale that we reduce annual CO_2 emissions to limit total additional emissions. This is consistent with the models presented in a recent report of the Intergovernmental Panel on Climate Change (2022). The continuation of current energy policies would result in a global temperature increase by 2100 of 2.7 °C to 3.1 °C.

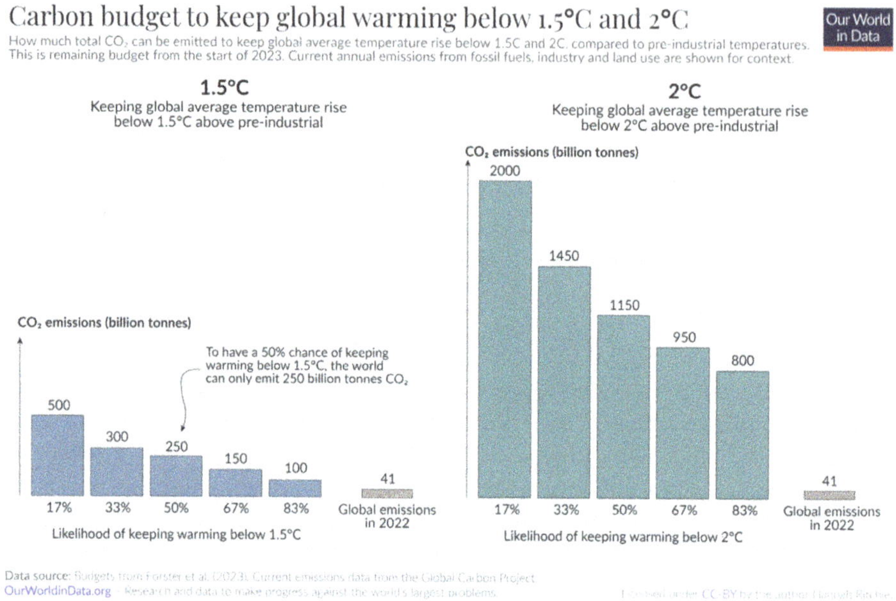

Figure 1.13. Total amount of CO_2 that can be emitted and keep global temperature increase below 1.5 °C and 2 °C compared top preindustrial levels. The annual CO_2 emissions (in 2022) are shown for comparison. Reproduced from Ritchie (2023). CC BY 4.0.

Pledges under agreements such as the Paris Agreement would limit the temperature increase to about 2.4 °C (see Dunlap 2023 and references therein). This falls short of the action needed to avoid serious environmental consequences. It is clear, therefore, that the accelerated development of low-carbon and carbon-free sources of energy is essential for the future sustainability of our environment. The next section reviews the major options for low-carbon energy.

1.5 Energy options for the future

A breakdown of world energy use in 2020 is shown in figure 1.14. This figure shows total energy use and energy used for electricity production. A comparison of these two graphs shows that low-carbon sources account for more than a third of electricity produced but only about 15% of total energy. The graphs also show that oil is a much greater portion of total energy than of electricity production. The major factor responsible for these differences is energy that is used for transportation. Oil (in the form of gasoline or diesel) accounts for the majority of transportation energy, while low-carbon sources contribute relatively little to transportation. The exceptions are small contributions from biofuels and electricity sources used for battery electric vehicles and plug-in hybrids. A breakdown of total primary energy use in the United States is illustrated in figure 1.15, which shows similar trends to overall world use.

Overall, the major contributions to low-carbon energy are hydroelectric, wind, solar, and nuclear. All of these are used more-or-less exclusively for electricity production. As we transition away from fossil fuels, many energy applications will

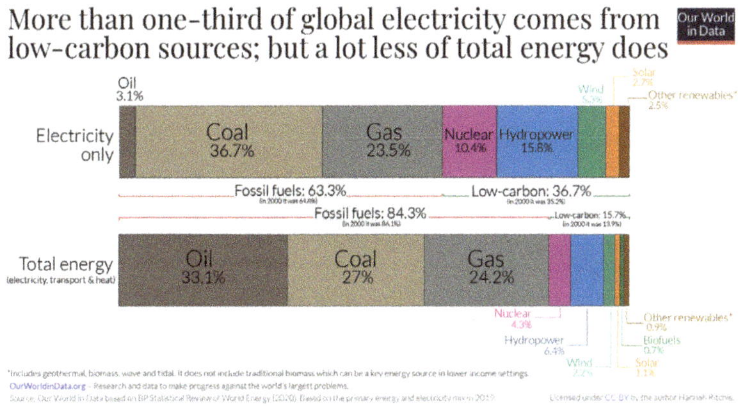

Figure 1.14. Breakdown of world primary energy used for electricity production (top) and all world primary energy (bottom) for 2020. Traditional biomass (e.g., wood), which is used primarily in developing countries as a source of heat, is not included in this analysis. Reproduced from Ritchie (2021). CC BY 4.0.

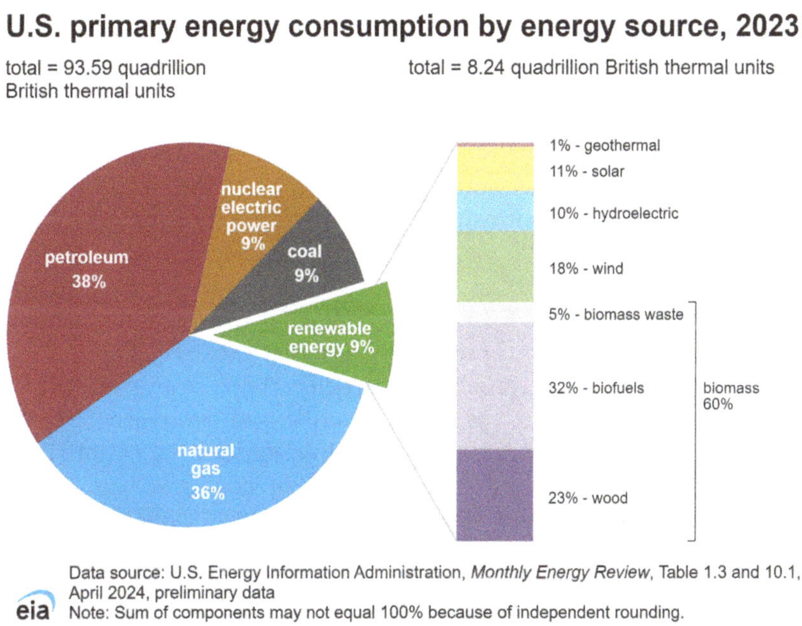

Figure 1.15. United States primary energy by source for 2022. Reproduced from U.S. Energy Information Administration (2023). Image stated to be in the public domain.

transition towards electricity. These include vehicles, where batteries (or other vehicle technologies that utilize electricity as an energy carrier) will become important, as well as heating applications that are currently based on fossil fuels such as fuel oil or natural gas.

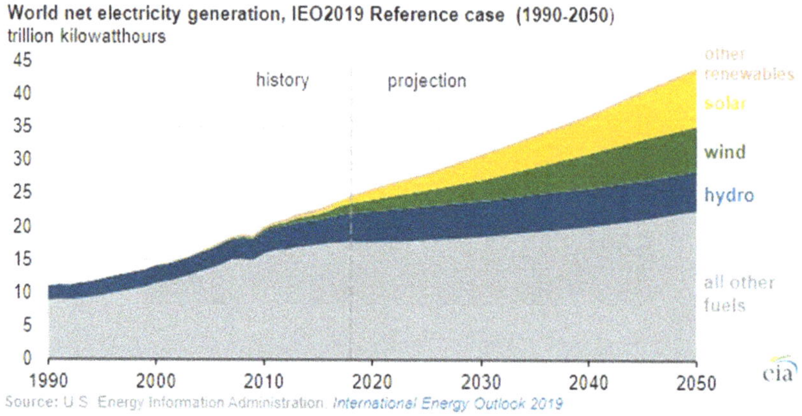

Figure 1.16. Past world sources of electricity generation and future predictions from United States Energy Information Administration. Reproduced from U.S. Energy Information Administration (2019). Image stated to be in the public domain.

The growth of major renewable energy sources (hydroelectric, wind and solar) used for electricity generation is shown in figure 1.16, along with predictions from the United States Energy Information Agency for future growth until 2050. More details concerning electricity produced from these sources is given below.

Figure 1.1 shows that the growth of hydroelectric energy began in the early part of the twentieth century. This corresponds to an increase in electricity use in developed countries. High head hydroelectric dams were a convenient approach to providing significant electrical capacity. In particular, hydroelectricity has a reasonably high capacity factor, i.e., the ratio of the actual electricity generated to the total that could be generated, and shows only moderate seasonal fluctuations. It is, therefore, a reliable means of providing base load capacity. Figure 1.17 shows details of anticipated future hydroelectric capacity. Aside from some growth over the next decade or so in China, the predicted growth of hydroelectric capacity up until 2050 is relatively small. This is because hydroelectricity is a mature technology and nearly all viable high head capacity in most countries has already been developed. Past about 2030, only minimal growth in low head capacity and some capacity in developing countries is expected.

Figures 1.18 and 1.19 show past and predicted future use of wind and solar generated electricity, respectively. Figure 1.18 shows that significant wind energy use began in the early-2000s, while figure 1.19 shows that significant solar energy use started a few years later. The increase in use of these two sources of energy can be directly related to the decrease in their cost, as shown in figure 1.20. In particular, it is clear that the growth of solar energy (i.e., photovoltaics) began in the 2010s when the cost to generated electricity became comparable to that of other technologies. Both sources of energy are expected to grow significantly between the present and 2050. Much of this growth is expected to occur in China and India, where significant increases in energy demand are anticipated.

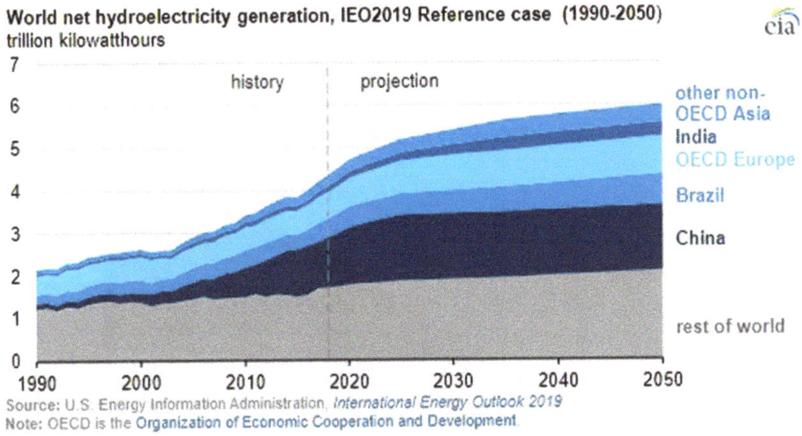

Figure 1.17. Past world hydroelectric electricity generation and future predictions from United States Energy Information Administration. Reproduced from U.S. Energy Information Administration (2019). Image stated to be in the public domain.

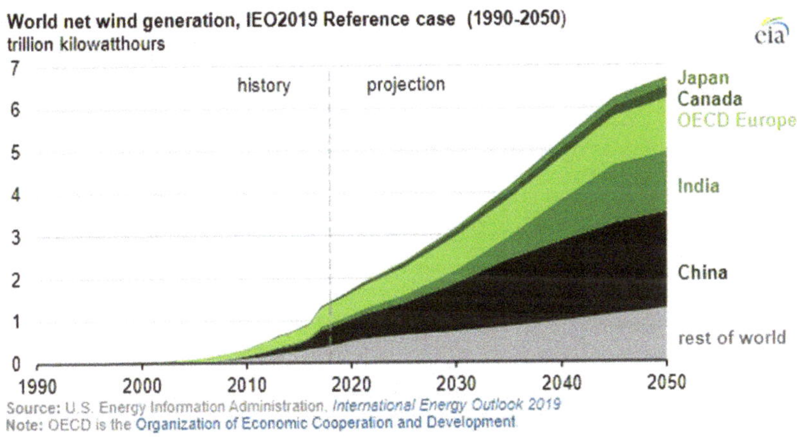

Figure 1.18. Past world electricity generation by wind and future and future predictions from United States Energy Information Administration. Reproduced from U.S. Energy Information Administration (2019). Image stated to be in the public domain.

The current use of nuclear power for electricity generation is summarized in figure 1.21. Several European countries rely heavily on nuclear power. For example, France, Slovakia, and Ukraine (prior to 2022) obtain more than half of their electricity from nuclear power. Further details on the development and growth of nuclear power are provided in section 5.9. Here we look briefly at a comparison of the benefits of different methods for electricity production.

There are several criteria that need to be considered in assessing the viability of different low-carbon energy sources. These factors include capacity factor, variations in output, levelized cost of electricity, power density, greenhouse gas emissions,

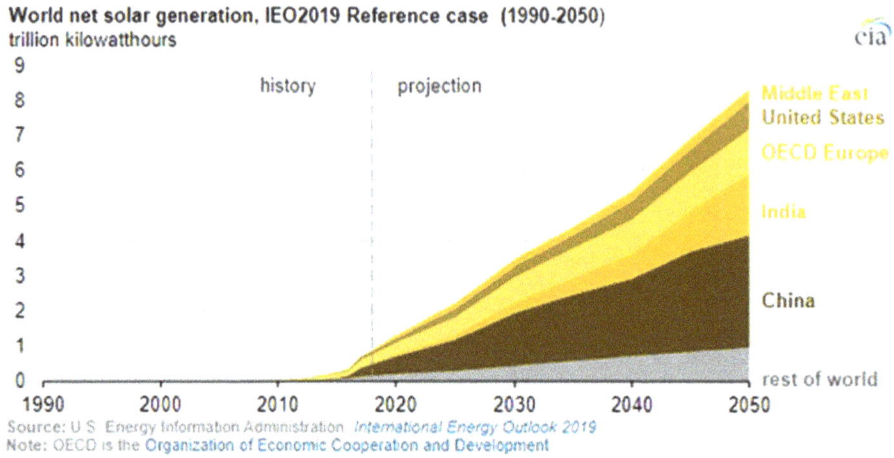

Figure 1.19. Past world solar electricity generation and future predictions from United States Energy Information Administration. Reproduced from U.S. Energy Information Administration (2019). Image stated to be in the public domain.

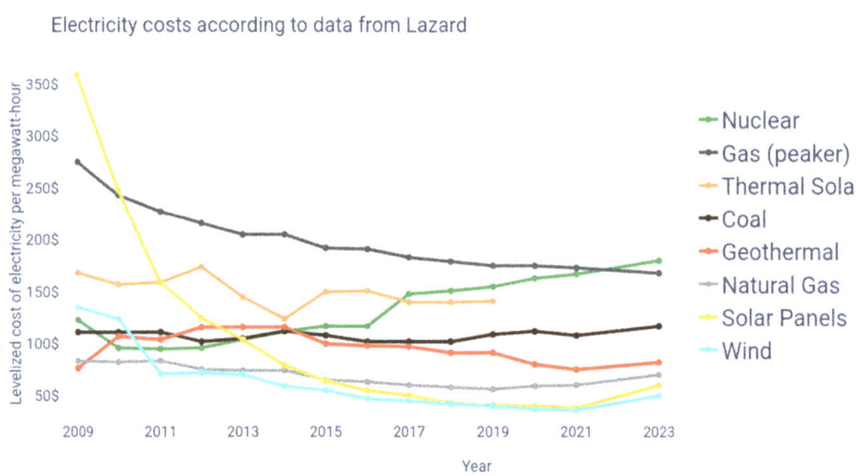

Figure 1.20. Levelized cost of electricity for major electricity sources between 2009 and 2023 in USD MWh^{-1}. Reproduced from Lazard (2023). CC BY 4.0.

and safety. We consider the first five factors below. Safety is considered, with specific reference to nuclear energy, in section 5.9.

Typical average capacity factors for some electricity generating technologies in the United States are given in table 1.1. There are a number of reasons for these values, and it is important to appreciate why no electricity source generates at a 100% capacity factor.

In recent years, some coal capacity has been replaced by natural gas capacity because natural gas produces less CO_2 per unit energy generated than coal. In addition, some coal capacity has been replaced by the increase in wind and solar

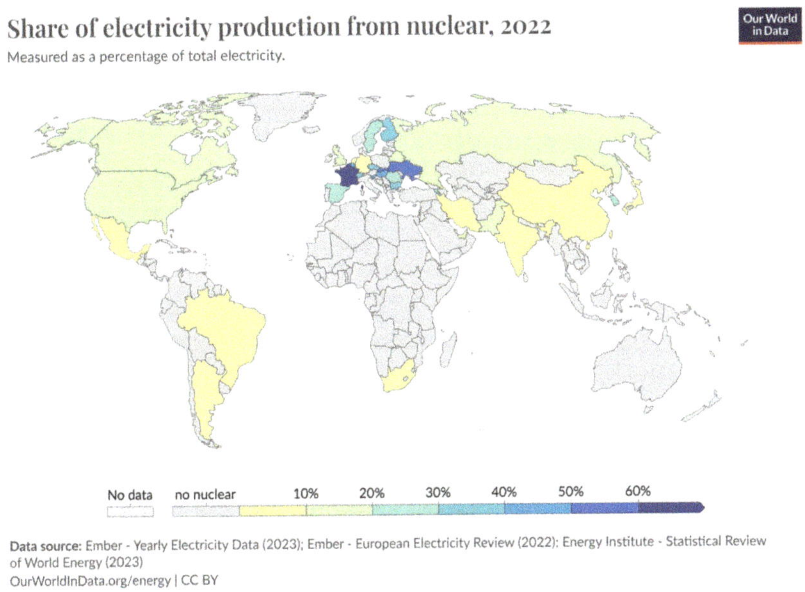

Figure 1.21. Share of nuclear energy for electricity production in different countries as of 2022. Reproduced from Ritchie and Rosado (2024). CC BY 4.0.

Table 1.1. Typical capacity factors for different electricity generating technologies. Data are average values for the United States for 2019. Data adapted from U.S. Energy Information Administration (2020a).

Energy source	Capacity factor (%)
Coal	47.5
Hydroelectric	41.2
Wind	34.3
Solar	24.1
Nuclear	93.4

capacity. The coal capacity factor also shows seasonal variations, as illustrated in figure 1.22. The reason for these fluctuations results from seasonal variations in electricity demand. In summer and winter, demand is higher due to the need of air conditioning and heating, respectively. Since the output of a coal fired generating station can be adjusted (up to its maximum capacity) in response to changes in demand, it serves to as a suitable means of providing base load capacity. Natural gas fired combustion turbines can respond more quickly to demand fluctuations and can supplement coal capacity.

Hydroelectric power shows seasonal fluctuations that peak in the spring when river flows are maximum. This typically does not correspond to times of maximum demand. Seasonal variations in hydroelectric generation and capacity factor are illustrated in figure 1.23.

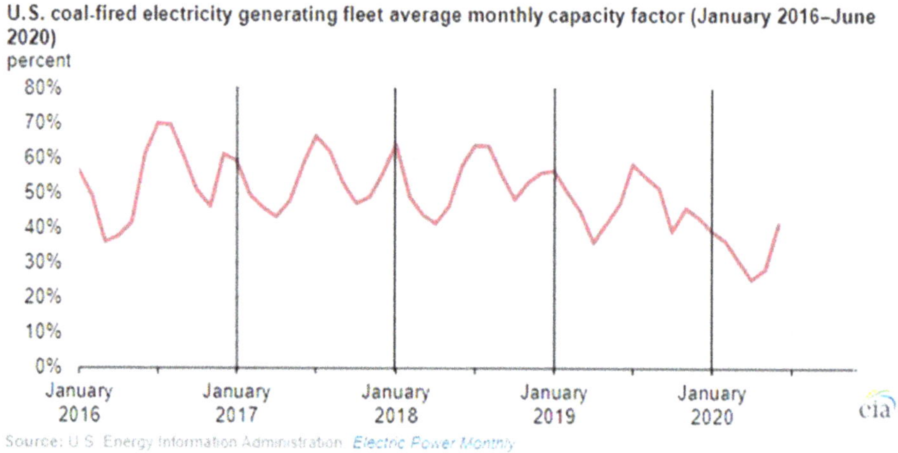

Figure 1.22. Variations in the coal fired generating station capacity factor in the United States from January 2016 to June 2020. Reproduced from U.S. Energy Information Administration (2020b). Image stated to be in the public domain.

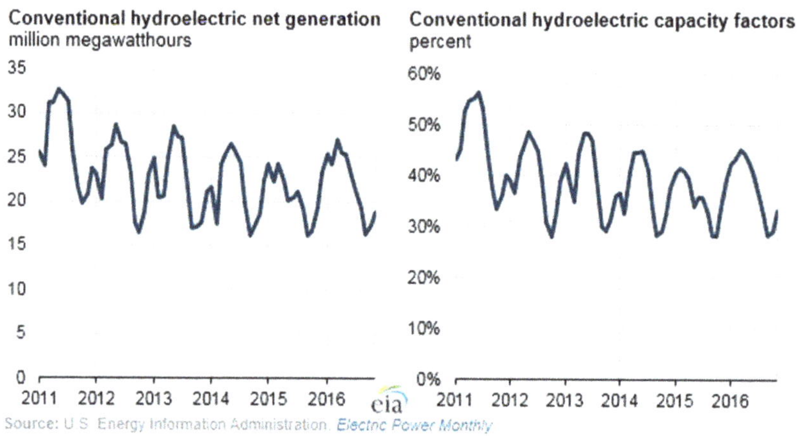

Figure 1.23. Seasonal variations in hydroelectric generation and capacity factor. Reproduced from U.S. Energy Information Administration (2017a). Image stated to be in the public domain.

Wind energy shows daily fluctuations but also shows reasonably well-defined seasonal fluctuations. As shown in figure 1.24, these trends depend on location but typically show maximum output in the late-spring and summer. These variations can be related to prevailing weather patterns. A comparison of figures 1.23 and 1.24 shows that the variations in the wind capacity factor are much greater than those for hydroelectric power.

Solar energy shows reasonably predictable daily and seasonal fluctuations. There are also some variations that result from short-term weather patterns. Daily variations in typical solar photovoltaic output are illustrated in figure 1.25.

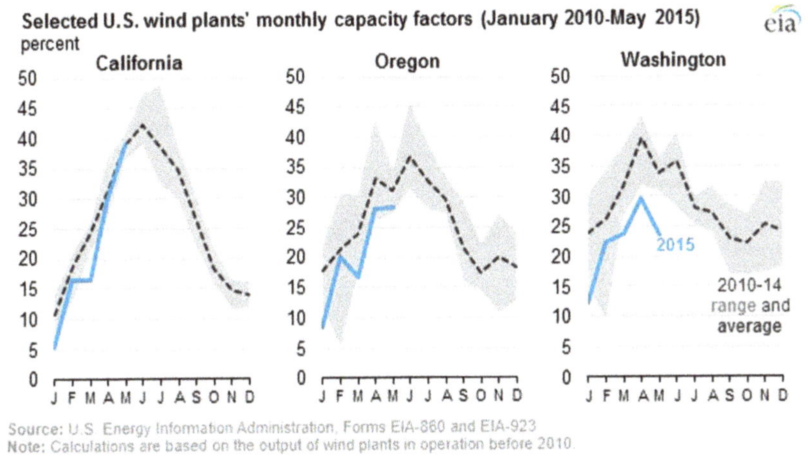

Figure 1.24. Typical average monthly capacity factors of wind turbines in California, Oregon, and Washington. Reproduced from U.S. Energy Information Administration (2015). Image stated to be in the public domain.

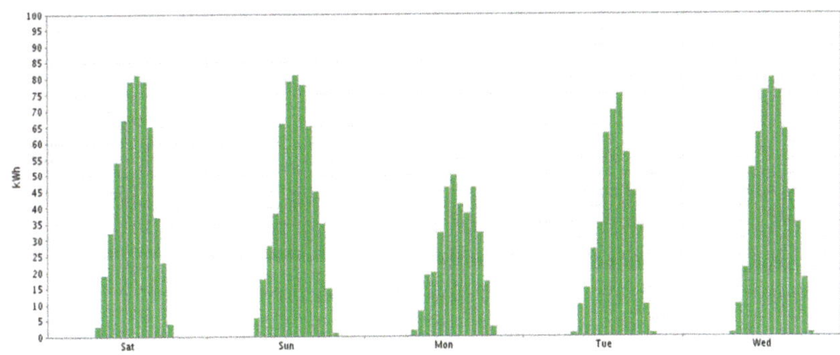

Figure 1.25. Hourly output from the PG&E solar panels installed at San Francisco Giants AT&T Park over the period from 16 to 20 August 2008. This hourly output from the PG&E solar panels installed at San Francisco Giants AT&T Park over the period from August 16-20, 2008 image has been obtained by the author from the Wikimedia website where it was made available by Delphi234 (2008). Image stated to be in the public domain. It is included within this chapter on that basis. It is attributed to Delphi234.

Table 1.1 shows that the capacity factor for nuclear energy is significantly greater than other technologies. Figure 1.26 compares the capacity factor for nuclear power with some other electricity generating methods. Although current nuclear fission reactors cannot be brought online quickly, as can (for example) natural gas fired combustion turbines, they can provide a consistently high capacity factor for base load generation. Some Generation IV nuclear reactor designs are more flexible and are capable of load following, either by reactors that are designed to respond to load fluctuations or by storage of thermal energy for subsequent electricity generation. Such reactors can be used to replace a larger fraction of renewable energy sources (see e.g., Jenkins *et al* 2018). Although individual reactors have downtime for

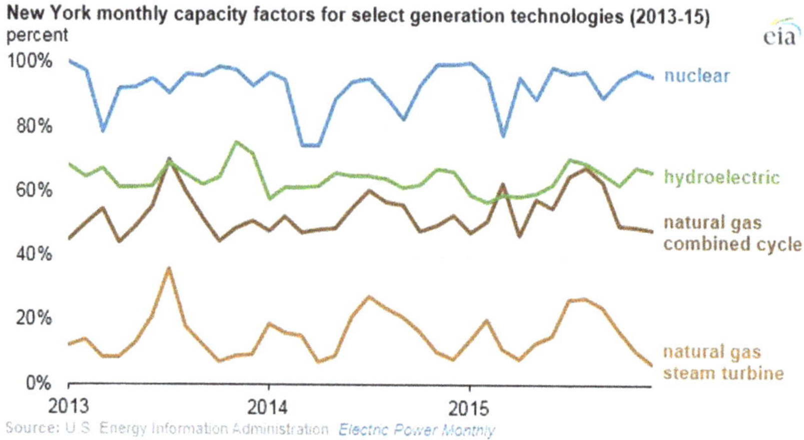

Figure 1.26. Monthly capacity factors for some generating technologies in New York from 2013 to the end of 2015. Reproduced from U.S. Energy Information Administration (2017b). Image stated to be in the public domain.

refueling and maintenance, the combined output of several reactors can provide consistently high output. It can be noted that New York has six reactors at four nuclear power plants and the variations in capacity factor shown in figure 1.26 are consistent with the temporary shutdown of individual reactors for maintenance or refueling.

The use of nuclear power or thermal (e.g., coal) generation for base load can be supplemented by renewables during peak daily demand. This approach is illustrated in figure 1.27, which shows the electricity generation in Germany over the course of one week. It can be noted that 5 July 2014 and 6 July 2014 are Saturday and Sunday, and the figure illustrates the reduced peak demand on weekends.

An analysis of capacity factor and, in particular, its daily and seasonal fluctuations, illustrates the advantage of low-carbon generating technologies that can provide base load capacity. This base capacity can be provided by hydroelectric and nuclear, while peak demand can be satisfied to a significant extent by more variable low-carbon sources such as wind and solar (see figure 1.27). The effective use of variable renewable energy sources requires the implementation of efficient energy storage, and this is particularly the case if solar and wind are to be used for base load capacity (see Dunlap 2020b, 2020c). The need for energy storage adds to the cost of renewable energy. This cost is not included in the levelized cost of electricity (LCOE), as discussed above. This additional levelized cost of storage (LCOS) depends on the details of the storage method. An analysis of LCOS for several battery types is summarized in figure 1.28. A comparison of these results with the LCOE illustrated in figure 1.20 for renewable sources shows that LCOS can be the major component of the cost of renewable energy and can push the cost of wind or solar close to that of nuclear.

A breakdown of world electricity generation in 2022 is shown in table 1.2. Fossil fuels account for about 17 400 TWh or slightly over 60% of the total. We can look at

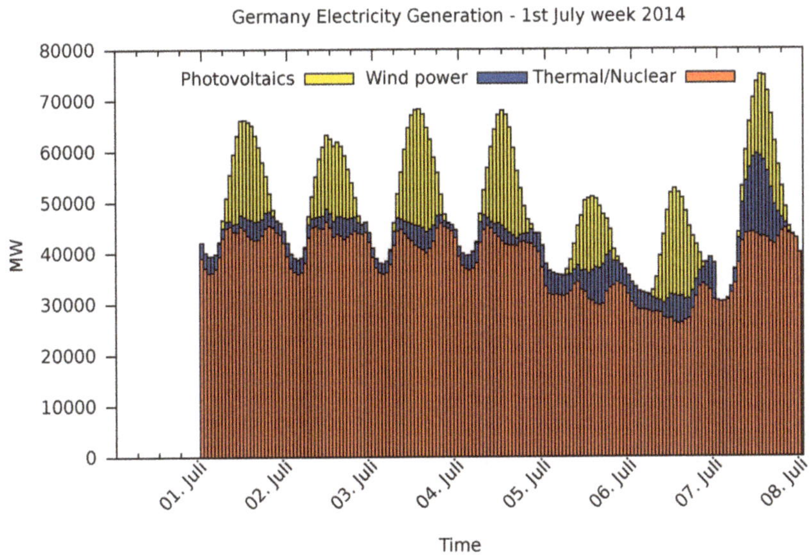

Figure 1.27. Hourly electricity generation in Germany over the first week in July 2014. Reproduced from Kopiersperre (2014). CC BY-SA 3.0.

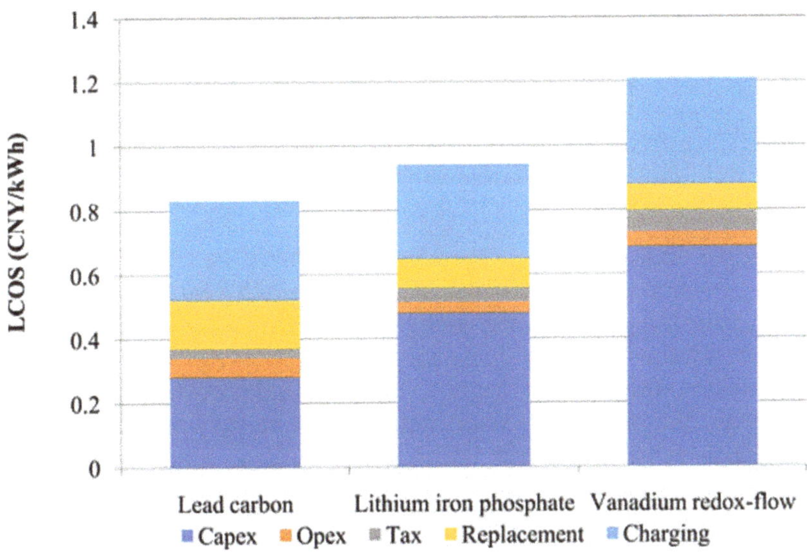

Figure 1.28. LCOS per kWh for different battery technologies with a breakdown of the components of the LCOS. Capex = capital expense, Opex = operating expense (1 CNY = 0.14 USD). Reproduced from Xu *et al* (2020). CC BY 4.0.

the requirements for various low-carbon electricity sources to replace this fossil fuel generated electricity. Using typical capacity factors from table 1.1, we obtain the result shown in table 1.3.

Table 1.2. Breakdown of world electricity sources for 2022. Data adapted from Ritchie and Rosado (2024).

Source	TWh	% of total
Coal	10 191	35.71
Natural gas	6309	22.12
Oil	885	3.10
Nuclear	2610	9.15
Hydroelectric	4327	15.17
Wind	2139	7.50
Solar	1289	4.52
Biomass	678	2.38
Other renewables	100	0.35
Total	28 528	100.00

Table 1.3. Required resources to replace 17 400 TWh of electricity generated annually from fossil fuels. The additional required capacity is (17 400 TWh)/(capacity factor × 8760 h).

Energy source	Additional required capacity (TW)
Wind	5.79
Solar	8.24
Nuclear	2.12

Based on the discussion above, we need to eliminate the vast majority of global greenhouse gas emissions on a time scale of about 25 years in order to have a reasonable probability of limiting global warming to 2 °C. This will require a substantial replacement of fossil fuel energy with low-carbon alternatives. As fossil fuel use for (e.g.) transportation, heating, and industrial processes transitions to low-carbon sources, the majority of final energy will be in the form of electricity (i.e., battery electric vehicles, electric heating, etc) with a small contribution from biofuels. This will increase the need for additional electricity generating capacity. That is, we have to both replace fossil fuel generated electricity and fossil fuel energy used for other purposes, with low-carbon electricity. On the other hand, this will be partially compensated for by the increased efficiency of (for example) battery electric vehicles compared to fossil fuel powered internal combustion engine vehicles. This situation has been considered in detail by Dunlap (2023). Since electricity accounts for about a third of end user energy (e.g., final energy in figure 1.2), the additional required generating capacity might, therefore, be of the order of two to three times that given in table 1.3 (see discussion in Dunlap 2023). Using 2.5 times the values in table 1.3, we can get a rough estimate of the necessary infrastructure that needs to be installed over the next 25 years or so in order to satisfy this energy requirement for wind, solar, and nuclear power.

Using the predictions for wind and solar generation for 2050 from figures 1.18 to 1.19, respectively, we can show that wind energy represents about 15% of the requirement and solar energy represents about 20% of the requirement.

Future nuclear energy development is somewhat more difficult to predict. However, we can approach the problem by looking at the infrastructure that would need to be installed and compare this to previous nuclear development. Using a requirement of 2.5 times the nuclear value in table 1.3 means that over the next 25 years we would require about 200 GW additional capacity per year. We can compare this to past nuclear development as discussed in detail in chapter 5. From figure 5.26 it can be seen that newly installed nuclear capacity was about 32 GW per year in the mid-1980s. This amounts to about 15% of the requirement.

The result of the above discussion is that our future energy mix will, out of necessity, consist of several low-carbon technologies, including wind, solar, and nuclear, as well as other technologies. In any case, the task of replacing fossil fuel generated electricity is very formidable. The inclusion of nuclear fission energy, to the extent that it is possible to develop it, is a necessary component of our future energy. Generation IV designs can provide reliable base load energy and, in some cases, are capable of effective load following. They can also supply a greater proportion of total energy demand.

Another way of looking at the capacity requirements is to consider power density, i.e., the land area that is required to supply the necessary capacity. Using the Alta Wind Energy Center in California, which has a capacity of 1548 MW and covers an area of 130 km^2, as an example, wind has a capacity per unit land area of 1548 MW/130 km^2 = 12 MW km^{-2}. For solar, we consider the Golmud Solar Park in Qinghai Province, China. This has a name plate capacity of 200 MW and covers an area of 5.64 km^2, giving a capacity per unit land area of 200 MW/5.64 km^2 = 35 MW km^{-2}. For nuclear, we consider the Bruce Nuclear Generating Station in Ontario, Canada (see further information in section 5.7) as an example. This station consists of eight CANDU pressurized heavy water-cooled nuclear with a total electrical capacity of 6610 MW. It occupies an area of 9.32 km^2 on the shore of Lake Huron, which it uses as a source of cooling water. The capacity per land area is 6610 MW/9.32 km^2 = 709 MW km^{-2}. The power density per unit land area is summarized in table 1.4 for these

Table 1.4. Estimated capacity per km^2 of land area for different low-carbon electricity generating technologies. The net capacity includes the estimated capacity factors from table 1.3.

Energy source	Total capacity per km^2 (MW km^{-2})	Net capacity per km^2 (MW km^{-2})
Wind	12	4.1
Solar	35	8.4
Nuclear	709	662

three energy sources where the net capacity includes a consideration of the capacity factor. While various factors must be included in an assessment of the feasibility of infrastructure development for different energy sources, it is clear that geographical resources need to be part of the analysis.

The final point to consider in this chapter is the extent to which so-called low-carbon energy sources are in fact low carbon. Life cycle analysis provides a measure of the greenhouse gas emissions per unit energy produced over the lifetime of the facility. The results of a recent analysis of net greenhouse emissions for different energy sources are summarized in figure 1.29. This figure also shows the results of a risk assessment, which are discussed further in section 6.9. Clearly, all renewable sources of energy, as well as nuclear, produce considerably less greenhouse gas emissions than any of the fossil fuels. There is, however, a substantial difference between biofuels and hydroelectric compared to wind, solar, and nuclear. It is of interest to look at the reasons for these differences.

In principle, biofuels are carbon neutral. Plant matter sequesters carbon dioxide from the atmosphere when it grows. When the biofuel produced from that plant matter is burned to produce energy, the same amount of carbon dioxide is released back into the atmosphere. In practice, however, the analysis of biofuel contributions to greenhouse gas emissions is very complex (see e.g., Takeuchi *et al* 2018). The use of fertilizers can emit nitrous oxide to the atmosphere, increasing agricultural land area can lead to deforestation and agricultural waste can lead to methane emission. Current biofuel production methods utilize fossil fuels, and these also contribute to CO_2 emissions. The details of the contribution of biofuels to greenhouse gas

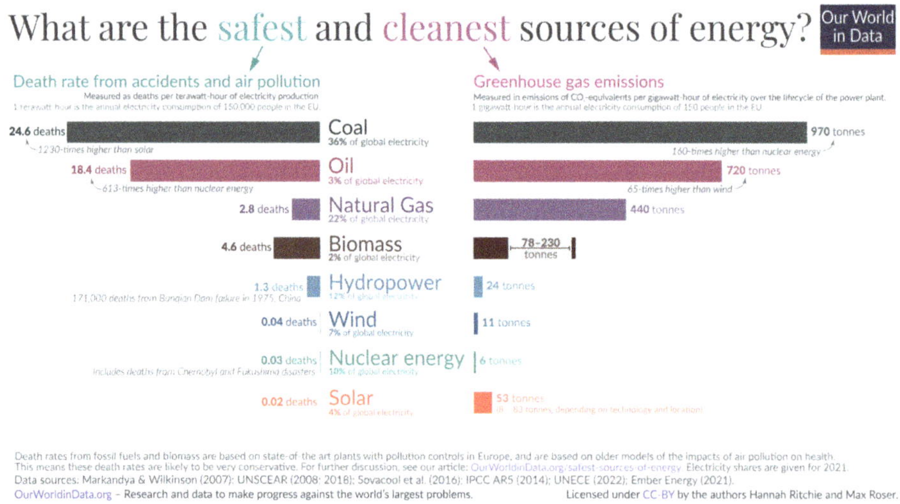

Figure 1.29. Death rate (in deaths per TWh) and greenhouse gas emissions (in CO_2 equivalent per GWh) for some common electricity producing technologies. Reproduced from Ritchie (2020b). CC BY 4.0.

emissions depends on the nature of the biofuel and the specifics of the agricultural and fuel production processes. Hence, figure 1.29 shows a wide range of values for the emissions from biomass energy.

Hydroelectric power contributes to greenhouse gas emissions in two ways. First, the construction of a hydroelectric dam requires considerable quantities on concrete. Concrete is made by mixing cement with various aggregate materials such as stone or sand. Cement is made by a variety of processes which typically begin by heating limestone ($CaCO_2$) to form lime (CaO) by the process,

$$CaCO_3 \rightarrow CaO + CO_2.$$

It is estimated that the production of 1 kg of concrete releases 0.93 kg of CO_2. Second, hydroelectric power produces greenhouse gas emissions by the construction of the reservoir. Reservoir construction can result in deforestation and decomposition of biomass. These factors are much more significant in tropical regions than in temperate regions.

In the case of wind and solar energy, greenhouse gas emissions are primarily the result of the manufacture of infrastructure. For nuclear energy there are greenhouse gas emissions associated with the manufacture of reactor components and during construction. There are also greenhouse gas emissions associated with the mining, transport, and processing of reactor fuel.

It is clear from the above discussions that nuclear fission energy must be seriously considered as a component of future low-carbon energy sources for the mitigation of global climate change.

References

Delorme 2019 Atmospheric carbon dioxide concentrations at Mauna Loa, Hawaii since 1958 https://commons.wikimedia.org/wiki/File:Mauna_Loa_CO2_monthly_mean_concentration.svg

Delphi234 2008 Hourly output from the PG&E solar panels installed at San Francisco Giants AT&T Park over the period from August 16–20 https://commons.wikimedia.org/wiki/File:ATTParksolarpaneloutput.png

Delphi234 2012 United States per capita energy use 1650–2010 https://commons.wikimedia.org/wiki/File:United_States_per_capita_energy_use_1650-2010.png

Delphi234 2014 World total primary energy production https://commons.wikimedia.org/wiki/File:World_total_primary_energy_production_chart_only.png

Dunlap R A 2020a *Renewable Energy: Volume 1: Requirements and Sources* (San Rafael, CA: Morgan & Claypool)

Dunlap R A 2020b *Renewable Energy: Volume 2: Mechanical and Thermal Energy Storage Methods Requirements and Sources* (San Rafael, CA: Morgan & Claypool)

Dunlap R A 2020c *Renewable Energy: Volume 3: Electrical, Magnetic and Chemical Energy Storage Methods Requirements and Sources* (San Rafael, CA: Morgan & Claypool)

Dunlap R A 2023 *Transportation Technologies for a Sustainable Future—Renewable Energy Options for Road, Rail, Marine and Air Transportation* (Bristol: IOP Publishing)

Energy Institute 2023 Statistical review of world energy www.energyinst.org/statistical-review

Fawzy S, Osman A I, Doran J and Rooney D W 2020 Strategies for mitigation of climate change: a review *Environ. Chem. Lett.* **18** 2069–94

Hansen J E *et al* 2023 Global warming in the pipeline *Oxford Open Climate Change* **3** kgad008
Hubbert M K 1956 *Nuclear Energy and the Fossil Fuels* Publication No. 95 (Houston, TX: Shell Development Company)) https://web.archive.org/web/20080527233843/http://hubbertpeak.com/hubbert/1956/1956.pdf
Intergovernmental Panel on Climate Change 2022 *Climate Change 2022 Mitigation of Climate Change Working Group III Contribution to the Sixth Assessment Report of the Intergovernmental Panel on Climate Change* ISBN 978-92-9169-160-9
Jenkins J D, Zhou Z, Ponciroli R, Vilim R B, Ganda F, de Sisternes F and Botterud A 2018 The benefits of nuclear flexibility in power system operations with renewable energy *Appl. Energy* **222** 872–84
Jouzel J *et al* 2007 Orbital and millennial antarctic climate variability over the past 800,000 years *Science* **317** 793–6
Kockmeyer 2013 Norwegian oil output and a fitted bell curve https://commons.wikimedia.org/wiki/File:NorwayoilproductionandHubbert2013.svg
Kopiersperre 2014 Germany wind and solar generation hourly for first week of July 2014 https://commons.wikimedia.org/wiki/File:German_Wind_and_Solar_2014-07-01_to_2014-07-07.svg
Lazard 2023 2023 Levelized cost of electricity www.lazard.com/research-insights/2023-levelized-cost-of-energyplus/
Lbeaumont 2011 Total world, annual primary energy consumption https://commons.wikimedia.org/wiki/File:Annual_world_primary_energy_consumption.svg
Luthi D *et al* 2008 High-resolution carbon dioxide concentration record 650,000–800,000 years before present *Nature* **453** 379–82
Mir-445511 2023 Lazard 2023 Levelized cost of electricity https://commons.wikimedia.org/wiki/File:Electricity_costs_in_dollars_according_to_data_from_Lazard.png
Ney E P 1959 Cosmic radiation and weather *Nature* **183** 451–2
NOAA National Centers for Environmental Information 2023 Monthly Global Climate Report for Annual 2022 www.ncei.noaa.gov/access/monitoring/monthly-report/global/202213
Our World in Data 2023 Global primary energy consumption by source https://ourworldindata.org/grapher/global-energy-consumption-source
Poynting M 2024 World's first year-long breach of key 1.5 °C warming limit *BBC* (8 February 2024) www.bbc.com/news/science-environment-68110310
Ritchie H 2020a Sector by sector: where do global greenhouse gas emissions come from? https://ourworldindata.org/ghg-emissions-by-sector
Ritchie H 2020b What are the safest and cleanest sources of energy? https://ourworldindata.org/safest-sources-of-energy
Ritchie 2021 Decarbonizing electricity is only one step towards a low-carbon energy system https://ourworldindata.org/low-carbon-electricity
Ritchie H 2022 Primary, secondary, final, and useful energy: why are there different ways of measuring energy? https://ourworldindata.org/energy-definitions
Ritchie H 2023 How much CO_2 can the world emit while keeping warming below 1.5 °C and 2 °C? https://ourworldindata.org/how-much-co2-can-the-world-emit-while-keeping-warming-below-15c-and-2c
Ritchie H and Rosado P 2024 Electricity mix https://ourworldindata.org/electricity-mix
Royal Society 2018 Greenhouse gas removal https://royalsociety.org/-/media/policy/projects/greenhouse-gas-removal/royal -society-greenhouse-gas-removal-report-2018.pdf

Smil V 2017 *Energy Transitions: Global and National Perspectives, 2nd edn* (Westport, CT: Praeger)

Sorrell S, Speirs J, Bentley R, Brandt A and Miller R 2009 *An Assessment of the Evidence for a Near-Term Peak in Global Oil Production* (London: UK Energy Research Centre)

Takeuchi K, Shiroyama H, Saito O and Matsuura M (ed) 2018 *Biofuels and Sustainability* (Tokyo: Springer)

U.S. Energy Information Administration 2015 West Coast wind patterns lead to below-normal wind generation capacity factors www.eia.gov/todayinenergy/detail.php?id=22452

U.S. Energy Information Administration 2017a Hydroelectric generators are among the United States' oldest power plants www.eia.gov/todayinenergy/detail.php?id=30312

U.S. Energy Information Administration 2017b Indian Point, closest nuclear plant to New York City, set to retire by 2021 www.eia.gov/todayinenergy/detail.php?id=29772

U.S. Energy Information Administration 2019 EIA projects that renewables will provide nearly half of world electricity by 2050 www.eia.gov/todayinenergy/detail.php?id=41533

U.S. Energy Information Administration 2020a *Electric power monthly with data for August 2020* www.eia.gov/electricity/monthly/archive/october2020.pdf

U.S. Energy Information Administration 2020b As U.S. coal-fired capacity and utilization decline, operators consider seasonal operation www.eia.gov/todayinenergy/detail.php?id=44976

U.S. Energy Information Administration 2023 U.S. primary energy consumption by energy source, 2023 www.eia.gov/energyexplained/us-energy-facts/

Weith A 2016 Glacier decrease on Svalbard in the years 1900–1960–2015 https://commons.wikimedia.org/wiki/File:Glacier_decrease_on_Svalbard_in_the_years_1900–1960-2015.jpg

Xu Y, Pei J, Cui L, Liu P and Ma T 2020 The levelized cost of storage of electrochemical energy storage technologies in China *Front. Energy Res.* **10** 873800

IOP Publishing

Generation IV Nuclear Reactors
Design, operation and prospects for future energy production
Richard A Dunlap

Chapter 2

Some introductory nuclear physics

Chapter 2 provides an overview of some basic nuclear physics that is required for an understanding of the operation of a nuclear fission reactor. This begins with a review of nuclear structure and the interactions between the particles in the nucleus. This is followed by a description of the nuclear binding energy and nuclear decay processes, including proton and neutron emission, alpha decay, and beta decay. Excited states are then described in the context of nuclear structure. Finally, nuclear reactions are considered, with particular emphasis of reactions involving neutrons. These neutron reactions include elastic and inelastic scattering along with (n, γ) reactions and induced fission.

2.1 Introduction

A description of the operation of a nuclear fission reactor relies on an understanding of the interactions between the protons and neutrons that make up the nucleus. The present chapter provides a brief overview of these interactions and the way in which they govern the properties of nuclei. In particular, those features of nuclei that are of particular relevance to the functioning of a nuclear fission reactor are emphasized. The topics discussed in this chapter are covered in more detail in any of the numerous nuclear physics texts that have been published, suitable references include Cottingham and Greenwood (2001), Dunlap (2023), Krane (1988), and Williams (1991).

2.2 Nuclear structure

The nucleus of an atom consists of a collection of nucleons, i.e., protons and neutrons, that are bound together. The element is determined by the number of protons, generally designated as Z, and this is referred to as the atomic number or proton number. The number of neutrons is generally designated as N. The total number of protons and neutrons, designated A, as

$$A = Z + N \tag{2.1}$$

doi:10.1088/978-0-7503-6069-2ch2

is referred to as the mass number. A nucleus with specific values of Z and N is referred to as a nuclide and is designated as ^{A}E, where the abbreviation for the name of the element, E, is determined by the value of Z and the neutron number, N, is determined from A and Z according to equation (2.1). For record keeping purposes when describing nuclear decays or reactions, it is sometimes convenient to write the name of the nuclide as $^{A}_{Z}\mathrm{E}^{N}$. It is often useful to consider families of nuclides with similar properties. Isotopes are a group of nuclides with common values of Z, i.e., they are of the same element. Isotones and isobars are, respectively, families of nuclides with common values of N and A.

While some combinations of protons and neutrons form stable nuclei, other combinations do not. Those that do not form stable nuclei can, in some cases, form unstable nuclei, while in other cases there is no experimental evidence that such combinations exist. The Segrè plot, named after the Italian American physicist Emilio Segrè (1905–89), is shown in figure 2.1. This plot illustrates the combinations of protons and neutrons that form stable and unstable nuclei.

A very significant feature illustrated in the Segrè plot is the fact that heavier nuclei have an increasingly greater number of neutrons compared to the number of

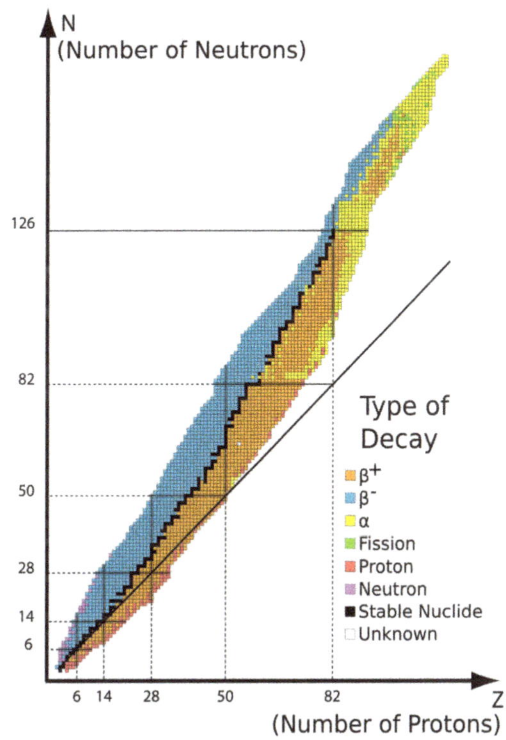

Figure 2.1. Segrè plot showing the combinations of protons and neutrons that form stable and unstable nuclei. Stable nuclides are shown by the black data points. Unstable nuclides are shown by the various colored data points where the decay processes as indicated are discussed in detail in section 2.4. Reproduced from Napy1kenobi (2009). CC BY-SA 3.0.

protons. The reasons for this behavior can be explained by a relatively simple quantum mechanical description of the particles and interactions in the nucleus and will be discussed in further detail in the next section.

2.3 Particles and interactions

The reasons why some combinations of protons and neutrons form nuclei and other combinations do not may be explained by looking in detail at the interactions between these particles. We begin here with an overview of the four known interactions in nature. These interactions and some of their relevant properties are summarized in table 2.1. Each interaction acts on objects with different characteristics. Gravity, which acts on masses, is the weakest of the interactions and acts primarily at large distances, and is not a factor in the description of nuclear behavior. From the standpoint of nuclear properties, the weak interaction acts on leptons, such as electrons and neutrinos, as well as hadrons, such as the proton and the neutron. The electromagnetic interaction acts on charges and in the present context acts on charged particles such as protons and electrons. The strong interaction acts on hadrons, such as protons and neutrons, but not on leptons, such as electrons. Each interaction can be described in terms of a meditating virtual particle as indicated in the table. The range of the interactions can be considered in the context of the typical size of a nucleus, which is a few fermis (fermi = fm = 10^{-15} m).

The properties of the nucleus can be understood by a consideration of the quantum mechanics of a bound system of nucleons. The application of the Schrödinger equation as described below can provide considerable insight into nuclear behavior. We can begin with the spherically symmetric three-dimensional time independent Schrödinger equation as

$$-\frac{\hbar^2}{2m}\nabla^2\psi + V(r)\psi = E\psi$$

where $\hbar$ is the reduced Planck constant, m is the particle mass, ψ is the wavefunction, $V(r)$ is the radially dependent potential, and E is the total energy. The application of this equation to (say) the electron in a hydrogen atom using the Coulomb potential allows for the determination of the relevant electron energy levels. The application of the Schrödinger to the nucleons in a nucleus is different than its application to the hydrogen atom for two reasons—there are two different types of particles to consider (protons and neutrons) and the appropriate potential is the result of all

Table 2.1. The four known interactions and some of their relevant properties. Adapted from Dunlap (2023).

Interaction	Acts on	Mediating boson	Range (m)	Relative strength
Gravity	Masses	Graviton	Infinite	10^{-39}
Weak	Leptons and hadrons	W^+, W^-, Z°	10^{-18}	10^{-5}
Electromagnetic	Charges	Photon	Infinite	10^{-2}
Strong	Hadrons	Gluons	10^{-15}	1

relevant interactions. In the case of the protons, the relevant interactions are the strong interaction and the Coulomb interaction, while in the case of neutrons only the strong interaction is relevant (as the neutrons are uncharged). The quantized energy levels as obtained from the Schrödinger equation for the protons and neutrons are illustrated schematically in figure 2.2. There are two features that are seen in the figure that are of particular relevance. First, the bottom of the potential well is slightly raised for the protons as compared to the neutrons. More importantly, however, the spacing of the proton energy levels is slightly greater than that of the neutrons. Both of these features are the result of the inclusion of the repulsive Coulomb interaction for the positively charged protons.

The information shown in figure 2.2 allows us to understand the features shown in the Segrè plot in figure 2.1. The component particles of the nucleus, i.e., the protons and neutrons, are fermions and are, therefore, governed by Fermi–Dirac statistics. This behavior includes the requirement that the protons and neutrons in the nucleus must obey the Pauli exclusion principle. As a result, there is a maximum number of nucleons in each of the energy levels as determined by the allowed quantum numbers for that level. In the ground state of the nucleus, the energy levels up to the so-called Fermi energy are occupied for both proton and neutron levels. Since the protons and neutrons in the nucleus interact strongly with each other, the equilibrium ground state configuration of the nucleus must have similar Fermi energies for these two species of nucleons. Since the neutron density of states is greater than the proton density of states, an equilibrium nuclear configuration can, in principle, accommodate more neutrons than protons. This potential excess of neutrons as compared to

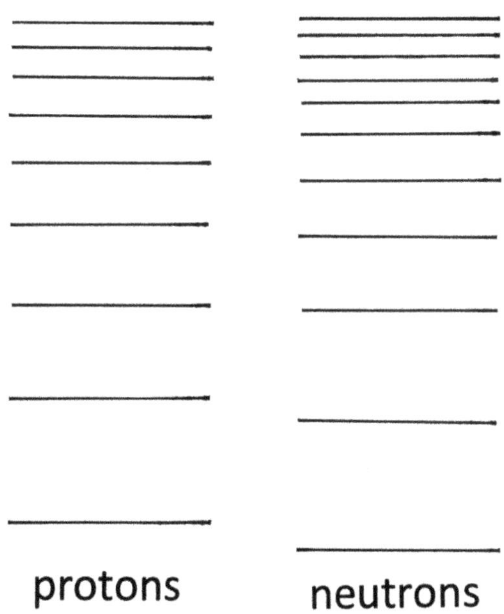

Figure 2.2. Schematic representation of the proton (left-hand side) and neutron (right-hand side) energy levels in square well potential.

protons becomes more significant as the total number of nucleons increases. This behavior explains the general shape of the curve shown in figure 2.1.

2.4 Nuclear binding energy

The above discussion provides some insight into the occurrence of stable and unstable nuclei as illustrated in figure 2.1. It can also, in a semiempirical way, explain the fate of unstable nuclei. Before we look at this behavior in more detail, it is useful to quantify the behavior of nuclei by defining the nuclear binding energy. Energy is expressed in terms of Einstein's mass–energy relation,

$$E = mc^2$$

where m is mass and c is the speed of light. The binding energy, B, which holds the nucleons together in the nucleus, is defined as a positive quantity and results in a decrease in the mass of the nucleus compared to the total mass of the constituent nucleons by the relation,

$$m_{\text{nuc}} = Zm_{\text{p}} + Nm_{\text{n}} - B/c^2 \tag{2.2}$$

where m_{p} and m_{n} are the proton mass and neutron mass, respectively. We can understand the signs of the quantities in equation (2.2) in the following way. The masses of the individual nucleons are measured when they are free particles, i.e., they are not interacting with each other. In order to totally isolate nucleons that are bound together in a nucleus, we need to add energy to the system. That energy is the total binding energy of the nucleus, B, and it represents an equivalent mass of B/c^2. Thus, separating bound nucleons adds mass, so the binding energy of the nucleus reduces the masses of the free particles as seen in equation (2.2). The binding energy is easily obtained from equation (2.2) as the difference between the mass of the free particles and the mass of the nucleus they form when bound together (times the speed of light squared). The more tightly bound the nucleons are, the lower the mass and the more likely the nucleus thus formed will be stable. Since the total nuclear binding energy will increase with the number of nucleons that are interacting, an important measure of nuclear stability is the binding energy per nucleon, defined as

$$\frac{B}{A} = \frac{(m_{\text{nuc}} - Zm_{\text{p}} - Nm_{\text{n}})c^2}{(Z + N)}.$$

A plot of the nuclear binding energy per nucleon as a function of the total number of nucleons is shown in figure 2.3. The plot shows data for all known nuclides, where the stable nuclides are shown by the black data points and these exhibit higher binding energies per nucleon than the unstable nuclides shown by the red data points.

There are two features shown in figure 2.3 that are of particular relevance to the operation of a nuclear fission reactor, the general shape of the binding energy curve for stable nuclei as a function of mass number and the features of the binding energy for individual nuclides as a function of the even or odd nature of Z and N. In the first case, the curve shows a rapidly increasing binding energy as a function of mass

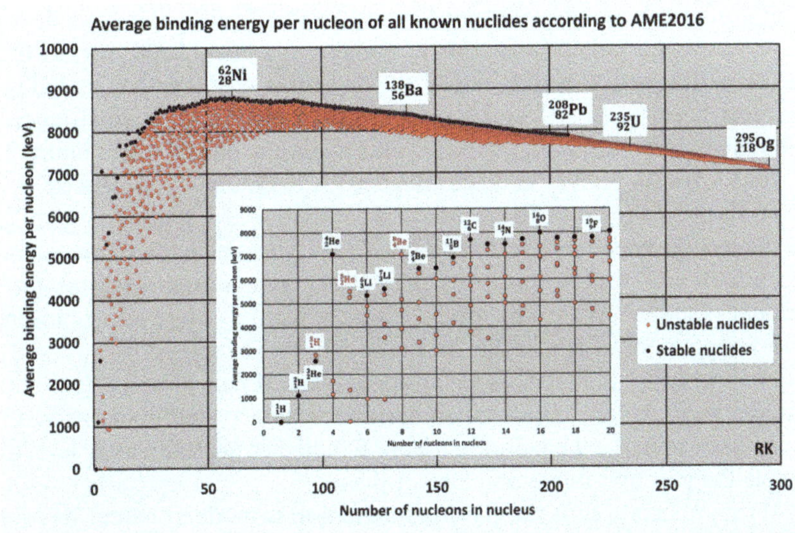

Figure 2.3. Binding energy per nucleon as a function of the total number of nucleons (mass number), A. Stable nuclei are represented by black data points and unstable nuclei are represented by red data points. The insert shows data for light nuclei. Reproduced from Kahn (2018). CC BY-SA 4.0.

number for light nuclei. There is a fairly broad peak centered at around $A = 55$ followed by a gradual decrease. The shape of this curve explains the basic operating principle of fusion and fission reactors. For a fusion reactor, two light nuclei are combined to form one heavier nucleus. Since the binding energy of the heavier nucleus is greater than the total binding energy of the two lighter nuclei, there is a net overall decrease in mass and this mass is converted into energy according to Einstein's relation. For a more detailed description of fusion energy see Dunlap (2021). For a fission reactor, a heavy nucleus is broken up into two lighter nuclei. This situation produces net energy as a result of the change in the binding energy per nucleon when the two lighter nuclei both have mass numbers that put them on the portion of the curve in figure 2.3 where the binding energy per nucleon decreases for large A.

The relationship between binding energy per nucleon and the evenness or oddness of Z and N can be seen in the insert in figure 2.3 by looking at the data for nuclei with $Z = N = 2n$, where n is an integer. These nuclides are ^{4}He, ^{8}Be, ^{12}C, ^{16}O, and ^{20}Ne. The sequence continues for larger values of n (not shown in the insert). Each of these nuclides is at a local maximum in the binding energy per nucleon. It is interesting to note that although ^{8}Be has a particularly large binding energy per nucleon, it is not stable. In fact, it decays by splitting into two ^{4}He nuclei with a half-life of 8×10^{-17} s. This happens because ^{4}He has an even greater binding energy per nucleon than ^{8}Be, and it is, therefore, energetically favorable for this decay to occur.

The reason for the features in the binding energy curve related to the evenness or oddness of Z or N is the strong pairing interaction between like nucleons in the nucleus. This causes an even number of like nucleons in the nucleus to form nucleon pairs of opposite spin. This pairing interaction plays a crucial role in the determination of the energies of excited nuclear states. In fact, it is typically easier to raise the energy of a pair of nucleons to a higher quantum level than to break the pair and raise the energy of only one of the nucleons. A macroscopic analog of this behavior has been proposed by Dunlap (2013) using the relative strength of the electromagnetic interaction and gravity. This analog is illustrated in figure 2.4. The ground state consists of two coupled magnets in a ground level defined by a zero vertical position as shown in figure 2.4(a). The first excited state consists of the two coupled magnets moving to a gravitationally higher potential, as shown in figure 2.4(b). The second excited state consists of one of the magnets moving to the higher gravitational potential. What this means is that it requires more energy to break the magnetic coupling of the pair of magnets and move one of them to a higher gravitational potential than it does to move the coupled pair of magnets to the higher gravitational level. An analogous situation can occur for paired nucleons in a nucleus where it is easier to move the pair of nucleons to a higher nuclear energy level than it is to decouple the pair and move only one of them.

Another manifestation of the pairing interaction is made apparent by looking at the number of stable nuclides with different nuclear configurations. Table 2.2 shows the number of known stable nuclides as a function of the evenness or oddness of the number of protons and neutrons. For Z and N both odd, there are both unpaired protons and neutrons. This configuration leads to only four known stable nuclides. If either the number of protons or the number of neutrons (but not both) is even,

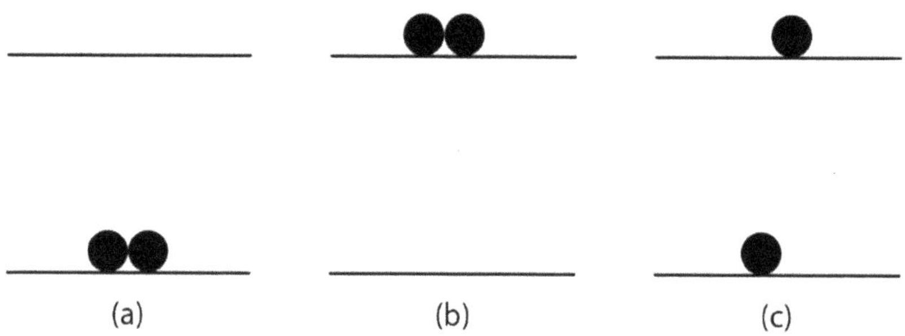

Figure 2.4. Analog of the nuclear pairing interaction using the strength of the magnetic interaction compared to gravity. (a) Analog of the nuclear ground state, (b) analog of the first excited nuclear state, and (c) analog of the second excited nuclear state.

Table 2.2. Number of known stable nuclides as a function of the evenness and oddness of the number of protons and neutrons. Adapted from Dunlap (2023).

Number of protons (Z)	Number of neutrons (N)	Number of known stable nuclides
Odd	Odd	4
Odd	Even	50
Even	Odd	55
Even	Even	165

then one of the nucleon species will form ground state pairs, and will, therefore, increase the binding energy and the corresponding nuclear stability. Thus, in these two cases about 50 stable nuclides are know. For both Z and N even, all nucleons are paired in the nuclear ground state and the corresponding nuclear stability is evidenced by the large number (165) of known stable nuclides. The relevance of the pairing interaction will be considered further below.

2.5 Nuclear decay processes

The properties of unstable nuclides have been alluded to in figure 2.1. This figure shows six common nuclear decay processes. Before discussing these processes in some detail, we consider some general properties of nuclear decay.

Nuclear decay is a statistical process that is described by a characteristic time constant, λ, referred to as the decay constant. The rate at which an unstable nuclear species (the parent nuclide) decays (to the daughter nuclide) in a sample containing $N(t)$ such nuclei is given by,

$$dN(t) = -\lambda N(t)dt. \tag{2.3}$$

The number of parent nuclei as a function of time is easily obtained from equation (2.3) by integration,

$$N(t) = N(0)e^{-\lambda t} \tag{2.4}$$

where $N(0)$ is the number of parent nuclei present at time $t = 0$. Two important parameters that describe the decay process can be obtained from equations (2.3) and (2.4). First, the half-life, $\tau_{1/2}$, of the decay process is the time required for the initial quantity of parent nuclei to be reduced by a factor of 2. From equation (2.4), this is obtained by calculating the time for which $N(\tau_{1/2}) = N(0)/2$,

$$\tau_{1/2} = \frac{\ln 2}{\lambda}.$$

Second, the decay rate as a function of time is found by combining equations (2.3) and (2.4) as

$$\left|\frac{dN(t)}{dt}\right| = N(0)\lambda e^{-\lambda t}.$$

These relations will be useful when assessing the properties of nuclear decays.

We can now look at the details of specific decay processes. Five of the decay processes shown in figure 2.1, i.e., proton emission, neutron emission, alpha decay, beta-minus decay (β^- decay), and beta-plus decay (β^+ decay), are discussed in some detail in the present section. Electron capture, which is analogous to β^+ decay, is also discussed. Fission will be considered in the next chapter.

2.5.1 Proton emission

As seen in figure 2.1, proton emission occurs for nuclei along the lower edge of the region of known nuclides. These nuclei have a much greater number of protons than that which is needed to form a stable nucleus. As illustrated by the energy levels for the neutrons and protons in figure 2.2, nuclei that undergo proton emission have a proton Fermi energy that is much greater than the neutron Fermi energy, meaning that the proton levels are filled to energies that are well above those of the filled neutron levels. In order to move the nucleus closer to stability, the most energetic proton can escape from the nucleus. The decay process corresponding to proton emission is

$$ {}^{A}_{Z}\mathrm{X}^{N} \rightarrow {}^{A-1}_{Z-1}\mathrm{Y}^{N} + \mathrm{p}. $$

Note that the name of the element has changed because the number of protons has changed. The nuclide on the left-hand side of the equation is the parent, while the nuclide on the right-hand side of the equation is the daughter. The energy associated with this decay process is the proton separation energy, S_{p}, and is given by the energy equivalence of the mass difference between the quantities on the left-hand side of the equation and the right-hand side of the equation, i.e.,

$$ S_{\mathrm{p}} = \left[m_{\mathrm{nuc}}\left({}^{A}_{Z}\mathrm{X}^{N}\right) - m_{\mathrm{nuc}}\left({}^{A-1}_{Z-1}\mathrm{Y}^{N}\right) - m_{\mathrm{p}}\right]c^2 \tag{2.5} $$

where it is important to note that the masses are nuclear masses. From a practical standpoint, masses that are measured experimentally, at least for heavier elements, are typically atomic masses not nuclear masses, i.e., they include the masses of the electrons (and also the mass equivalent of the electronic binding energies). Thus, the nuclear mass, as defined in equation (2.2), can be written as

$$ m_{\mathrm{nuc}} = m - Zm_{\mathrm{e}} - b/c^2 \tag{2.6} $$

where m is the atomic mass, m_{e} is the electron mass, and b is the total electronic binding energy of the atom. Substituting equation (2.6) into (2.5) gives

$$ S_{\mathrm{p}} = \left[m\left({}^{A}_{Z}\mathrm{X}^{N}\right) - m\left({}^{A-1}_{Z-1}\mathrm{Y}^{N}\right) - m_{\mathrm{p}} - m_{\mathrm{e}} - \frac{b\left({}^{A}_{Z}\mathrm{X}^{N}\right) - b\left({}^{A-1}_{Z-1}\mathrm{X}^{N}\right)}{c^2}\right]c^2 \tag{2.7} $$

where the masses are now the masses of neutral atoms. Although equation (2.7) includes a term for the electronic binding energies, this term is generally ignored. The mass equivalent of the total electronic binding energy of a heavy atom can be fairly significant. However, equation (2.7) includes a term related to the difference in electronic binding energies of the parent and the daughter. This difference in electronic binding energies is of the order of the binding energy of the least bound electron and is negligible. Thus, we can write the proton separation energy as

$$ S_{\mathrm{p}} = [m({}^{A}_{Z}\mathrm{X}^{N}) - m({}^{A-1}_{Z-1}\mathrm{Y}^{N}) - m_{\mathrm{p}} - m_{\mathrm{e}}]c^2 \tag{2.8} $$

It is important to note that decay processes can occur spontaneously if they are energetically favorable, i.e., proton emission can, in principle, occur when the energy in equation (2.8) is positive. However, as is the case for decays involving charged particles, a consideration of Coulomb effects is also relevant. This is discussed further below for other decay processes, particularly alpha decay and for fission in the next chapter.

2.5.2 Neutron emission

Neutron emission occurs for nuclei where the neutron Fermi energy is much greater than the proton Fermi energy and these nuclei are on the upper side of the upper side of the region of known nuclides in figure 2.1. Neutron emission is defined as

$$ {}^{A}_{Z}\mathrm{X}^{N} \rightarrow {}^{A-1}_{Z}\mathrm{X}^{N-1} + \mathrm{n} $$

where the name of the daughter element is the same as the name of the parent element because the number of protons in the nuclei has not changed. The derivation of the neutron separation energy follows along the lines of the derivation of the proton separation energy as shown above. It is important, however, to be careful with electron masses because the number of atomic electrons does not change in the decay. The decay energy is, therefore, given as

$$ S_{\mathrm{n}} = \left[m\left({}^{A}_{Z}\mathrm{X}^{N}\right) - m\left({}^{A-1}_{Z}\mathrm{Y}^{N-1}\right) - m_{\mathrm{n}}\right]c^{2} \tag{2.9} $$

where, once again, the electronic binding energies have been ignored.

2.5.3 Alpha decay

An alpha particle (α-particle) is the nucleus of a ^{4}He atom and consists of a bound system of two neutrons and two protons. It has a charge of $+2e$ from the presence of the two protons and is a particularly stable configuration of nucleons, as indicated by its high binding energy (shown in the insert in figure 2.3). Alpha decay (α-decay) corresponds to the emission of an α-particle from a nucleus according to the process,

$$ {}^{A}_{Z}\mathrm{X}^{N} \rightarrow {}^{A-2}_{Z-2}\mathrm{Y}^{N-2} + {}^{4}_{2}\mathrm{He}^{2} \tag{2.10} $$

Sometimes written as

$$ {}^{A}_{Z}\mathrm{X}^{N} \rightarrow {}^{A-2}_{Z-2}\mathrm{Y}^{N-2} + \alpha. $$

As seen in figure 2.1, α-decay most commonly occurs for heavy nuclides with an excess of protons. The energy associated with the α-decay process, Q_α, follows from the discussion above,

$$ Q_{\alpha} = [m({}^{A}_{Z}\mathrm{X}^{N}) - m({}^{A-4}_{Z-2}\mathrm{Y}^{N-2}) - m({}^{4}_{2}\mathrm{He}^{2})]c^{2} $$

where, because all masses are atomic masses, the electron masses are all properly counted. This energy is sometimes written in terms of the α-particle mass as

$$ Q_{\alpha} = \left[m\left({}^{A}_{Z}\mathrm{X}^{N}\right) - m\left({}^{A-4}_{Z-2}\mathrm{Y}^{N-2}\right) - m_{\alpha} - 2m_{\mathrm{e}}\right]c^{2} \tag{2.11} $$

The significance of the high binding energy for the ^{4}He nucleus can be seen by writing equation (2.11) in terms of binding energies as defined by equation (2.2). This gives,

$$Q_\alpha = \left[B\left({}^{A-4}_{Z-2}\mathrm{Y}^{N-2}\right) + B\left({}^{4}_{2}\mathrm{He}^{2}\right) - B\left({}^{A}_{Z}\mathrm{X}^{N}\right)\right].$$

Thus, the large value of $B(^4\mathrm{He})$ optimizes the energy associated with the decay process. While a positive value of the energy in equation (2.11) is a requirement for the occurrence of α-decay, a proper consideration of this process requires a detailed look at the effect of Coulomb interactions.

Figure 2.5 shows the potential experienced by an α-particle. Inside the nucleus (of radius R), the α-particle, which exists with an energy E_α from equation (2.11), is bound to the other nucleons by the strong interaction. Outside the nucleus the strong interaction with the nuclear nucleons is negligible (because of the strong interaction's short range) and the α-particle experiences the repulsive long-range Coulomb interaction with the positively charged nuclear protons. In order for α-decay to occur, the α-particle must tunnel quantum mechanically through the Coulomb barrier to become a free particle (at radius r_α).

It is obvious that the tunneling probability (and hence the half-life of the decay process) will be a sensitive function of E_α and the relative height and width of the Coulomb barrier. The relationship between half-life and decay energy (the Geiger–Nuttall law) was first pointed out by Johannes Wilhelm 'Hans' Geiger (1882–1945) and John Mitchell Nuttall (1890–1958) (Geiger and Nuttall 1911, 1912). The quantum theory of barrier tunneling as it pertains to α-decay was first presented

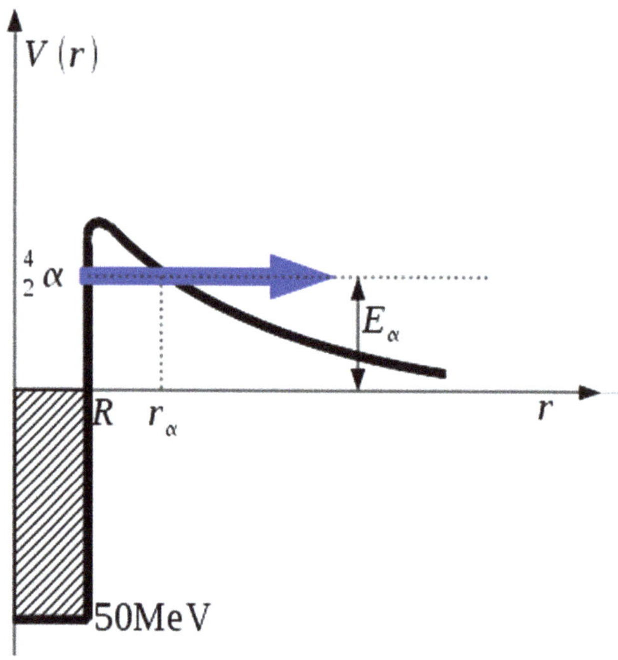

Figure 2.5. Radial dependence of the potential for an α-particle in a nucleus. This Alpha particle tunneling image has been obtained by the author from the Wikimedia website where it was made available by Persino (2010) under a CC BY 3.0 licence. It is included within this chapter on that basis. It is attributed to Persino.

by George Gamow (1904–68) (Gamow 1928). This theory shows that the half-life of the α-decay process may be expressed as

$$\tau = \tau_0 e^G$$

where τ_0 is a constant with a value of the order of 10^{-24} s and the factor G is a function of the nuclear charge, E_α, and the ratio R/r_α. The application of Gamow theory to the Geiger–Nuttall law is illustrated in figure 2.6. Each line represents different isotopes of the same element, and hence the same nuclear charge. Angular momentum (i.e., spin) also plays a role in determining α-decay half-lives, but this effect is not relevant to the graph shown in figure 2.6 because only data for even Z and even N nuclei are illustrated.

Since energy is shown on a linear scale and half-life is shown on a logarithmic scale in figure 2.6, the sensitivity of the decay process to energy is clear. In addition, the range of half-lives is significant. The shortest half-lives are in the range of milliseconds for nuclides with low atomic numbers and large decay energies. The longest half-lives are in the range of 10^{18} s for nuclides with the smallest decay energies. (By comparison the age of the Universe is about 4×10^{17} s.) The features in figure 2.6 illustrate the importance of the height and width of the Coulomb barrier that the α-particle needs to tunnel through in determining the half-life of the decay.

In the analysis of experiments, such as that shown in figure 2.6, it is important to realize that in equation (2.10) the energy associated with the decay process is shared between the daughter nucleus and the α-particle. A simple application of the laws of

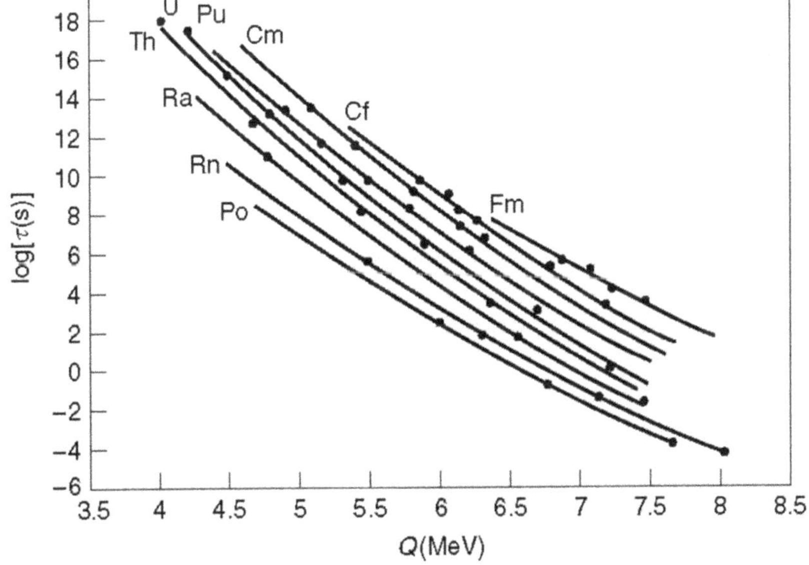

Figure 2.6. The Geiger–Nuttall relationship as described by Gamow theory for some even Z and even N nuclides. Reproduced from Dunlap (2023).

conservation of energy and momentum shows that the measured kinetic energy of the α-particle, T_α, in terms of the total decay energy, Q_α, is

$$T_\alpha = \frac{Q_\alpha}{1 + \frac{m_{\text{nuc}}\left({}^{4}_{2}\text{He}^{2}\right)}{m_{\text{nuc}}\left({}^{A-2}_{Z-2}\text{Y}^{N-2}\right)}}.$$

Typically, for the α-decay of a heavy nucleus the α-particle kinetic energy is about 98% of the decay energy.

2.5.4 Beta decay

As illustrated in figure 2.1, beta-decay (β-decay) processes are the most common forms of decay, with β^- decay occurring for nuclides with excess neutrons and β^+ decay occurring for nuclides with excess protons (compared to similar mass stable nuclei). The details of each of these processes are described below.

2.5.4.1 β^- Decay

The β^- decay process converts a neutron to a proton and emits an electron and antineutrino by the process

$$\text{n} \rightarrow \text{p} + \text{e}^- + \bar{\nu}$$

where $\bar{\nu}$ is an antineutrino (specifically an electron antineutrino). The details of this process can be understood on the basis of the quark structure of the proton and neutron, but such details are not needed for the present discussion. In terms of parent and daughter nuclei, the β^- decay process is written as

$${}^{A}_{Z}\text{X}^{N} \rightarrow {}_{Z+1}^{A}\text{Y}^{N-1} + \text{e}^- + \bar{\nu}.$$

The energy released in the β^- decay can be written (in terms of atomic masses) as

$$Q_{\beta^-} = \left[m\left({}^{A}_{Z}\text{X}^{N}\right) - m\left({}_{Z+1}^{A}\text{Y}^{N-1}\right)\right]c^2 \tag{2.12}$$

where the mass of the antineutrino is assumed to be zero (or at least negligible) and the differences in the electronic binding energies are assumed to be small. In the derivation of equation (2.12) it is important to carefully account for all electron masses.

In the β^- decay process, the decay energy is shared between the electron and the antineutrino (the nuclear recoil energy is negligible). Since it is fairly straightforward to measure the energy spectrum of the emitted electrons (see figure 2.7), it was known, even in the early days of experimental nuclear physics, that the electron did not always acquire all of the decay energy but showed a distribution with a maximum energy, $E_{\max}$, equal to the decay energy in equation (2.12). In fact, it was on the basis of experimental results such as those shown in the figure that, in 1930, Wolfgang Pauli (1900–58) hypothesized the existence of the neutrino. In 1933, Enrico Fermi (1901–54) proposed the theory of beta decay that explained the shape of the observed electron energy spectrum.

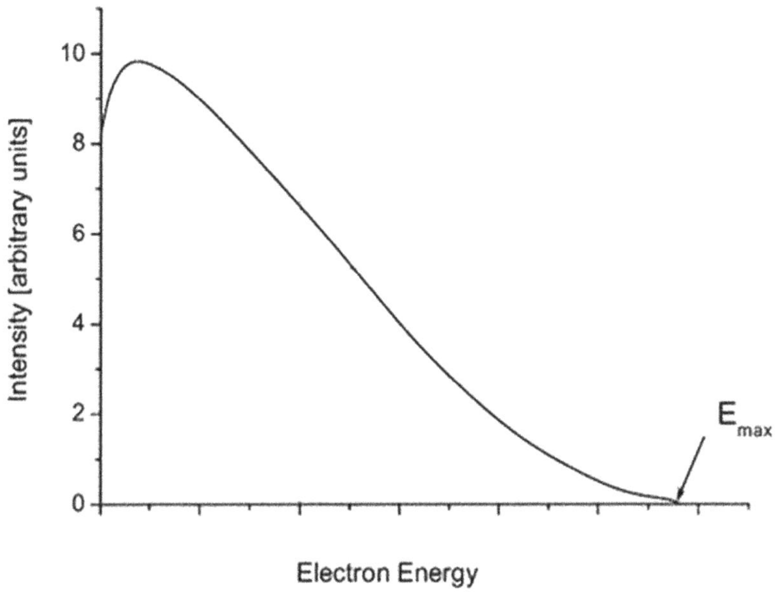

Figure 2.7. Typical electron energy spectrum from β^- decay. Reproduced from HPaul (2020). CC BY-SA 4.0.

2.5.4.2 β^+ Decay and electron capture

Nuclei with an excess number of protons (compared to similar mass stable nuclei) decay by β^+ decay, and thereby convert a proton to a neutron, a positron (antielectron), and a neutrino (electron neutrino),

$$\mathrm{p} \rightarrow \mathrm{n} + \mathrm{e}^+ + \nu \tag{2.13}$$

In terms of parent and daughter nuclei, this is written as

$$^A_Z\mathrm{X}^N \rightarrow {}_{Z-1}^{\ A}\mathrm{Y}^{N+1} + \mathrm{e}^+ + \nu.$$

The energy released in the β^+ decay can be written (in terms of atomic masses) as

$$Q_{\beta^+} = \left[m\left(^A_Z\mathrm{X}^N\right) - m\left({}_{Z-1}^{\ A}\mathrm{Y}^{N+1}\right) - 2m_\mathrm{e}\right]c^2 \tag{2.14}$$

Again, since this is expressed in terms of atomic masses, it is important to carefully account for all electron masses.

Electron capture is a process that is analogous to β^+ decay. In this case, a proton in the nucleus interacts with an atomic electron that has a wavefunction that overlaps the nucleus to produce a neutron. Such electrons are from the inner shell (e.g. K-shell). The process is,

$$\mathrm{e}^- + \mathrm{p} \rightarrow \mathrm{n} + \nu$$

or in terms of the nuclei involved

$$\mathrm{e}^- + {}^A_Z\mathrm{X}^N \rightarrow {}_{Z-1}^{\ A}\mathrm{Y}^{N+1} + \nu \tag{2.15}$$

The decay energy can be calculated to be (again in terms of atomic masses)

$$Q_{ec} = \left[m\left({}^{A}_{Z}X^{N}\right) - m\left({}_{Z-1}{}^{A}Y^{N+1}\right)\right]c^2 - b_e \tag{2.16}$$

where b_e is the binding energy of the electron that is captured. Unlike our previous assumption that electron binding energies could be ignored, this is not necessarily the case for electron capture. In the case of electron capture, the electronic bonding energy is the binding energy of the captured inner shell electron. This can be in the range of many 10 s of keV and can be important for the overall energy calculation. A comparison of equation (2.16) with (2.14) shows that the energy for electron capture is greater than that for β^+ decay by a factor of $2m_ec^2 - b_e$. Since $2m_ec^2 = 1.022$ MeV, electron capture is always more energetically favorable. The factor of $2m_ec^2$ can be understood by adding an electron to both sides of equation (2.13). This will give equation (2.15) with an additional electron–positron pair on the right-hand side which provides the mass equivalent of the additional $2m_ec^2$ of energy.

An important characteristic of β^+ decay/electron capture that is seen from the discussion above is, that in the case of decays which have a decay energy of less than about 1 MeV, β^+ decay is not energetically favorable and electron capture is the only possible decay route. This is the case, for example, for the beta decay of ^{37}Ar to ^{37}Cl, where the decay energy is 0.81 MeV. For larger decay energies, both β^+ decay and electron capture can be energetically favorable and can occur. Decays with significantly large energies are typically only β^+ decay. For example, the beta decay of ^{37}Kr to ^{37}Ar, with an energy from equation (2.14) of 5.14 MeV, is known to decay only by β^+ decay.

2.6 Excited nuclear states

Nuclei can exist in excited states, as illustrated in figure 2.4. Excited states are formed by moving nucleons to higher energy levels and/or by breaking nucleon pairs. In some nuclei, excited states can also correspond to degrees of freedom associated with rotational or vibrational modes. Dunlap (2023) has provided a brief introduction to the formation and properties of nuclear excited states. For the present discussion, we consider common methods by which excited states can be populated and the ways in which they decay.

2.6.1 Population of excited states

Two common reasons for the population of excited nuclear states are nuclear decays and nuclear reactions. In the present section we deal primarily with the population of excited states during decay processes. Nuclear reactions are considered in detail in the next section.

Figure 2.8 shows two decay processes that result in the population of excited states of the daughter nucleus. Figure 2.8(a) shows the β^- decay of ^{137}Cs and figure 2.8(b) shows the α-decay of ^{235}U. Both of these decays have particular relevance to nuclear fission reactors, as will be discussed in the next chapter.

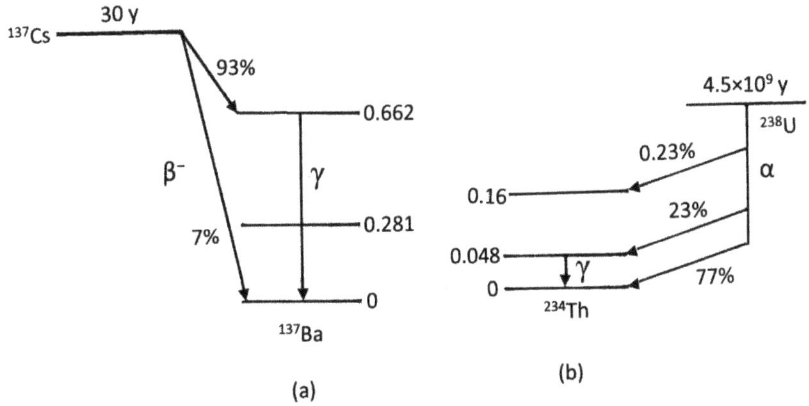

Figure 2.8. (a) β^- Decay of ^{137}Cs as described by equation (2.17), and (b) α-decay of ^{238}U as described by equation (2.18). Excited state energies are in MeV.

The β^- decay of ^{137}Cs is described by the process,

$$^{137}\text{Cs} \rightarrow {}^{137}\text{Ba} + e^- + \bar{\nu} \tag{2.17}$$

As seen in the figure, the β^- decay of ^{137}Cs can populate the 662 keV state or the ground state of ^{137}Ba. The relative fraction of these two decays, known as the branching ratios or branching fractions, are 93.5% to the 662 keV state and 6.5% to the ground state. Meanwhile, in general, decay energies are an important factor in determining the probability of a specific decay process, with greater energies giving rise to greater transition probabilities, angular momentum conservation considerations are also very important and are a determining factor in the branching ratios for the β^- decay of ^{137}Cs.

The α-decay of ^{238}U is described by the process,

$$^{238}\text{U} \rightarrow {}^{234}\text{Th} + \alpha \tag{2.18}$$

As seen in figure 2.8(b), this decay populates a variety of different excited states of the daughter ^{234}Th nucleus.

Another feature that is illustrated in figure 2.8 deals with the spacing of the energy levels in the two diagrams. It is common that the excited state energy level spacing is greater for light (i.e., smaller) nuclei than for heavy nuclei. This feature is roughly analogous to the spacing of the quantized energy level of a particle in a box, which decreases with increasing box size. The small spacing of nuclear energy levels for heavy nuclei plays an important role, as described in the next chapter, in the description of fission reactions.

2.6.2 Decay of excited states

The most common method of decay for excited nuclear states is through a so-called isomeric transition. This transition occurs between the excited state and a lower energy (either another excited state or the ground state) of the same nuclide.

This decay process emits a gamma ray (high-energy photon) and is referred to as gamma-decay (γ-decay). Figure 2.8 shows these decays for the excited states that have been populated as a result of α-decay or β-decay. The gamma-decay process, as shown in figure 2.8(a), for example, can be written as

$$^{137}\text{Cs}^* \rightarrow {}^{137}\text{Cs} + \gamma$$

where the asterisk following the name of the nuclide on the left-hand side of the equation is used to designate that the nucleus is in an excited state. The emission of gamma radiation during α-decay and β-decay processes is significant for the shielding requirements of radioactive material, as will be considered in further detail in chapter 6.

It is also possible for excited nuclear states to decay by one of the modes discussed above. Figures 2.9(a) and (b) show examples of two possible processes, neutron emission and alpha decay, respectively.

In the case of neutron emission, the process shown in the figure is,

$$^{9}\text{Be}^* \rightarrow {}^{8}\text{Be} + \text{n} \tag{2.19}$$

and this has an energy, as calculated from equation (2.9), of −1.67 MeV. This means that the decay shown in equation (2.19) cannot occur for the ground state of ^{9}Be. However, if a ^{9}Be nucleus is in an excited state with an energy greater than 1.67 MeV, then there is sufficient energy for the process in equation (2.19) to occur. As figure 2.9(a) shows, the 2.43 MeV state of ^{9}Be decays by neutron emission to the ground state of ^{8}Be.

Alpha decay from excited nuclear states follows along the same lines as neutron emission and the process shown in figure 2.9(b) is,

$$^{12}\text{C} \rightarrow {}^{8}\text{Be} + \alpha.$$

From equation (2.19) the energy associated with this process is −7.37 MeV. Thus, alpha decay of ^{12}C is only permitted from an energy standpoint for excited states above 7.37 MeV. As illustrated in figure 2.9(b) this decay is permitted and is observed from the 7.66 MeV excited state of ^{12}C.

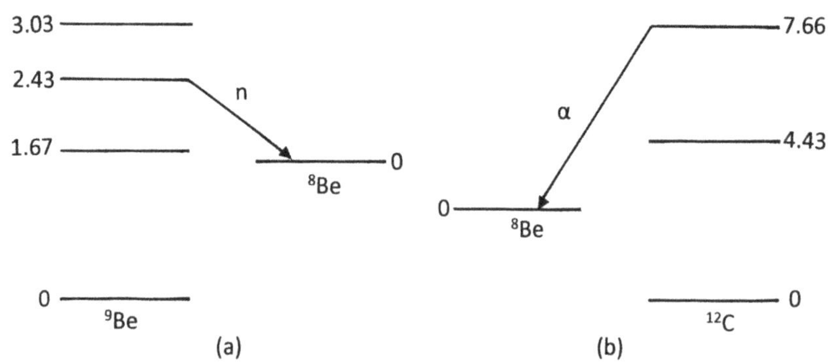

Figure 2.9. Simplified energy level diagrams showing (a) neutron emission from an excited state of ^{9}Be and (b) α-decay of an excited state of ^{12}C. Excited state energies are in MeV.

2.7 Nuclear reactions

Nuclear reactions differ from typical decay processes because they involve the interaction of two quantities. A reaction may be expressed as

$$a + A \rightarrow B + b \tag{2.20}$$

where typically a particle, a, is incident on a nucleus, A, yielding a particle, b, and a nucleus, B. The process in equation (2.20) is often written in the shorthand notation A(a,b)B. Here we consider some of the most common types of reactions.

2.7.1 Elastic scattering

The simplest nuclear processes that can be described by equation (2.20) are not true reactions but are scattering processes where a and b are the same particle. Scattering processes can be either elastic or inelastic. In an elastic scattering process, defined as A(a,a)A, kinetic energy and momentum are conserved. A simple nonrelativistic one-dimensional example considers a particle, a, for example a proton or neutron, with mass m_a and an initial velocity v_a (in the laboratory frame) incident on a nucleus, A, with mass m_A at rest ($v_A = 0$, in the laboratory frame). After the scattering event the particle has velocity v_a' and the nucleus has velocity v_A'. Conservation of momentum requires,

$$m_a v_a = m_a v_a' + m_A v_A'$$

and conservation of kinetic energy requires,

$$\frac{1}{2} m_a v_a^2 = \frac{1}{2} m_a v_a'^2 + \frac{1}{2} m_A v_A'^2.$$

These two equations can be solved for the final velocities of the two masses as

$$v_a' = \frac{m_a - m_A}{m_a + m_A} v_a$$

and

$$v_A' = \frac{2m_a}{m_a + m_A} v_a.$$

These equations can be solved to give the kinetic energy of the mass, A, after the scattering as

$$E_A' = \frac{4 m_A m_a}{(m_A + m_a)^2} E_a \tag{2.21}$$

and the kinetic energy of the scattered particle as

$$E_a' = \left[\frac{m_A - m_a}{m_A + m_a}\right]^2 E_a.$$

Equation (2.21) gives the kinetic energy gained by the nucleus and also the kinetic energy lost by the incident particle due to recoil of the nucleus. The relative loss of energy of the incident particle during the elastic scattering event is,

$$\frac{E'_{\mathrm{A}}}{E_{\mathrm{a}}} = \frac{4m_{\mathrm{A}}m_{\mathrm{a}}}{(m_{\mathrm{A}} + m_{\mathrm{a}})^2}.$$

As an example, we can consider a neutron ($m_{\mathrm{a}} \approx 1$ u) incident on a ^{238}U nucleus ($m_{\mathrm{A}} \approx 238$ u). In this case, the neutron experiences an energy loss of about 1.7%.

2.7.2 Inelastic scattering

Inelastic scattering corresponds to the process A(a,a)A*, i.e., some of the energy of the incident particle is given to the nucleus and leaves the nucleus in an excited state (designated as A*). A detailed analysis of the energy available to the nucleus, ΔE, has been given by Dunlap (2023). This energy is given in terms of the scattering angle, θ, as

$$\Delta E = E_{\mathrm{a}}\left[1 - \frac{m_{\mathrm{a}}}{m_{\mathrm{A}^*}}\right] - E'_{\mathrm{a}}\left[1 + \frac{m_{\mathrm{a}}}{m_{\mathrm{A}^*}}\right] + 2\frac{m_{\mathrm{a}}}{m_{\mathrm{A}^*}}[E_{\mathrm{a}}E'_{\mathrm{a}}]^{1/2}\cos\theta$$

where m_{A^*} is the mass of the nucleus in the excited state.

2.7.3 Nuclear reactions

Nuclear reactions are typically defined by equation (2.20) when a $\neq$ b. Some examples of low-energy nuclear reactions are as follows:

(1) A nucleon is incident on a nucleus. This nucleon is absorbed by the nucleus and a different nucleon is emitted. Examples of such a reaction are the (n, p) reaction and the (p, n) reaction.

(2) A compound particle is incident on a nucleus and one component of the compound particle is absorbed. An example is the (d, p) reaction where a deuteron (bound neutron–proton pair) is incident on a nucleus, the neutron is absorbed, and the proton is emitted. This type of reaction is referred to as a *stripping reaction.*

(3) A particle that is incident on a nucleus gains one or more nucleons from the nucleus. This is referred to as a *pickup reaction*, and a typical example is the (d, α) reaction.

(4) An incident particle is absorbed by a nucleus and the excess energy is used to elevate the nucleus to an excited state. The most common example of this type of reaction is the (n, γ) reaction where an incident neutron is absorbed, and the excited state thus formed decays by gamma emission. This process is often referred to as radiative capture.

The (n, γ) reaction is of particular relevance to the operation of a fission reactor. This reaction may be written as

$$n + {}^{A}X \rightarrow {}^{A+1}X^{*} \rightarrow {}^{A+1}X + \gamma \tag{2.22}$$

and is described graphically in figure 2.10. Here, E_1 and E_2 are excited states of the nucleus ${}^{A+1}X$. If the state E_2 is populated by the (n, γ) reaction and subsequently decays by gamma emission to a state below the neutron separation energy, S_n, as given by the equation (2.9), then the neutron cannot escape and has been captured.

It is possible that the ${}^{A+1}X^{*}$ nucleus that is formed by neutron absorption, as illustrated in figure 2.10, may undergo processes other than γ-emission. Neutron emission is one possibility, in which case the nucleus decays back to the original nucleus by the process,

$$n + {}^{A}X \rightarrow {}^{A+1}X^{*} \rightarrow {}^{A}X + n.$$

This is referred to as a decay through the incident channel. Another possibility is induced fission, where the ${}^{A+1}X^{*}$ nucleus formed by neutron absorption breaks up into two smaller nuclei (along with some left-over neutrons). This process is described by,

$$n + {}^{A}X \rightarrow {}^{A+1}X^{*} \rightarrow \text{fission products.}$$

In assessing the possibility of this process, it is important to consider the nucleon pairing energy, as described above. In general, if the original ${}^{A}X$ nucleus is an even–even nucleus, then it is particularly stable and the ${}^{A+1}X^{*}$ nucleus that is formed by neutron absorption is in a relatively low excited state. On the other hand, if the original ${}^{A}X$ nucleus is an odd–even or even–odd nucleus, then the ${}^{A+1}X^{*}$ nucleus that is formed by neutron absorption is left in a higher excited state. In this case, there is a greater probability of induced fission. This process is discussed in detail in the next chapter. The greater stability of even–even nuclei, compared to odd–even or

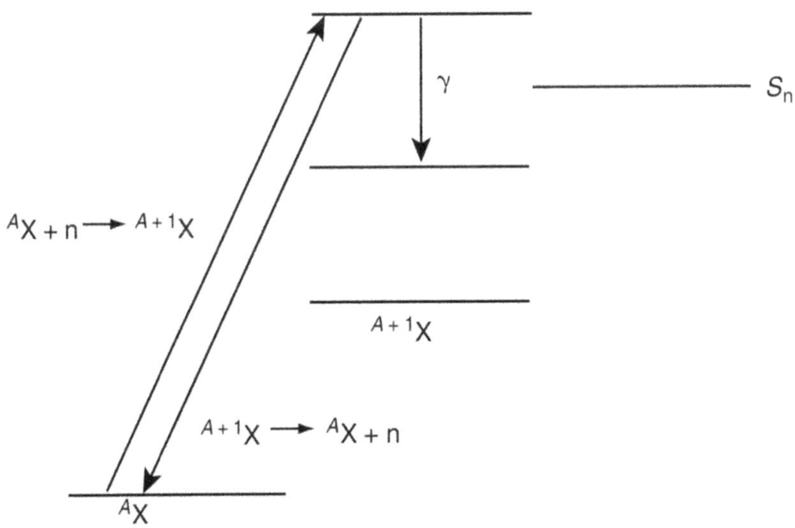

Figure 2.10. The (n, γ) reaction described by equation (2.22). Reproduced from Dunlap (2023).

even–odd nuclei, is also evidenced by a comparison of the half-lives of the two naturally occurring uranium isotopes; 7.038×10^8 years and 4.468×10^9 years for ^{238}U ($N = 146$, $Z = 92$) and ^{235}U ($N = 143$, $Z = 92$), respectively. Induced fission is discussed in further detail in the next chapter.

References

Cottingham W N and Greenwood D A 2001 *An Introduction to Nuclear Physics* 2nd edn (Cambridge: Cambridge University Press)

Dunlap R A 2013 A macroscopic analogue of the nuclear pairing potential *Eur. J. Phys. Educ.* **4** 25–30 (CC BY 3.0)

Dunlap R A 2021 *Energy from Nuclear Fusion* (Bristol: IOP Publishing)

Dunlap R A 2023 *An Introduction to the Physics of Nuclei and Particles*— 2nd edn (Bristol: IOP Publishing)

Gamow G 1928 Zur quantentheorie des atomkernes *Z. Phys.* **51** 204–12

Geiger H and Nuttall J M 1911 The ranges of the α particles from various radioactive substances and a relation between range and period of transformation *Phil. Mag. Ser. 6* **22** 613–21

Geiger H and Nuttall J M 1912 The ranges of α particles from uranium *Phil. Mag. Ser. 6* **23** 439–45

HPaul 2020 A beta spectrum https://commons.wikimedia.org/wiki/File:Beta_spectrum.png

Kahn R 2018 Average binding energy per nucleon as a function of the number of nucleons in the nucleus https://commons.wikimedia.org/wiki/File:Nuclear_binding_energy_RK01.png

Krane K S 1988 *Introductory Nuclear Physics* (New York: Wiley)

Napy1kenobi 2009 Graph of isotopes by type of nuclear decay https://commons.wikimedia.org/wiki/File:Table_isotopes_en.svg

Persino 2010 Alpha particle tunnelling https://commons.wikimedia.org/wiki/File:Przej%C5%9Bcie_tunelowe_cz%C4%85stki_alpha.png

Williams W S C 1991 *Nuclear and Particle Physics* (Oxford: Oxford University Press)

Chapter 3

Principles of nuclear fission reactors

In chapter 3, the basic operation of a nuclear fission reactor is considered in the context of the nuclear physics described in chapter 2. This begins with a detailed description of the nuclear fission process, with particular emphasis on the fission of the naturally occurring isotopes of uranium. The basic design of a thermal neutron reactor is described, including an analysis of the material requirements for its basic components, reactor fuel, neutron moderator, control rods, and coolant. The requirements for the production of a controlled nuclear fission reaction are considered along with the utilization of delayed neutrons for reactor control. Finally, the basic operational principles of fast neutron reactors, including breeder reactors, are discussed.

3.1 Introduction

Chapter 2 overviewed the properties of nuclei and has provided the basis for understanding the production of useful energy from nuclear fission. The present chapter provides an overview of the operation of traditional nuclear power reactors. Further technical details of these reactors can be found in Bodansky (2004), and Murray and Holbert (2015).

3.2 Fission processes

Fission is the process by which a heavy nucleus breaks up into two lighter nuclei. Although alpha decay is, in a sense, a fission process where one of the fission products is a ^{4}He nucleus, the term fission is generally used to describe processes where the two product nuclei are both sizeable fractions of the original nuclear mass. A generic fission process may be described as

$$^{A}\mathrm{X} \rightarrow {}^{A'}\mathrm{Y} + {}^{A-A'}\mathrm{Z}.$$

The energy associated with such a process is given by

doi:10.1088/978-0-7503-6069-2ch3

$$E = \left[m({}^{A}\mathrm{X}) - m({}^{A'}\mathrm{Y}) - m({}^{A-A'}\mathrm{Z})\right]c^2$$

where the masses can be either atomic masses or nuclear masses because the number of electrons is conserved. From equation (2.2), this expression may be written in terms of binding energies as

$$E = B({}^{A'}\mathrm{Y}) + B({}^{A-A'}\mathrm{Z}) - B({}^{A}\mathrm{X}) \tag{3.1}$$

In the simple case of symmetric fission, where $A' = A/2$, we can write equation (3.1) as

$$E = A\left[\frac{B({}^{A/2}\mathrm{Y})}{A/2} - \frac{B({}^{A}\mathrm{Y})}{A}\right]$$

where the term in the square brackets is the difference in the binding energy per nucleon for the product nuclides minus the binding energy per nucleon for the original nuclide. An inspection of figure 2.3 clearly shows that fission will be energetically favorable for all processes where the fission fragments have A greater than about 55.

It is important to consider why nuclei with A greater than about 110 can be stable and not spontaneously undergo fission (which would be energetically favorable). The reason is the same as that for alpha decay. The decay products (i.e., fission fragments) must tunnel through the Coulomb barrier in order to be free from the strong nuclear potential well. Figure 3.1 shows the Coulomb barrier height for fission as a function of the total number of nucleons as calculated on the basis of the so-called liquid drop model. The broken line shows a more detailed calculation that takes into account nuclear shell structure. These models are discussed in detail in Krane (1988). Clearly, the Coulomb barrier is very large for nuclei with $A \approx 110$; so even though fission would be energetically favorable, it cannot occur. The Coulomb barrier only becomes compatible with fission for nuclei with A in excess of 220 or so. In fact, for $A > 300$ the Coulomb barrier becomes small, indicating that this represents the size limit to fission-stable nuclei. For nuclides that are of interest as fission reactor fuels, it is the behavior around $A \approx 230$–240 that is of relevance, and this will be considered in more detail below.

The above description of fission has taken a rather simple approach. There are three additional important points that we need to consider when dealing with actual fission processes.

First, actual fission processes are not symmetric, and the distribution of product nuclei sizes is characteristic of the original nucleus. This is referred to as the fission yield and is discussed in further detail for specific fission processes below.

Second, as illustrated in the Segrè plot in figure 2.1, heavy stable nuclei have a greater N/Z ratio than light stable nuclei. This means that if a heavy nucleus undergoes fission, then there will be too many neutrons to be accommodated in the fission fragments. This means that an actual fission process (in this case spontaneous fission) is described by the relation,

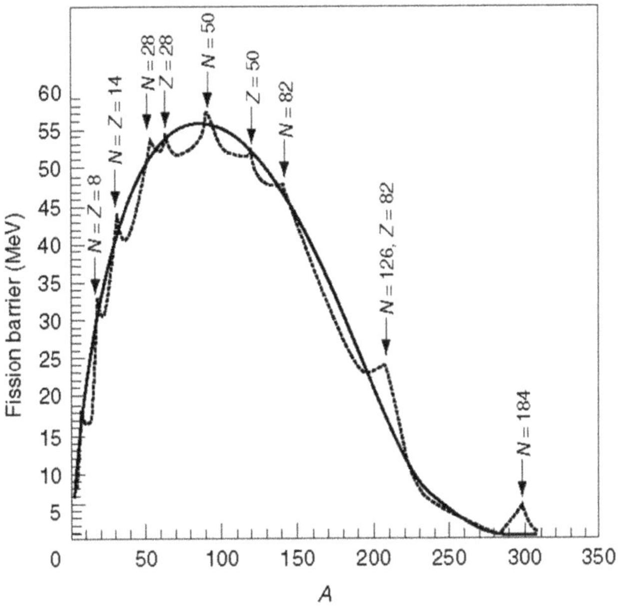

Figure 3.1. Coulomb barrier for spontaneous fission of heavy nuclei as a function of the total number of nucleons. Reprinted from Myers and Swiatecki (1966), Copyright (1966), with permission from Elsevier.

$$^{A}X \rightarrow {}^{A'}Y + {}^{A-A'-\delta}Z + \delta n \quad (3.2)$$

where δ excess neutrons are released during the fission. Even though these excess neutrons are released, the fission products on the right-hand side of equation (3.2) are still on the neutron-rich side of the stability line in figure 2.1 and will undergo decay processes (typically β^-) to reach stability.

The emission of excess neutrons brings up the third point that needs to be considered here. The excess neutrons that are released can interact with other nuclei and provide energy that can help to overcome the fission Coulomb barrier. This process is referred to as induced fission (rather than spontaneous fission, as in equation (3.2)) and corresponds to the process,

$$n + {}^{A}X \rightarrow {}^{A+1}X^* \rightarrow {}^{A'}Y + {}^{A+1-A'-k}Z + kn$$

where $k = \delta + 1$. This equation shows that the nucleus ^{A}X absorbs a neutron yielding an excited state of the nucleus $A+1$, which then undergoes fission. The details of the energetics of this process for specific situations are considered in detail in the next section.

3.3 The fission of uranium

Naturally occurring uranium consists of two isotopes, ^{235}U and ^{238}U, with natural abundances of 0.72% and 99.274%, respectively. There is also a small amount of

^{234}U (0.0055%) and trace amounts of several other isotopes, but these are not relevant to the operation of a fission reactor. As we will see below, it is only because of the existence of the small fraction of ^{235}U that nuclear fission reactors are possible.

Both ^{235}U and ^{238}U are β-stable and their primary mode of decay is through the α-decay processes,

$$^{235}\mathrm{U} \rightarrow {}^{231}\mathrm{Th} + \alpha$$

and

$$^{238}\mathrm{U} \rightarrow {}^{234}\mathrm{Th} + \alpha.$$

Both ^{235}U and ^{238}U will also undergo spontaneous fission. Although with half-lives in the range of 10^{16}–10^{17} years, this process does not contribute directly to any significant energy production. However, the excess neutrons that are produced by spontaneous fission can cause induced fission events. As long as δ in equation (3.2) is greater than zero, there is a possibility of a chain reaction that can provide useful energy output.

It is important to look at the induced fission processes for the two isotopes of uranium in some detail. We begin with the neutron capture process for these two uranium isotopes,

$$\mathrm{n} + {}^{235}\mathrm{U} \rightarrow {}^{236}\mathrm{U} + 6.54\ \mathrm{MeV} \tag{3.3}$$

and

$$\mathrm{n} + {}^{238}\mathrm{U} \rightarrow {}^{239}\mathrm{U} + 4.81\ \mathrm{MeV} \tag{3.4}$$

where the excess energy associated with the product nucleus is for neutrons that have low kinetic energy (generally referred to as thermal neutrons). The importance of neutron energy is discussed further below. The fairly small difference in excess energy for the two reactions in equations (3.3) and (3.4) is highly significant. If we look closely at the graph of the fission Coulomb barrier in figure 3.1 for the region around $A = 235$ to 238, we see that the Coulomb barrier is around 6.2 MeV. This means that the excess energy that is produced when a thermal neutron is absorbed by a ^{235}U nucleus is greater than the Coulomb barrier, while the excess energy in the case of ^{238}U is less than the Coulomb barrier. This means that thermal neutrons can induce fission in a ^{235}U nucleus but not in a ^{238}U nucleus.

The induced fission process for ^{235}U may be written as

$$\mathrm{n} + {}^{235}\mathrm{U} \rightarrow {}^{236}\mathrm{U}^* \rightarrow {}^{A'}\mathrm{Y} + {}^{236-A'-k}\mathrm{Z} + k\mathrm{n} \tag{3.5}$$

and is illustrated graphically in figure 3.2. In equation (3.5), k is on average typically around 2.5. This means that there is, on average, a gain of about 1.5 neutrons for every induced fission event and these are given off on a timescale that is characteristic of the fission process, i.e., about 10^{-14} s. These excess neutrons can go on to induce further fissions in ^{235}U and can be responsible for a continuing chain reaction.

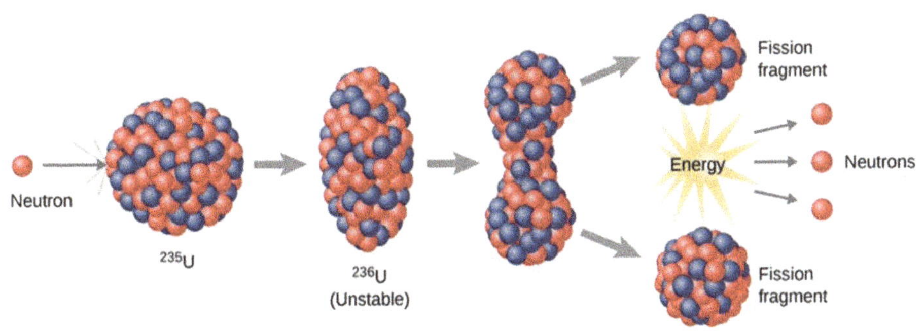

Figure 3.2. Schematic representation of neutron induced fission of ^{235}U. Reproduced from Ling *et al* (2016). CC BY 4.0.

For low-energy neutrons, induced fission is highly unlikely in ^{238}U. However, if the incident neutron provides more than about 1.4 MeV of energy (i.e. 6.2–4.81 MeV), then induced fission can occur by the process

$$n + {}^{238}U \rightarrow {}^{239}U^* \rightarrow {}^{A'}Y + {}^{239-A'-k}Z + kn \tag{3.6}$$

Nuclei that can undergo induced fission when interacting with a thermal neutron are referred to as fissile. Nuclei that require additional energy from the kinetic energy of an incident neutron to undergo induced fission are referred to as non-fissile.

An interesting feature in the properties of uranium, and which is a general feature of most heavy nuclei, is that fissile ^{235}U has an even number of protons ($Z = 92$) and an odd number of neutrons ($N = 143$), while non-fissile ^{238}U has an even number of both protons and neutrons ($Z = 92$, $N = 146$). The reason for this feature can be readily explained in terms of the discussion in chapter 2 on the pairing interaction (see figure 2.4 and accompanying discussion). Thus, the pairing of nucleons makes fission less energetically favorable, and this explains the difference between fissile nuclei with an unpaired nucleon (as for ^{235}U) and non-fissile nuclei with only paired nucleons (as for ^{238}U).

Two important features of induced fission, such as in equation (3.5), require careful consideration in an analysis of nuclear power: first, that all fission processes do not yield the same fission products; and second, that the fission products are not of equal size. Figure 3.3 shows the so-called fission yield, i.e., the distribution of fission fragment mass for induced ^{235}U fission and two other potentially useful fissile nuclides. This figure shows that for induced ^{235}U fission, one fission fragment has a mass number around 95 and the other fission fragment has a mass number around 140. A typical induced ^{235}U fission might, therefore, be

$$n + {}^{235}U \rightarrow {}^{97}Y + {}^{137}I + 2n \tag{3.7}$$

where the number of neutrons and protons is conserved.

From figure 2.3, we can make an estimate of the actual energy release in a fission process as given in equation (3.7). ^{235}U has a binding energy per nucleon of about 7.6 MeV. Nuclei with $A \approx 97$ have a binding energy per nucleon of about 8.8 MeV

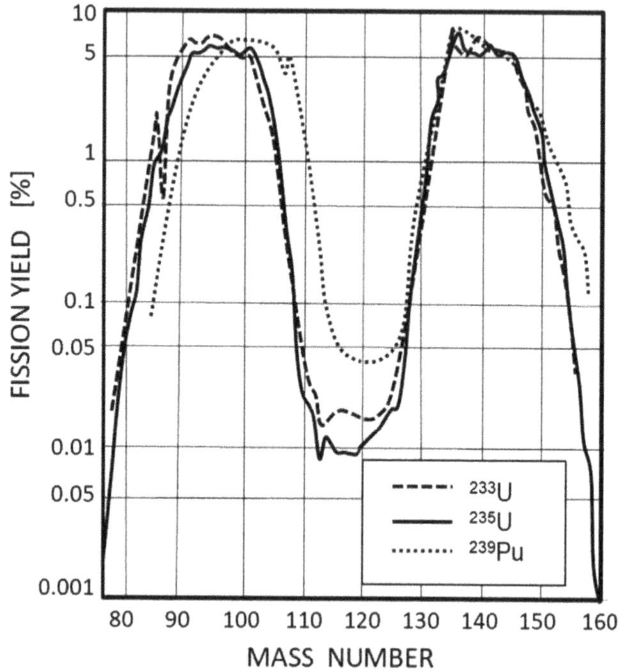

Figure 3.3. Fission yield for ^{233}U, ^{235}U and ^{239}Pu. Reproduced from Woźnicka (2019). CC BY 4.0.

and nuclei with $A \approx 137$ have a binding energy per nucleon of about 8.3 MeV. So, the total energy release from equation (3.7) is

$$(97 \times 8.8 \text{ MeV}) + (137 \times 8.3 \text{ MeV}) - (235 \times 7.6 \text{ MeV}) = 205 \text{ MeV}.$$

The actual measured average energy per induced fission in ^{235}U is 202.79 MeV (Kopeikin *et al* 2004). This energy is manifested in several forms. Most of the fission energy is released as kinetic energy of the fission fragments. Some kinetic energy is also associated with the released neutrons. The fission fragments are typically left in excited states, which decay very quickly by γ-emission. Although typically one or two additional neutrons are released during the fission, the fission fragments are still neutron rich and will decay over some period of time by β^- decay. These decays (referred to as delayed) emit electrons, additional γ-rays, and antineutrinos. The significance of these decays and other processes in the fission fragments are discussed further in section 3.7 and in chapter 6. The approximate breakdown of the energy release from nuclear fission is given in table 3.1. While most of the prompt energy can be recovered in the form of thermal energy that can be used to generate electricity, most of the delayed energy is lost.

3.4 Neutron interactions in uranium

In order to better understand how nuclear fission can be used as a source of energy, it is necessary to look in more detail at the interaction between neutrons and uranium nuclei, particularly as a function of neutron energy. The first thing that is important

Table 3.1. Typical distribution of energy from the fission of ^{235}U. Adapted from Dunlap (2023).

Energy source	Energy (MeV)	Percent total
Fission fragments	167	81.5
Prompt neutrons	5	2.4
Prompt γ-rays	6	2.9
Delayed electrons	8	3.9
Delayed γ-rays	7	3.4
Delayed antineutrinos	12	5.9
Total	205	100

to consider is the energy of the neutrons that are emitted in fusion processes such as equations (3.5) and (3.6). On average, the neutrons that are produced in such fusion processes have kinetic energies around 2 MeV. This means that they have, at least in principle, sufficient kinetic energy to induce fission in either ^{235}U or ^{238}U. If, on the average, one (or about 40%) of these emitted neutrons goes on to induce another fission process, then a sustained fission chain reaction can occur. In order to determine if this situation will occur, we have to look at all of the possible processes that a neutron can be involved in. As an example, we consider a neutron incident on a sample of ^{235}U. All of the possible interactions have been discussed above or in chapter 2, so we can summarize them here.

Elastic scattering: This is the ^{235}U(n, n)^{235}U process and serves only to reduce the kinetic energy of the neutron. As noted in the previous chapter, this reduction is of the order of about 2% of the neutron's kinetic energy before the scattering. This process can occur for neutrons of any energy.

Inelastic scattering: This is the ^{235}U(n, n)^{235}U* process and results is a more significant decrease in the neutron kinetic energy and also leaves the ^{235}U nucleus in an excited state. For this process to occur, the neutron kinetic energy (as measured in the center of mass frame) must be at least equal to the energy of the first excited nuclear state. For ^{235}U, this threshold energy is 14 keV. (For ^{238}U, the threshold energy for inelastic scattering is 44 keV.)

Radiative capture: This is the ^{235}U(n, γ)^{236}U* reaction, which results in the neutron being absorbed by the ^{235}U nucleus, thereby creating ^{236}U in an excited state. If the exited ^{236}U nucleus decays by gamma emission to a state with energy less than the ^{236}U neutron separation energy,

$$E_\mathrm{n} = [m(^{235}\mathrm{U}) + m_\mathrm{n} - m(^{236}\mathrm{U})]c^2,$$

then the neutron cannot escape and has been captured. From equation (3.3), the separation energy is 6.54 MeV. (For ^{238}U the neutron separation energy from equation (3.4) is 4.81 MeV.) The threshold neutron energy for radiative capture is the energy difference between the neutron separation energy and the energy of the next available energy level of the daughter nucleus. Since the nuclear energy levels in heavy nuclei such as ^{235}U and ^{238}U are closely spaced, particularly at energies of a few MeV above the ground state, the threshold neutron energy for radiative capture is small—of the order of 1–10 eV for uranium isotopes.

Induced fission: Induced fission is defined in equation (3.5) for ^{235}U (and in equation (3.6) for ^{238}U). Since ^{235}U is fissile, this process can occur for neutrons of any energy and, as noted above, a threshold neutron energy of about 1.4 MeV is needed to induce fission in ^{238}U.

On the basis of the above neutron reactions, we can see that elastic and inelastic scattering will merely reduce the energy of the neutron, but the neutron will still be available to undergo other reactions, such as fission, as long as it remains in the sample and above the threshold energy. Radiative capture, on the other hand, captures the neutron and prevents it from undergoing further reactions. So, ultimately a neutron that is incident on a sample of uranium either gets captured, induces fission, or passes through the sample before one of these two processes occurs. The relative cross sections for the processes as a function of neutron energy ultimately determines which of these possibilities occurs. Table 3.2 summarizes the dominant and less likely neutron processes for the two uranium isotopes as a function of different energy regions.

It is interesting to look at the conditions that are necessary for a sample of uranium to undergo a fission chain reaction. As an example, we consider the behavior of fission neutrons emitted in a sample of pure ^{235}U. Each neutron that induces fission will produce k neutrons in accordance with equation (3.5). However, not all neutrons will induce additional fission events because some will get lost due to radiative capture or by exiting the sample. If, on average, each neutron has a probability, η (from 0 to 1) of inducing a fission reaction, then each fission will, on average, result in a gain of $(\eta k - 1)$ neutrons. If there are n neutrons present in the sample at time $t = 0$, then at a time $t + dt$ the number of neutrons will be given as

Table 3.2. Dominant and less likely neutron processes in different energy regions for ^{235}U and ^{238}U. Adapted from information in Dunlap (2023).

	^{235}U		^{238}U	
Energy	Dominant process	Other processes	Dominant process	Other processes
<1 eV	Fission	Elastic scattering	Elastic scattering	
1–100 eV	Radiative capture	Fission elastic scattering	Radiative capture	Elastic scattering
100 eV–1 MeV	Radiative capture inelastic (>14 keV)	Fission	Radiative capture	Elastic scattering inelastic (>44 keV)
>1 MeV	Inelastic	Fission	Inelastic	Radiative capture fission (>1.4 MeV)

$$n(t + dt) = n(t)\left[1 + (\eta k - 1)\frac{dt}{t_0}\right] \tag{3.8}$$

where t_0 is the average time required for a neutron to induce fission. The rate of change of the number of neutrons in the sample will be given from equation (3.8) as

$$\frac{dn}{dt} = \frac{n(t + dt) - n(t)}{dt} = \frac{(\eta k - 1)}{t_0}n(t).$$

This differential equation is readily integrated to give

$$n(t) = n(0)\exp\left[\frac{(\eta k - 1)t}{t_0}\right] \tag{3.9}$$

Thus, if $(\eta k - 1) > 0$, then the number of neutrons in the sample and hence the energy produced will grow exponentially with a time constant t_0. If $(\eta k - 1) < 0$, and then the number of neutrons will decrease exponentially. A controlled chain reaction would require $(\eta k - 1) = 0$. Since $k \approx 2.5$ (a detailed analysis gives $k = 2.42$), then a controlled chain reaction would imply $\eta \approx 0.4$.

For neutrons in ^{235}U, table 3.2 shows that in the MeV energy range the processes that can occur are inelastic scattering and induced fission. Radiative capture does not become significant until the energy falls below about 10 keV. Thus, η will be less than unity because some of the neutrons will exit the sample before inducing fission. We can make an estimate of η as a function of the sample size by looking in more detail at the interaction probabilities for neutrons in the MeV energy range. At 1–2 MeV, the fission cross section for ^{235}U is about 1 b (b = barn = 10^{-28} m^2 (100 fm^2)), as shown in figure 3.4. This figure also shows that the radiative capture cross section is small. At these same energies, the inelastic scattering cross section is about 6 b. Thus, induced fission accounts for about 1/7 of the total neutron interaction cross section; meaning that, on average, a neutron will undergo seven interactions to yield one induced fission (the remaining interaction being inelastic scattering).

The distance that a neutron travels between interactions is given by the mean free path as

$$d = \frac{1}{\sigma\rho}$$

where σ is the total interaction cross section and ρ is the number density of nuclei. For ^{235}U, $\rho = 4.83 \times 10^{28}$ m^{-3} and using $\sigma = 7$ b gives a mean free path of $d = 0.03$ m. Since the path of the neutron in the sample can be described as a random walk, the average total distance traveled between fissions will be given by (0.03 m) × $(7)^{1/2} = 0.08$ m. This means that a sphere of ^{235}U that is less than 0.08 m in radius will not sustain a chain fission reaction, while a sphere with a radius greater than 0.08 m will sustain a chain reaction. For the uranium density of 1.9×10^4 kg m^{-3}, this radius corresponds to a critical mass of 41 kg. A more detailed analysis of fission in a sample of ^{235}U, gives a slightly greater critical mass (47 kg).

This analysis shows that for a sample of ^{235}U with a mass less than the critical mass, a sustained fission chain reaction will not occur, and the number of neutrons

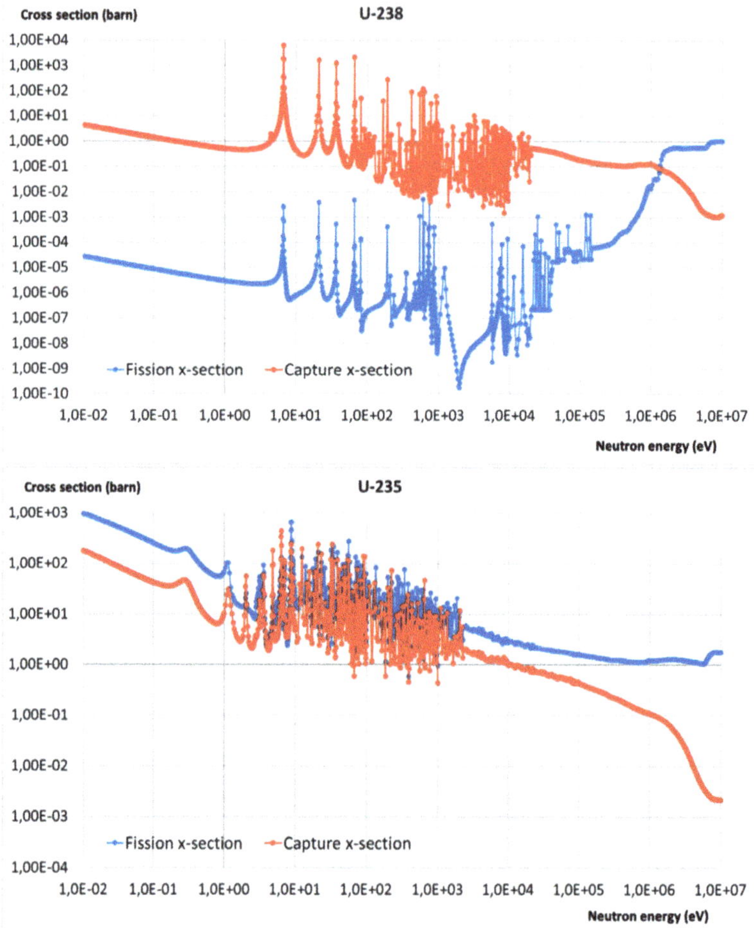

Figure 3.4. Neutron induced fission cross sections (blue) and radiative capture cross sections (red) for ^{235}U and ^{238}U as a function of incident neutron kinetic energy. [b = barn = 10^{-28} m^2]. Reproduced from Ripani (2018). CC BY 4.0.

produced will remain at a small value from the occasional spontaneous or induced fission event. In a sample of ^{235}U with a mass greater than the critical mass, the chain reaction in the sample will continue in an uncontrolled manner. We can use the results given in equation (3.9) to get an idea of the timescale for such a reaction. The timescale for the actual fission process is very short. However, the chain reaction is governed by the timescale of the movement of neutrons through the sample. Since fission neutrons have a typical energy of around 2 MeV, we can easily calculate the neutron velocity as

$$v = \sqrt{\frac{2E}{m_{\mathrm{n}}}}.$$

This gives $v = 2 \times 10^7$ m s^{-1}. Since the neutron, on average, travels a distance of 7×0.03 m $= 0.21$ m, the fission timescale is about 10^{-8} s. A rough analysis based on equation (3.9) gives the time for the entire critical mass to fission be around 5×10^{-7} s. For an average energy release of 200 MeV per fission (see table 3.1 above), this represents the release of about 1.2×10^{16} J for a critical mass of ^{235}U. This is equivalent to the electrical output of a large coal-fired generating station over a period of about 6 months. Obviously, this chain reaction releases a large amount of energy over a short period of time in an uncontrolled way and is the basis of the operation of a fission weapon. In a common weapon design (see chapter 4), two subcritical masses of fissile material are brought together quickly in order to produce a mass above criticality that will rapidly release a substantial amount of energy. The ^{235}U fission weapon dropped on Hiroshima in 1945 contained a total of 51 kg ^{235}U, just about the critical mass. Only about 2% of the total available fission energy was actually released during the detonation because the uranium fuel blew apart before all of the nuclei could undergo fission.

The design of a fission reactor in which a controlled fission reaction can be sustained requires a careful consideration of the energy dependence of the neutron cross section of both ^{235}U and ^{238}U. We can see from table 3.2 that the most significant fission reactions occur for ^{235}U nuclei for low-energy neutrons. This feature is better quantified by looking at the energy dependence of the fission cross section, as shown in figure 3.4. Clearly, the fission cross section for low-energy neutrons in ^{235}U is almost four orders of magnitude greater than the fission cross section for either uranium isotope at high energy, i.e., around 2 MeV. It is also apparent from the figure that the neutron cross section for fission is much greater for ^{235}U than for ^{238}U. This is generally true for other heavy nuclei, where fissile nuclei have larger thermal neutron cross sections than non-fissile nuclei. This feature is illustrated in figure 3.5, which shows the thermal neutron absorption cross section as a function of neutron number for uranium nuclei from ^{230}U to ^{238}U. Since uranium has an even number of protons ($Z = 92$), the uranium nuclei all have paired protons and nuclei with an odd neutron number have an unpaired neutron while nuclei with an even neutron number have all paired neutrons. The large cross section for fissile uranium nuclei results from the propensity for an incident neutron to pair with the unpaired uranium neutron.

3.5 Basic thermal neutron reactor design

Figure 3.4 shows the regions of radiative capture (i.e., (n, γ) resonances) for ^{235}U and ^{238}U. These occur between a fraction of an eV and a few keV for ^{235}U, and between a few eV and 100 keV for ^{238}U. Radiative capture is the only significant reaction that competes with induced fission because other neutron processes, i.e., elastic and inelastic scattering, do not consume neutrons but allow them to go on to be involved in other reactions. Therefore, to effectively control fission reactions in a reactor, it is important to avoid the (n, γ) resonances and to control the rate at which induced fission occurs. Figure 3.6 shows the basic design of the core of a thermal neutron reactor. (Fast neutron reactors will be discussed below.) Thermal neutron reactors primarily produce

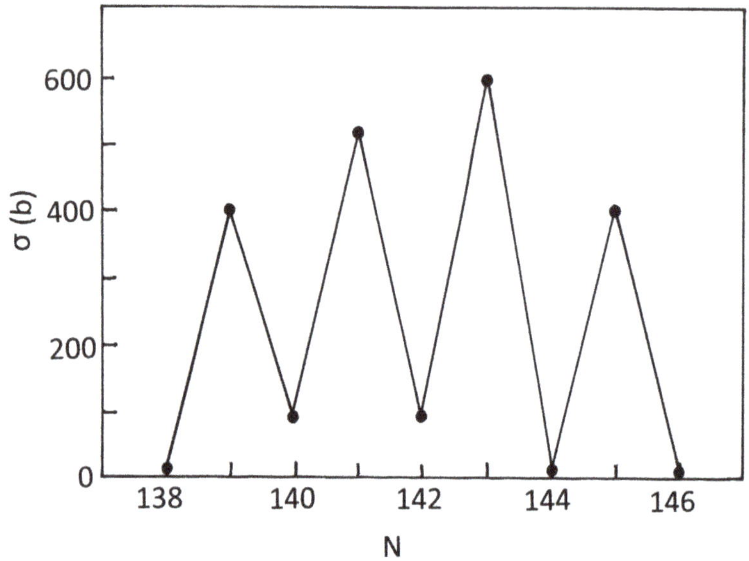

Figure 3.5. Thermal neutron absorption cross section for uranium isotopes as a function of neutron number, N. Data are from Lederer and Shirley (1978) [b = barn = 10^{-28} m^2].

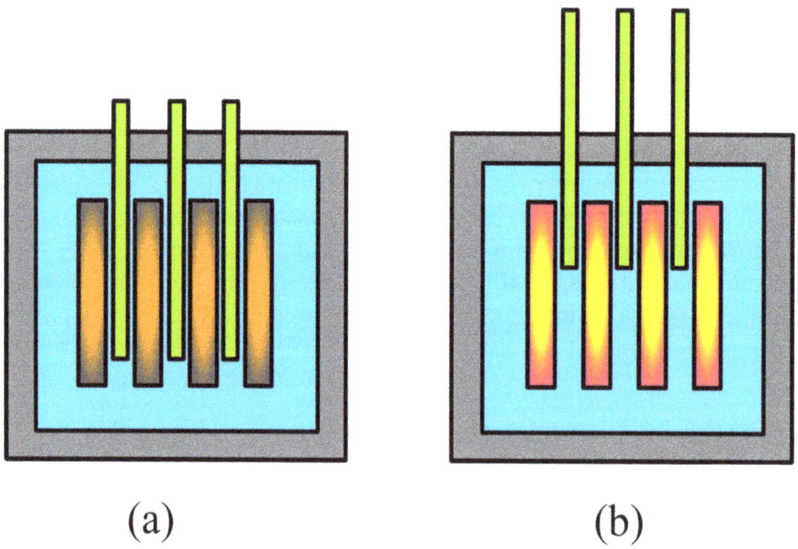

Figure 3.6. Control rods in a thermal neutron fission reactor. (a) Control rods inserted between fuel elements to block the flow of neutrons and (b) control rods withdrawn to allow flow of neutrons between fuel rods (green = control rods, red/yellow = fuel elements, blue = moderator, gray = shielding/enclosure vessel). This Schematic drawing of control rods in a nuclear reactor image has been obtained by the author from the Wikimedia website where it was made available by Pbech (2009) under a CC0 licence. It is included within this chapter on that basis. It is attributed to Pbech.

energy as a result of fusion induced by neutrons with typical energy less than about 1 eV in fissile ^{235}U. In the diagram, each fuel element (sometimes referred to as a fuel rod or fuel assembly, depending on the specific design, see chapter 5) contains a mixture of ^{235}U and ^{238}U (not necessarily in their relative natural abundances) that includes much less than the critical mass of fissile material. These fuel elements are surrounded by a material (known as the moderator) that is designed to reduce the energy of neutrons that are emitted from the fuel elements. Specific reactor designs, such as those discussed in detail in chapter 5, may use the moderator material to reduce neutron energy, cool the core, and transfer heat to a turbine for electricity generation. The mass and geometry of the fuel rods are designed so that essentially all of the MeV energy neutrons produced within the element escape into the moderator. Both ^{235}U and ^{238}U have neutron cross sections in the MeV range that are dominated by inelastic scattering. This means that the neutrons escape from the fuel element, generally before causing induced fission or being absorbed in an (n, γ) process (which occurs at considerably lower energy in both isotopes).

The moderator is a material that interacts with neutrons over a wide range of energies (i.e., <1 eV to >1 MeV) primarily by elastic or inelastic scattering. These scattering events reduce the energy of the neutrons (typically to an energy below the (n, γ) resonances) without absorbing them. Therefore, the neutrons that travel through the moderator enter into an adjacent fuel element with low energy. From table 3.2, it is seen that for neutrons below 1 eV, the dominant neutron process in ^{235}U is induced fission (with some elastic scattering), while in ^{238}U it is elastic scattering. Therefore, these low-energy neutrons will continue to lose kinetic energy through elastic scattering, or they will induce fission. Using this approach, we can ensure that the neutrons that enter an adjacent fuel element are not likely to be lost to (n, γ) processes and that the majority will ultimately induce fission in the ^{235}U component of the fuel.

In order to ensure that the fission chain reaction that is created within the fuel elements proceeds in a controlled manner, we need to control the flow of neutrons between the fuel elements. This is done, as shown in figure 3.6, by inserting control rods between the fuel elements. The control rods are comprised of a material that has a large radiative capture cross section over a wide range of energies. The reactor is started with the control rods entirely blocking the fuel elements from their neighboring elements, as in figure 3.6(a). The control rods are retracted, as in figure 3.6(b), while monitoring the neutron density in the reactor. The control rods are maintained in a position that provides an induced fission rate that is appropriate for the operation of the reactor and maintains a suitable operating temperature. The details of how this is done are described in more detail in section 3.7. In the next section, we consider the material requirements for the important components of the reactor, e.g. the fuel elements, the moderator, and the control rods.

3.6 Reactor materials

3.6.1 Uranium fuel

The design of the fuel elements in a thermal neutron reactor varies considerably. The present section describes a fairly generic fuel element design that applies to many current model power reactors. More specific details of current fuel element design are presented in the relevant sections of chapter 5. These specific details include specific composition of the fuel, including the level of isotopic enrichment, size and number of fuel rods in a fuel assembly, and the overall size of the assembly.

Uranium, which is used as fuel for thermal neutron reactors, is obtained from uranium ore, which is mined in various locations, or is obtained from reprocessed spent reactor fuel. Details of uranium production in different countries and fuel reprocessing are discussed in chapter 5. Here we provide an overview of the production of uranium fuel from ore. Uranium, which is extracted from mined ore is generally in the form of an oxide with a composition of 70%–90% U_3O_8 (triuranium octoxide). Other uranium oxides, such as uranium hydroxide, as well as uranium sulfate, are also commonly present. This material is often referred to as 'yellowcake'. Although it is sometimes yellow in color, this is not always the case, depending on the exact composition. Uranium used as fuel in thermal neutron reactors may be unenriched (natural abundances of ^{235}U and ^{238}U) or enriched in ^{235}U, depending on the specific design of the reactor.

Various chemical processes are used to process uranium for reactor use. A typical process for converting U_3O_8 into usable fuel begins with its conversion to UO_2 (uranium dioxide) by the reaction

$$U_3O_8 + 2H_2 \rightarrow 3UO_2 + 2H_2O. \quad (3.10)$$

For uranium that is used in the unenriched state, the UO_2 can be used directly as a reactor fuel. For uranium that is used in reactors requiring enriched fuel, the U_3O_8 is typically converted to UF_6 (uranium hexafluoride) for enrichment and further processing. A typical process for this conversion is

$$UO_2 + 4HF \rightarrow UF_4 + 2H_2O$$

followed by

$$UF_4 + F_2 \rightarrow UF_6.$$

While a variety of different enrichment techniques have been used in the past, gas centrifuges are used nearly exclusively at present for the separation of ^{235}U and ^{238}U for fission reactor fuel. UF_6 is a solid at room temperature, which sublimes (at atmospheric pressure) at 56.5 °C. UF_4 cannot be used directly in the processes described below because of its high boiling point, 1417 °C. UF_6, which is heated above its sublimation temperature, is fed into a centrifuge, as illustrated in figure 3.7. The centrifuge rotates at high-speed (typically >50 000 rpm) causing the heavier gas molecules (that is the UF_6 molecules containing ^{238}U) to move towards the outside of the cylinder, while lighter gas molecules (i.e., the UF_6 molecules containing ^{235}U, which are about 0.85% lighter than the heavy molecules) move towards the center of the cylinder. This creates a concentration gradient across the radius of the cylinder where enriched uranium (i.e., uranium with more than the natural abundance of ^{235}U) collects near the cylinder axis while depleted uranium (i.e., uranium with less the natural abundance of ^{235}U) collects near the outside edge of the cylinder. In the design illustrated in figure 3.7, there is a so-called countercurrent set up in the

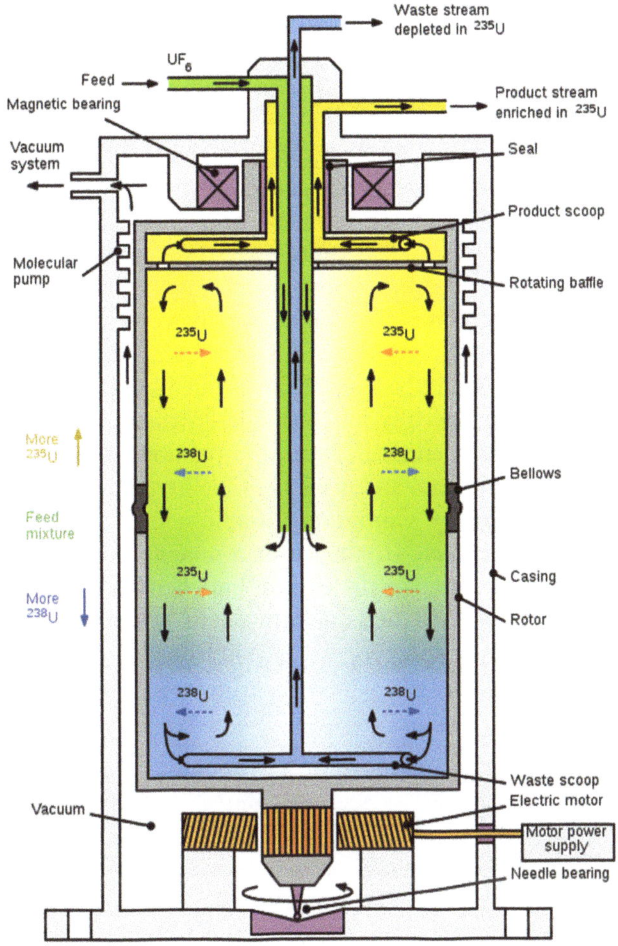

Figure 3.7. Diagram of a gas centrifuge used to separate ^{235}U and ^{238}U. This centrifuge design uses countercurrent circulation as described in the text for increased separation. This Simplified diagram of a gas centrifuge used for uranium isotope separation with a countercurrent circulation for increased separation image has been obtained by the author from the Wikimedia website where it was made available by Inductiveload (2020). Image stated to be in the public domain. It is included within this chapter on that basis. It is attributed to Inductiveload.

cylinder to enhance the separation effect. This countercurrent represents a gas flow along the axial direction of the cylinder, as indicated by the arrows in the figure, and results in a vertical concentration gradient in the cylinder. The countercurrent may be introduced by a vertical temperature gradient or by mechanical movement using a rotor. As illustrated in the figure, for this particular design, the enriched uranium (referred to as the product) is extracted from the top of the cylinder and the depleted uranium (referred to as the waste) is extracted from the bottom of the cylinder.

The degree to which the uranium is enriched in ^{235}U is referred to as the separation factor and is defined as

$$\alpha = \frac{\left(\frac{x_p}{1 - x_p}\right)}{\left(\frac{x_W}{1 - x_W}\right)}$$

where x_p and x_w are the ^{235}U concentrations in the product and waste streams, respectively. A typical countercurrent gas centrifuge produces a separation factor of about 1.3. This is not sufficient for practical applications, so a large number of centrifuges in series are utilized. The product stream from one centrifuge is fed into the next higher stage, while the waste stream is returned to the next lower stage to be added to the input stream. Generally, numerous series configurations of centrifuges are arranged in parallel to increase output. Such an arrangement is shown in figure 3.8. The devices shown in the figure from about 40 years ago are about 12 m in length, current devices are typically much smaller, around 4 m in length.

Fuel for fission reactor use, either in natural form, as extracted from uranium ore, or in enriched form, as UF_6 from an enrichment facility, must be processed to make fuel that is suitable for a reactor. Commonly, uranium in the form of UO_2 is used for the fabrication of reactor fuel. In the case of unenriched fuel, UO_2 from ore processing can be used directly. In the case of enriched fuel, enriched UF_6 can be converted to uranium dioxide by heating with steam by the reaction

$$UF_6 + 2H_2O \rightarrow UO_2F_2 + 4HF$$

Figure 3.8. Array of gas centrifuges used to enrich uranium. This 1984 photograph is of the gas centrifuge plant in Piketon, OH. Reproduced from Brouwers (2023). CC BY 4.0.

followed by

$$3UO_2F_2 + 2H_2O + H_2 \rightarrow U_3O_8 + 6HF.$$

The resulting triuranium octoxide is converted to uranium dioxide by the reaction given in equation (3.10).

Uranium dioxide is a desirable chemical form for fission reactor fuel for several reasons, e.g., it has a high melting point (2865 °C) compared to elemental uranium (1132 °C) and it is chemically stable because it is already fully oxidized, and therefore cannot burn. Other chemical forms of uranium have been used in specific cases. Some of these are described in the later chapters of this book that deal with specific Generation IV reactor designs.

For reactor fuel, powdered uranium oxide is pressed into centimeter-sized pellets and sintered at high temperature. A typical fuel pellet produced in this manner is illustrated in figure 3.9. These fuel pellets are stacked in tubes made from a suitable metal with high thermal and chemical stability that has a very low neutron absorption cross section to produce fuel rods. Stainless steel has sometimes been used for fuel rod tubes, although current reactor designs most commonly use tubes made from zirconium alloy. Figure 3.10 illustrates the structure of a typical reactor fuel rod. These rods are, most commonly, combined in an array to produce the fuel assembly (sometimes called a fuel bundle), as shown in the figure. Typically, bundles consist of 100 or more rods and reactor cores are made up of hundreds of bundles. The number of rods in each bundle and the number of bundles in the reactor core, as well as the length of the fuel assembly, depends on the specific design of the reactor, and is discussed in more detail for the common current reactor types in chapter 5 and for Generation IV reactors in subsequent chapters.

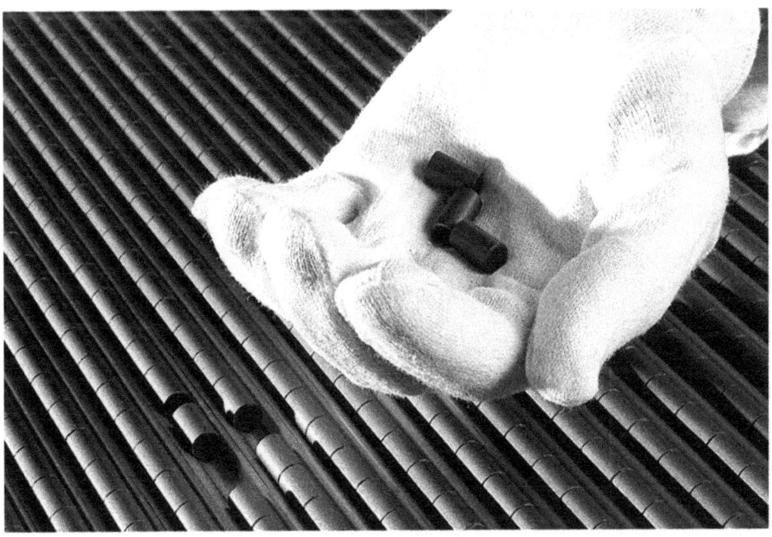

Figure 3.9. Typical uranium fuel pellets for a thermal neutron power reactor. Reproduced from U.S. Nuclear Regulatory Commission (2023). U.S. Nuclear Regulatory Commission (2023). Image stated to be in the public domain.

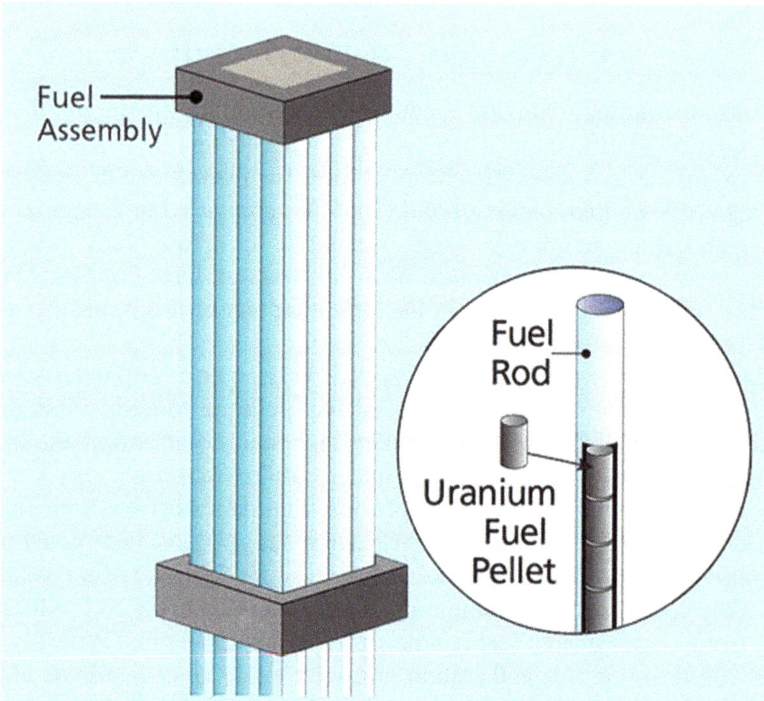

Figure 3.10. Design of a typical thermal neutron reactor fuel assembly. Reproduced from U.S. Nuclear Regulatory Commission (2016). CC BY 2.0.

3.6.2 Moderator materials

The moderator reduces the energy of the neutrons as they traverse the distance between adjacent fuel elements. Neutrons that are produced in one fuel element may exit the element into the moderator with minimal reduction of energy, i.e., at around 2 MeV. It is important for the operation of the reactor as described above that when these neutrons enter another fuel element their energy has been reduced to a value where the neutron cross section is dominated by the fission cross section of ^{235}U. Although true thermal neutrons are in thermal equilibrium at a temperature of 300 K and have an energy of about 0.25 eV, the actual degree of moderation (or thermalization) in a reactor depends on the nature of the moderator (as discussed below) and the details of the reactor design. A suitable moderator must satisfy several important criteria. The most important of these, for the purpose of the construction of a thermal neutron reactor, are described below. Other factors are of relevance for other types of reactors and are discussed later in the book.

Small nuclear mass: Since the neutrons lose energy primarily through elastic scattering with nuclei in the moderator, the use of a moderator containing light nuclei will increase its efficiency. Following the discussion in chapter 2 for one-dimensional scattering, the final energy, E_a', of a particle of mass m_a, elastically scattered from a stationary particle of mass m_A is given by

$$E_a' = \left[\frac{m_A - m_a}{m_A + m_a}\right]^2 E_a \quad (3.11)$$

where E_a is the particle's initial energy. In a two-dimensional case, equation (3.11) refers to the situation where the incident particle is backscattered, i.e. the scattering angle is 180°. In the forward scattering limit, the scattering angle is 0° and the final energy of the scattered particle will be

$$E_a' = E_a. \quad (3.12)$$

The average scattered particle final energy for the two-dimensional case is the average of equations (3.11) and (3.12) or

$$E_a' = \frac{1}{2}\left(1 + \left[\frac{m_A - m_a}{m_A + m_a}\right]^2\right)E_a \quad (3.13)$$

Although a simple analysis of neutrons in a moderator based on equation (3.13) would provide an estimate of the average neutron energy after a known number of elastic scattering events, this does not provide the most useful approach to the determining the suitability of a moderator. The reason for this is the fact that as the neutrons are moderated, their energy distribution becomes skewed, where a few high energy neutrons remain in the distribution and cause a mean energy that is higher than that which is actually relevant to the reactor operation. A measure of the effectiveness of neutron moderation, ξ, that is often used is the logarithm of the energy ratio before and after the scattering, i.e.,

$$\xi = \ln\left(\frac{E_a}{E_a{}'}\right) \quad (3.14)$$

Since the overall goal of the moderator is to reduce the neutron energy from its initial value of around 2 MeV to a value that is below about 1 eV, the results shown in equations (3.13) and (3.14) can be used to determine the mean number of scattering events, n, that are necessary to accomplish this energy reduction, i.e.,

$$n = \frac{1}{\xi}\ln\left(\frac{E_{max}}{E_{min}}\right)$$

where typically $E_{max} = 2$ MeV and $E_{min} = 1$ eV. As noted in the previous chapter, the energy loss of a neutron scattering from a uranium nucleus is less than 2%. A simple inspection of equation (3.13) shows that the most efficient moderators are those with nuclei of low mass number.

Small (n, γ) *cross section:* A small radiative capture cross section over a wide range of energies (~1 eV–2 MeV) is important because the role of the moderator is to lower the energy of the neutrons and not to remove them. The important criterion for assessing the relevance of radiative capture is the ratio σ_{el}/σ_c, where σ_{el} is the elastic scattering cross section and σ_c is the neutron capture cross section. The effect of neutron cross sections on the desirability of a material as a moderator can be assessed by combining the effects of nuclear mass, as discussed above, with the relative neutron cross sections for elastic scattering and radiative capture. For this purpose, it is convenient to define an overall moderating ratio, M, as

$$M = \xi \frac{\sigma_{el}}{\sigma_c}.$$

An additional disadvantage of a moderator with a significant (n, γ) cross section is that radioactive by-products can be produced, and this can require additional safety measures and/or create waste disposal problems. Large neutron absorption cross sections make the light elements lithium and boron unacceptable for use as moderators, at least for thermal neutron reactors. However, the neutron absorption properties of boron make it useful for control rods, as noted below.

High density: The number of collisions per unit time is proportional to the number density of moderator nuclei. For this reason, moderator materials must be either liquids or solids. Gases, even under pressure, are not effective moderators. This condition eliminates helium as a possible moderator.

Chemical and thermal stability: It is important that the moderator material is chemically stable so that it does not deteriorate over time and that it does not interact adversely with other reactor components. It is also essential that the moderator material retains its integrity at the reactor operating temperature.

Low toxicity: A moderator material with a low toxicity is desirable. Toxic materials can complicate reactor construction and there is a possibility of environmental contamination in the event of a reactor malfunction. This is particularly the case for liquid moderators where moderator leaks are a concern. Toxicity makes beryllium an undesirable moderator.

Economic viability: Cost must be considered as an important factor in the design of a commercial power reactor. Since a large amount of moderator material is typically needed for a thermal fission reactor, economic factors must be a consideration. Beryllium, in addition to its toxicity, is an unacceptable moderator material due to its cost.

It is unfortunate, when all of the factors above are taken into consideration, that there really is no ideal moderator material. So, the materials that are used as moderators in thermal reactors must inevitably have some undesirable properties that must be taken into account in the reactor design. There are three materials that are currently used as moderators in thermal fission reactors—water, heavy water, and graphite (carbon). The properties of these materials as moderators are summarized in table 3.3. The properties of these three moderator materials are discussed below.

Water (H_2O): Water provides neutron moderation by virtue of its hydrogen content. The oxygen in H_2O has a small neutron cross section and is more-or-less inert, and the liquid form of water ensures a suitable moderator density. Since hydrogen is the lightest nucleus (consisting of only a single proton), it would seem to be the ideal moderator providing a large energy transfer per scattering as given by

Table 3.3. Summary of neutron moderating properties of some commonly used moderators. Data adapted from Duderstadt and Hamilton (1976).

Moderator	ξ	n	M
H_2O	0.92	16	71
D_2O	0.51	29	5670
Graphite	0.16	91	192

Table 3.4. Thermal neutron absorption cross sections, (n, γ). Data are from Lederer and Shirley (1978).

Nuclide	σ_c (b)
^{1}H	0.332
^{2}H	0.0005
^{12}C	0.0034

equation (3.13) and a corresponding large value of the parameter ξ in equation (3.14). This large value of ξ is given in table 3.3 and the small number of scattering events necessary for thermalization is also shown in the table. Hydrogen, however, suffers from a significant thermal neutron cross section, as shown in table 3.4, that leads to neutron absorption via the (n, γ) process,

$$n + p \rightarrow d + \gamma$$

where d is the deuteron, and the energy release is 2.22 MeV (the deuteron binding energy). While significantly smaller than the cross section of control rod materials, as discussed below, the cross section of hydrogen is sufficiently large that it diminishes its ability to act as a moderator material. The moderating ratio for water, as seen in table 3.3, is the lowest of the three commonly used moderator materials. The fact that neutrons are lost in the moderator as a result of radiative capture, means that the amount of moderator that can be used is limited in order to preserve the neutron number at a level that can maintain a chain reaction. The result of this feature of water as a moderator is that the nuclear fuel has to be enriched above the 0.72% natural abundance of ^{235}U. Typical light water reactors use fuel enriched to about 3% ^{235}U to compensate for neutron loss in the moderator.

Heavy water (D_2O): As for light water, the oxygen content of heavy water is basically inert. The greater mass of the deuteron (compared to the proton) results in a smaller value of ξ and a larger value of n, as shown in table 3.3. However, the most significant difference between light water and heavy water as a moderator is the difference in the neutron absorption cross section. This is illustrated by the data in table 3.4, where it is seen that the thermal neutron capture cross section for ^{2}H is about three orders of magnitude smaller than for ^{1}H. The result of this difference is a greatly improved moderating ratio for heavy water, as shown in table 3.3. This improved moderator efficiency results in the ability to use natural (unenriched) uranium as a fuel in reactors that are heavy water-moderated. While heavy water may seem to be an ideal moderator, it does have two drawbacks. First, it is expensive because deuterium represents only 0.015% of naturally occurring hydrogen and must be artificially separated from the remaining 99.985% of hydrogen atoms. The second concern is with the small neutron absorption cross section shown in table 3.4. Although radiative capture is much less likely than in light water and does not measurably influence the use of D_2O as a moderator, the small number of reactions that do occur produce a radioactive by-product. The neutron absorption reaction for deuterium is

$$\mathrm{n} + \mathrm{d} \rightarrow \mathrm{t} + \gamma$$

where t is the triton (^{3}H nucleus) and the energy release is energy 6.26 MeV. Tritium, ^{3}H, decays by β^- decay with a half-life of 12.3 years. Therefore, over time, the heavy water moderator will become radioactive, and this must be a consideration from a safety and waste disposal standpoint.

Graphite (C): The much greater mass of carbon compared to hydrogen and deuterium results in a significantly smaller value of ξ and a larger value of n, as shown in table 3.3. However, the small neutron absorption cross section, as shown in table 3.4, provides a good overall moderating ratio. The reasonable value of M allows some graphite-moderated reactors to operate using unenriched uranium as a fuel, although some graphite-moderated reactors utilize slightly (~2%) enriched uranium. The small neutron absorption cross section of carbon results in reaction

$$\mathrm{n} + {}^{12}\mathrm{C} \rightarrow {}^{13}\mathrm{C} + \gamma$$

with 4.95 MeV produced. The ^{13}C by-product is nonradioactive and there are no concerns such as those for tritium production in heavy water-moderated reactors. Potential safety risks associated with using graphite as a moderator in a fission reactor are discussed in chapter 6.

3.6.3 Control rods

Control rods are made from materials that have a large (n, γ) cross section because their purpose is to stop neutrons from passing from one fuel element to another. It is important that neutron absorption in the control rod is due to the (n, γ) reaction and not due to induced fission because the latter would produce additional neutrons. Control rods are typically made from a combination of materials that will insure a large neutron absorption cross section over a wide range of energies. Two materials that are often components of control rods are boron and cadmium. These materials cannot undergo induced fission by neutron irradiation. In the case of boron, it is not energetically favorable for this reaction to occur, and in both cases the Coulomb barrier, as shown in figure 3.1, is prohibitive. We look at the neutron related properties of these two elements in some detail.

Naturally occurring boron consists of two isotopes, ^{10}B and ^{11}B, in proportions of 20% and 80%, respectively. The energy dependence of the neutron absorption cross

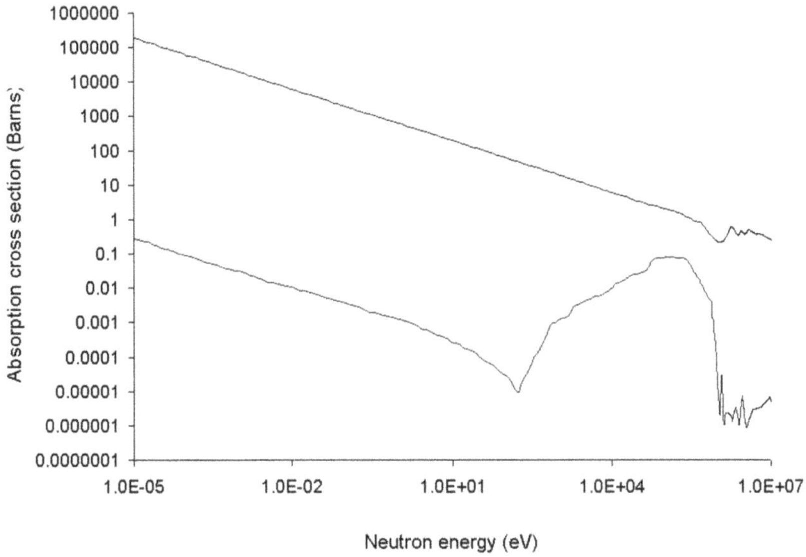

Figure 3.11. Neutron absorption cross sections as a function of neutron energy for ^{10}B and ^{11}B. This Neutron absorption cross sections as a function of neutron energy for B-10 and B-11 image has been obtained by the author from the Wikimedia website where it was made available by Cadmium (2006). Image stated to be in the public domain. It is included within this chapter on that basis. It is attributed to Cadmium.

section of these two isotopes is illustrated in figure 3.11 The figure shows that ^{10}B, despite its lower natural abundance, is responsible for virtually all of the neutron absorption due to a cross section that is more than five orders of magnitude greater at nearly all energies than for ^{11}B. The (n, γ) reaction for ^{10}B is

$$n + {}^{10}B \rightarrow {}^{11}B^{*} \rightarrow {}^{11}B + \gamma$$

where 11.45 MeV of energy is released. The data in figure 3.11 also show the characteristic $1/E$ dependence of the cross section at low energies as predicted by the Breit–Wigner formula (see Dunlap (2023)).

Naturally occurring cadmium consists of eight isotopes with $106 \leqslant A \leqslant 116$. The neutron absorption cross section is primarily from ^{113}Cd, which has a natural abundance of about 12.2%. The (n, γ) reaction for ^{113}Cd is

$$n + {}^{113}Cd \rightarrow {}^{114}Cd^{*} \rightarrow {}^{114}Cd + \gamma$$

where 9.04 MeV of energy is released. Figure 3.12 shows the neutron absorption cross section for elemental cadmium. The region above about 20 eV is characterized by a series of sharp (n, γ) resonances. There is a lower energy (n, γ) peak at about 0.2 eV. Note that the energy scale is logarithmic, so the 0.2 eV peak is actually quite sharp. Below this peak, the cross section follows the characteristic $1/E$ energy dependence. This figure shows the similarity of the cadmium cross section with that of boron.

The large cross section for neutron absorption of natural boron and cadmium (along with a number of other elements, such as, silver, hafnium and indium, which are sometimes used as components of control rods) ensures that neutrons are

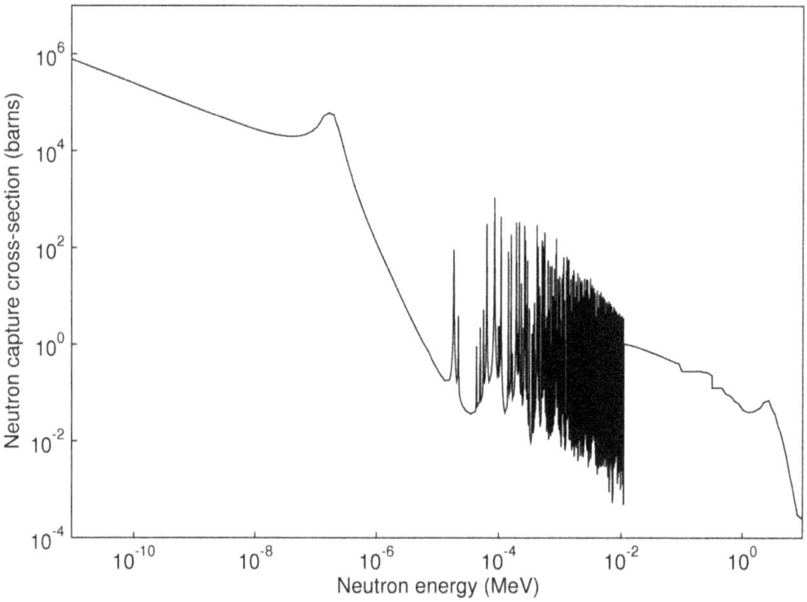

Figure 3.12. Total neutron absorption cross section for elemental cadmium as a function of energy. Reproduced from ThorEA (2017). CC BY-SA 3.0.

effectively prevented from traveling from one fuel element to another. The large low-energy neutron absorption cross section for these elements is best utilized when the neutron energy is significantly reduced from its initial ~2 MeV by traveling through a region of moderator prior to entering into the control rod. A typical arrangement of control rods in the fuel rod bundle is shown in figure 3.13.

3.6.4 Coolant

The coolant is used to maintain the temperature of the reactor core at an acceptable level and to transfer the heat which is produced by the fission reactions to a system for generating electricity. Thermal neutron reactors which utilize a liquid moderator, i.e., H_2O or D_2O, can also use the moderator liquid as a coolant. A simple water-moderated thermal neutron reactor schematic showing the means of transferring heat from the reactor core to a turbine connected to a generator to produce electricity is illustrated in figure 3.14. This is a boiling water reactor (described in more detail in chapter 5) where water, which circulates through the reactor core, serves as the moderator and is heated by the fission reactions in the fuel elements to produce steam. The steam is circulated through a turbine which turns a generator to produce electricity. The steam, after passing through the turbine, transfers its remaining excess heat through a heat exchanger (condenser) to cooling water that dissipates the heat in a large body of water or through a cooling tower. Further details of these cooling methods are discussed in chapter 5. Finally, the cold water is pumped back into the reactor core. Reactors that utilize a solid moderator material, i.e. graphite, need an additional material to serve as coolant.

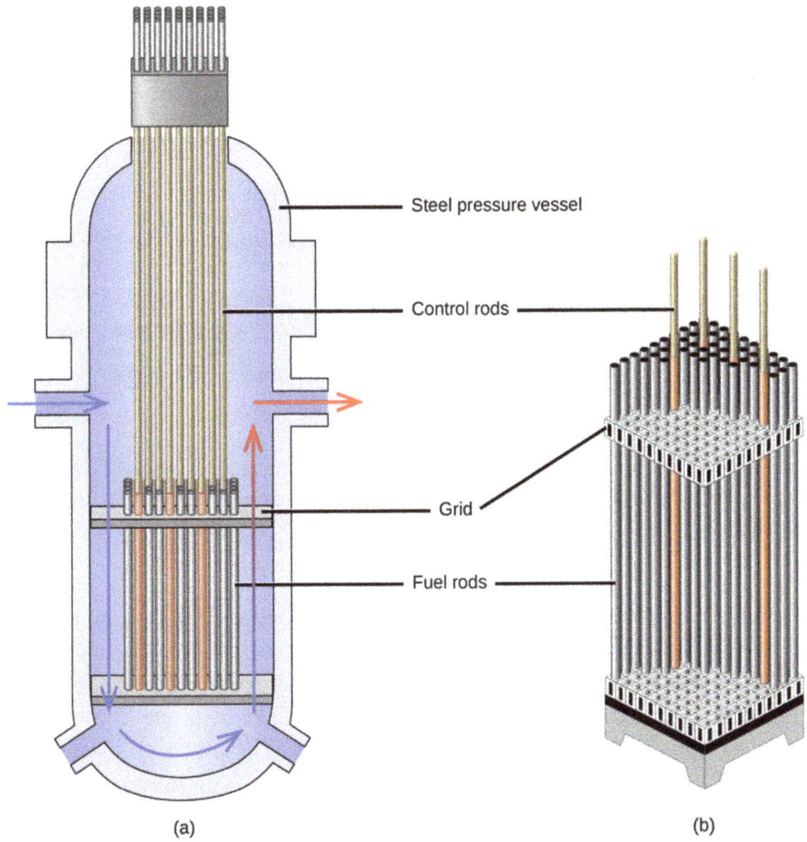

Figure 3.13. Diagram of the typical relationship between fuel rods and control rods in a thermal neutron fission reactor. Reproduced from Flowers *et al* (2015). CC BY 4.0.

The graphite-moderated reactors that are in current use are either gas-cooled, using carbon dioxide as a coolant, or water-cooled. Other coolants may be suitable for fast neutron reactors or other advanced reactor designs, and are discussed in the appropriate sections later in this book.

3.7 Delayed neutrons and reactor stability

As noted previously, the sustained operation of a thermal neutron reactor is achieved by regulating the position of the control rods in order to maintain a controlled chain reaction in the fuel elements where the reactor core temperature and neutron density are held at appropriate levels. However, the basic physics of this process are not so simple. The main point concerning this process that needs to be understood in detail is the timescale on which fission processes occur. As noted in section 3.4, the timescale for the fission process is of the order of 10^{-8} s. This represents the mean time between the time when a fission neutron is released by a fission event and the time it induces a subsequent fission in a sample of ^{235}U of

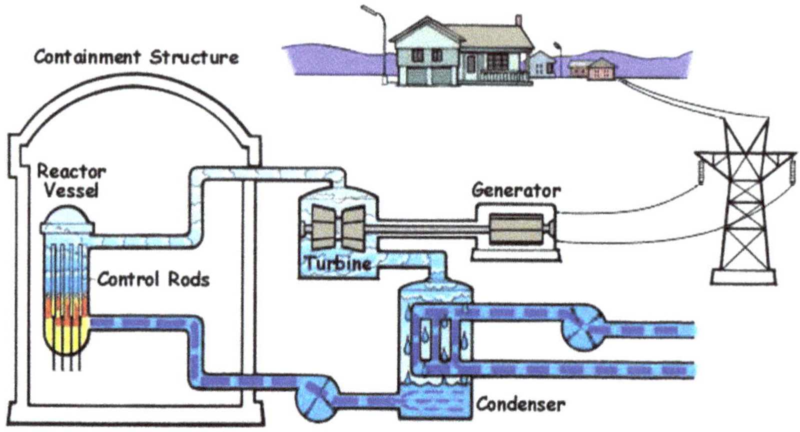

Figure 3.14. Diagram of a simple boiling water reactor. This boiling water reactor image has been obtained by the author from the Wikimedia website where it was made available by Internovice (2010). Image stated to be in the public domain. It is included within this chapter on that basis. It is attributed to Internovice.

critical mass. This, however, is not the actual process that occurs in a thermal neutron reactor. In the reactor case, the fission neutron exits a fuel rod and travels through the moderator (where it loses energy) and into another fuel rod, where it induces fission. Thus, the neutron travels a much longer distance than in the simple example in section 3.4, and at a diminishing velocity. Therefore, the fission timescale is substantially longer, probably of the order of 10^{-4} s. This timescale, however, is still too short to allow the control rods, which might be of the order of 2 m in length, to be moved (by some mechanical means) in response to changes in the core neutron density in order to maintain the reactor's operating conditions within safe limits.

In order to understand how thermal neutron reactors are actually controlled, we have to look in more detail at the properties of the fission fragments. As noted, these have an excess of neutrons and decay primarily by β^- decay, producing the delayed energy release indicated in table 3.1. A typical fission fragment decay scheme following from ^{137}I, as in equation (3.7), is illustrated in figure 3.15. This sequence of β^- decays is discussed further in chapter 6. Occasionally, ^{137}I decays to an excited state of ^{137}Xe, as shown in the figure. If this excited state is above the neutron separation energy (about 4 MeV), then the ^{137}Xe can decay by neutron emission,

$$^{137}\mathrm{Xe} \rightarrow {}^{136}\mathrm{Xe} + \mathrm{n}.$$

This neutron emission will occur (in this case), on the average, 23 s after the fission process and is referred to as a delayed neutron. Neutron emission occurs in various fission fragments and on the average contributes 0.0158 neutrons per fission.

We can now return to equation (3.9) to understand how delayed neutrons contribute to reactor stability. From equation (3.9) it can be shown that a controlled chain reaction requires

$$\eta k = 1 \tag{3.15}$$

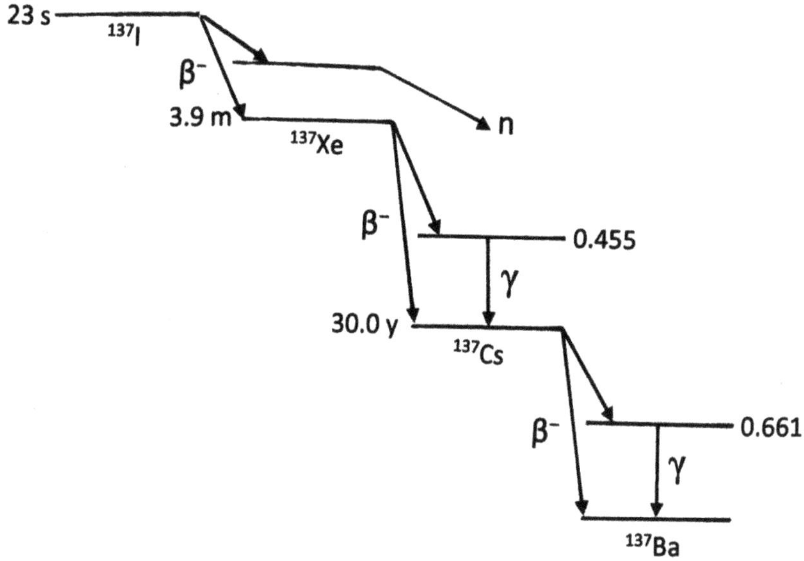

Figure 3.15. Energy level diagram for $A = 137$ β^- decays.

If we now include the average number of delayed neutrons per fission, k', then equation (3.15) becomes,

$$\eta k + \eta k' = 1 \tag{3.16}$$

If we establish a base operating condition where $\eta k = 0.99$ (for example) then the control rods can be used to adjust the flow of neutrons between fuel elements on the basis of delayed neutron to maintain equation (3.16). It is, therefore, the timescale for the delayed neutrons, tens of seconds, that allows the control rods to be adjusted in accordance with feedback from the reactor.

A final point that requires consideration is thermal stability. From a safety standpoint it is essential that the core of a reactor does not exceed a reasonable temperature limit. For this reason, it is important that the number of neutrons that induce fission decreases as a function of increasing reactor temperature, T. This is referred to as a negative coefficient of reactivity and means that

$$\frac{d\eta}{dT} < 0 \tag{3.17}$$

While most of the fission neutrons exit the fuel element where they were created and travel through the moderator, a small number of these neutrons will lose energy through interactions with nuclei in the fuel element. Some of these interactions will be (n, γ) reactions, which result in the neutron being absorbed. Figure 3.4 shows the radiative capture cross sections for ^{235}U and ^{238}U as a function of energy. The figure shows that there are (n, γ) resonance peaks in the range of about 1 eV to 10 keV. These are most notable for ^{238}U because this is the predominant isotope in the fuel and the magnitude of the resonant peaks is greater than for ^{235}U. These resonance peaks are the most significant contribution to neutron absorption within the fuel element and represent a small reduction in the factor η.

The degree to which temperature affects the value of η can be determined on the basis of the shape of the (n, γ) resonance peaks. The resonant peaks of nuclei that are in motion are Doppler shifted and a collection of nuclei that are in thermal motion have their resonant peaks broadened by the Doppler effect. This broadening in most significant for the lower energy peaks where the thermal energy is a more significant fraction of the neutron kinetic energy. A particularly relevant example is the lowest lying (n, γ) resonance in ^{238}U, which occurs at $E = 6.67$ eV, as seen in figure 3.4. This resonance corresponds to the reaction

$$n + {}^{238}\mathrm{U} \rightarrow {}^{239}\mathrm{U}^*$$

and the energy is the difference between the neutron separation energy of ^{239}U and the energy of the next highest excited state. Figure 3.16 shows the effects of temperature on the shape of this resonance peak. This behavior can be described quantitatively in terms of the additional linewidth, ΔE, that results from thermal motion as (see Dunlap 2023)

$$\Delta E \approx \left[\frac{4m_n E k_B T}{m({}^{238}\mathrm{U})}\right]^{1/2} \tag{3.18}$$

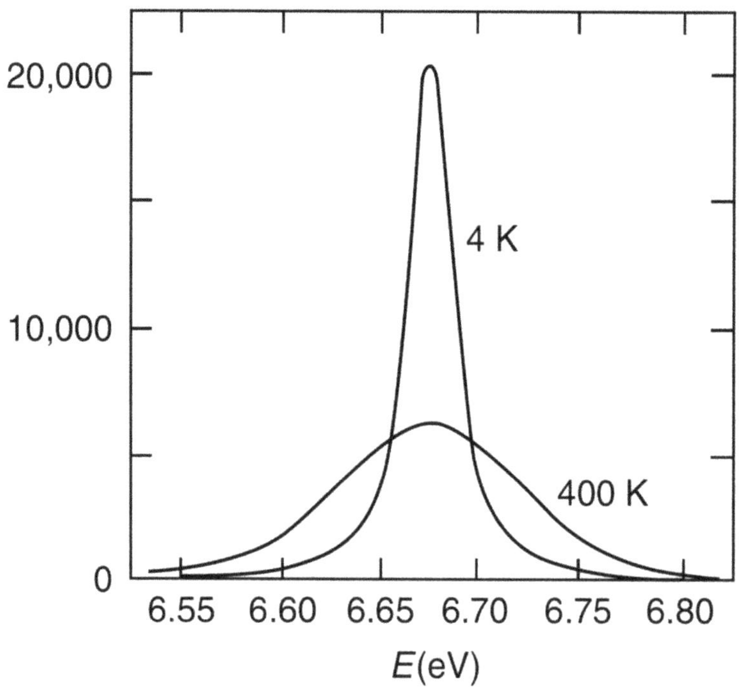

Figure 3.16. Shape of the 6.67 eV (n, γ) resonance in ^{238}U at 4 K and at 400 K. Reproduced from Dunlap (2023).

where k_B is the Boltzmann constant and T is the temperature in Kelvin. The energy in equation (3.18) is actually the center of mass energy, but the large difference in mass of the neutron and the uranium nucleus means that this is virtually the same as the neutron kinetic energy in the laboratory frame. This broadening makes it more difficult for neutrons, as they lose energy through elastic collisions, to avoid having the correct energy to be radiatively captured. This behavior gives rise to a decrease in the number of neutrons that escape from the fuel element and to a corresponding decrease in η that becomes more significant with increasing temperature. The mechanism as described above ensures that equation (3.17) is satisfied and contributes to the overall safety of the reactor. Details of the stability conditions for the Generation IV reactor designs are described in the relevant sections in the latter part of this book.

3.8 Principles of fast neutron reactors

Thermal neutron reactors, as described above, utilize energy that is extracted from the fission of ^{235}U and the potential energy content of the ^{238}U component of natural uranium is (for the most part) not utilized. Fuel reprocessing may make some of this energy available and is considered further in chapter 5. Fast neutron reactors, sometimes called fast breeder reactors, are a way of utilizing the energy associated with non-fissile materials such as ^{238}U, which in this case is referred to as a fertile material. In theory, the energy output of uranium could be increased by a factor approaching 100. In the future, this approach may also be utilized (as described below) to produce fissile ^{233}U from non-fissile (but fertile) ^{232}Th.

The basic principle of the fast breeder reactor is to utilize radiative capture of fast neutrons by non-fissile ^{238}U to produce a fissile material by the reaction

$$\mathrm{n} + {}^{238}\mathrm{U} \rightarrow {}^{239}\mathrm{U} + \gamma. \tag{3.19}$$

Although ^{239}U is fissile, it cannot be used as a reactor fuel because it is not stable against β^- decay. This decay process is

$${}^{239}\mathrm{U} \rightarrow {}^{239}\mathrm{Np} + \mathrm{e}^- + \overline{\nu} \tag{3.20}$$

and has a half-life of 23.5 min. The decay in equation (3.20) is followed by the β^- decay of ^{239}Np

$${}^{239}\mathrm{Np} \rightarrow {}^{239}\mathrm{Pu} + \mathrm{e}^- + \overline{\nu} \tag{3.21}$$

with a half-life of 2.4 days. The ^{239}Pu produced in equation (3.21) is also not β stable but has a half-life of about 24 000 years. This is sufficiently long for the fissile ^{239}Pu produced through neutron capture by ^{238}U to be used as a fuel for a nuclear reactor. In fact, ^{239}Pu produced in the fuel rods of conventional uranium-fueled thermal neutron reactors by ^{238}U capture of fast neutrons before they escape from the fuel element into the moderator contributes to the energy output of the reactor by induced fission,

$$\mathrm{n} + {}^{239}\mathrm{Pu} \rightarrow {}^{240}\mathrm{Pu}^* \rightarrow {}^{A'}\mathrm{Y} + {}^{240-A'-k}\mathrm{Z} + k\mathrm{n}$$

A fast breeder reactor is specifically designed to optimize the production of fissile ^{239}Pu from fertile ^{238}U. The basic design of a typical fast breeder reactor is illustrated in figure 3.17. While the design shown in the figure appears quite similar to that of a pressurized water reactor, there are some significant differences.

The coolant in a pressurized water reactor acts as both a coolant and a moderator. The latter by virtue of the low mass (and acceptable density) of the protons in H_2O (or the deuterons in D_2O in a heavy water reactor). A fast breeder reactor may be cooled by gas, molten salt, or liquid metal. The details of fast breeder reactor coolants are discussed in later chapters. In the case of gas-cooled reactors, the low density of the gas does not provide any significant neutron energy moderation. Nuclei in molten salt or liquid metal coolants have a much larger mass than the protons or deuterons in pressurized water reactors, and therefore have a relatively small effect on the neutron energy through elastic scattering. Induced fission reactions are therefore the result of fast neutrons incident on ^{235}U (or sometimes ^{239}Pu) nuclei. The choice of coolant is governed by considerations of factors such as heat capacity, chemical reactivity, toxicity, (n, γ) cross sections, and melting temperature.

Following along the discussion above concerning the relationship between the degree of neutron energy moderation and fuel enrichment for light water and heavy water-moderated thermal neutron reactors, we can see that the fast breeder reactor will require fuel with a high degree of enrichment. Typically, 20%–25% uranium enriched in ^{235}U is needed for a sustained fission reaction to be possible.

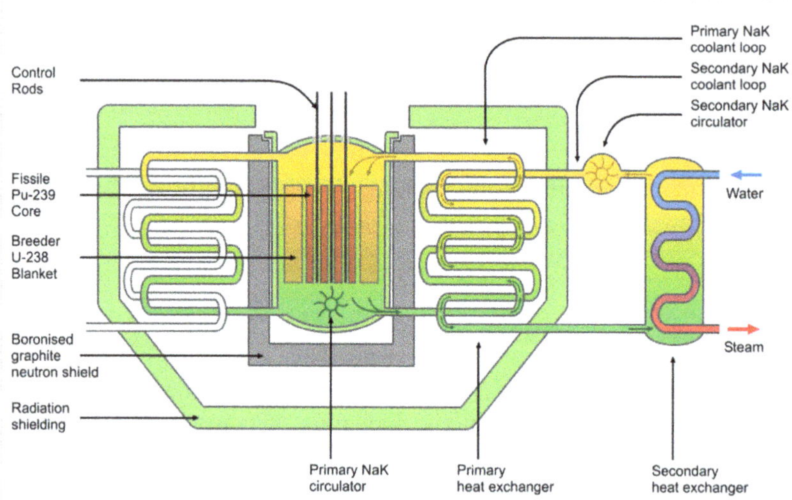

Figure 3.17. Schematic diagram showing the operation of the Dounreay Fast Reactor. Reproduced from Emoscopes (2005). CC BY-SA 3.0.

Finally, as seen in figure 3.17, a breeder blanket surrounds the reactor core. This blanket in a typical fast breeder reactor contains a large amount of ^{238}U, which is converted into fissile ^{239}Pu by the reactions shown above.

Thus, the fast breeder reactor converts fertile ^{238}U into fissile ^{239}Pu, which can be used as a fuel in a thermal neutron reactor or in a fast neutron reactor. An important factor in the effectiveness of this approach is the breeding ratio (BR). The BR is defined as

$$\mathrm{BR} = \frac{\text{Number of fissile nuclei produced}}{\text{Number of fissile nuclei consumed}}.$$

The concept of a breeder reactor requires that BR > 1, i.e., there is a net gain in the number of fissile nuclei. In the above discussion, the number of fissile nuclei produced is the number of new ^{239}Pu nuclei and the number of fissile nuclei consumed is the number of ^{235}U nuclei (extracted from natural uranium) combined with the number of ^{239}Pu nuclei (produced in previous breeding cycles) that undergo fission. It is sometimes convenient to define the breeding gain (BG), which is defined as

$$\mathrm{BG} = \mathrm{BR} - 1.$$

In experimental breeder reactors, a value of BG = 0.2 is acceptable, while a value of 0.4 is considered exceptional. The value of BG is, therefore, a measure of the effectiveness of a breeder reactor program and is a measure of how quickly an expansion of the number of breeder reactors can be accomplished. This effectiveness can be expressed quantitatively as the doubling time. The doubling time is defined as the time required for a breeder reactor to breed as much new fissile material as the fissile material content of its core. This is a measure of how long one breeder reactor needs to produce enough fuel for a new breeder reactor.

Previous work on breeder reactors has followed the approach described above by breeding fissile ^{239}Pu from fertile ^{238}U. As noted above, this approach can also be used to breed fissile ^{233}U from fertile naturally occurring ^{232}Th. Since thorium has no naturally occurring fissile isotope, the process would begin with the use of ^{235}U or ^{239}Pu to breed ^{233}U from ^{232}Th, which could then be used to breed more ^{233}U. The breeding reactions are as follows:

$$\mathrm{n} + {}^{232}\mathrm{Th} \rightarrow {}^{233}\mathrm{Th} + \gamma.$$

^{233}Th then decays by β^- decay,

$$^{233}\mathrm{Th} \rightarrow {}^{233}\mathrm{Pa} + \mathrm{e}^- + \bar{\nu}$$

with a half-life of about 22 min. This β^- decay is followed by

$$^{233}\mathrm{Pa} \rightarrow {}^{233}\mathrm{U} + \mathrm{e}^- + \bar{\nu}$$

with a half-life of about 27 days. ^{233}U is β-stable and decays by α-decay,

$$^{233}\mathrm{U} \rightarrow {}^{229}\mathrm{Th} + \alpha$$

with a half-life of 1.6×10^5 years. Thus, the ^{233}U is sufficiently stable to serve as a fission reactor fuel.

References

Brouwers J J H 2023 Innovative methods of centrifugal separation *Separations* **10** 181

Bodansky D 2004 *Nuclear Energy—Principles, Practices and Prospects* 2nd edn (New York: Springer)

Cadmium 2006 Neutron absorption cross sections as a function of neutron energy for B-10 and B-11 https://commons.wikimedia.org/wiki/File:Neutroncrosssectionboron.png

Duderstadt J J and Hamilton L J 1976 *Nuclear Reactor Analysis* (New York: Wiley)

Dunlap R A 2023 *An Introduction to the Physics of Nuclei and Particles* 2nd edn (Bristol: IOP Publishing)

Emoscopes 2005 Schematic diagram showing the operation of the DFR (Dounreay Fast Reactor) https://commons.wikimedia.org/wiki/File:DFR_reactor_schematic.png

Flowers P, Robinson W R, Langley R and Theopold K 2015 *Chemistry* (Houston: OpenStax) https://openstax.org/books/chemistry/pages/21-4-transmutation-and-nuclear-energy

Inductiveload 2020 Simplified diagram of a gas centrifuge https://commons.wikimedia.org/wiki/File:Countercurrent_Gas_Centrifuge.svg

Kopeikin V, Mikaelyan L and Sinev V 2004 Reactor as a source of antineutrinos: thermal fission energy *Phys. At. Nucl* **67** 1892–9

Krane K S 1988 *Introductory Nuclear Physics* (New York: Wiley)

Lederer C M and Shirley V S 1978 *Table of Isotopes* 7th edn (New York: Wiley)

Ling S J, Sanny J and Moebs W 2016 *University Physics* **vol 3** (Houston, TX: OpenStax)) https://openstax.org/books/university-physics-volume-3/pages/10-5-fission

Murray R L and Holbert K E 2015 *Nuclear Energy—An Introduction to the Concepts, Systems, and Applications of Nuclear Processes* 7th edn (Waltham, MA: Butterworth-Heinemann)

Myers W D and Swiatecki W J 1966 Nuclear masses and deformations *Nucl. Phys.* **81** 1–60

Pbech 2009 Schematic drawing of control rods in a nuclear reactor https://commons.wikimedia.org/wiki/File:Control_rods_schematic.svg

Ripani M 2018 Energy from nuclear fission *EPJ Web Conf.* **189** 00013

ThorEA 2017 *Thermal, epithermal and fast neutron spectra* https://thorea.fandom.com/wiki/Thermal,_Epithermal_and_Fast_Neutron_Spectra

U.S. Nuclear Regulatory Commission 2016 *Fuel rod* https://flickr.com/photos/69383258@N08/26072384441

U.S. Nuclear Regulatory Commission 2023 *Pellet, fuel* www.nrc.gov/reading-rm/basic- ref/glossary/pellet-fuel.html

Wikipedia 2010 Boiling water reactor https://commons.wikimedia.org/wiki/File:BoilingWaterReactor.JPG

Woźnicka U 2019 Review of neutron diagnostics based on fission reactions induced by fusion neutrons *J. Fusion Energy* **38** 376–85

IOP Publishing

Chapter 4

The early history of nuclear fission energy

This chapter provides a review of the development of our understanding of nuclear fission and the history of the development of early nuclear reactors. This chapter begins with an overview of the development of nuclear physics, and particularly with the discovery of nuclear fission in the early half of the twentieth century. The utilization of nuclear fission for the construction of nuclear weapons during World War II is discussed. The history of the development of early nuclear reactors for power production in the 1950s and 1960s is presented. This discussion includes the Obninsk nuclear power plant, which was the first nuclear reactor to produce electrical power, as well as the first nuclear submarine, the USS Nautilus, and early power reactors built in the United Kingdom and the United States.

4.1 Introduction

The concept of the atomic nature of matter dates back to the early Greek philosophers Leucippus of Miletus (fifth century BCE) and Democritus (c.460–c.370 BCE). It is generally believed that this theory and other early atomic theories viewed each material as being comprised of a unique type of atom that endowed the material with its unique physical properties. In the seventeenth century, Robert Boyle (1627–91) developed the concept of elements and the idea that they could be combined to form different chemical compounds. This idea was further developed in the early-nineteenth century by John Dalton (1766–1844). Dalton proposed an atomic model where the atoms of different elements were distinguished on the basis of their weight, with hydrogen as the element with the lightest atoms. Shortly after Dalton's work, William Prout (1785–1850) suggested that all elements had atoms that were comprised of varying numbers of hydrogen atoms. While this approach seemed valid on the basis of Dalton's measured atomic masses, more accurate measurements showed that this hypothesis was not strictly correct. By the late-nineteenth century evidence began to suggest that atoms had internal structure, and this led to our understanding of the atomic nucleus. This chapter provides an

doi:10.1088/978-0-7503-6069-2ch4

overview of the discoveries up to the mid-twentieth century that provided an accurate description of the nucleus and how this knowledge led to the understanding and utilization of nuclear fission reactions.

4.2 The development of nuclear physics

In 1896, Antoine Henri Becquerel (1852–1908) first observed radiation resulting from the radioactive decay of an unstable nucleus. Subsequent work by Becquerel along with that of Ernest Rutherford (1871–1937) showed that three types of radiation were produced by the decay of different materials and that these types of radiation could be distinguished on the basis of three different electric charges—positively charged, negatively charged, and neutral.

Another important step in our understanding of atomic structure came in 1897 when J J (Joseph John) Thomson (1856–1940) discovered that the cathode rays (referred to as Lenard rays at the time) produced in a vacuum tube were negatively charged particles. Thomson measured the charge to mass ratio of these particles and estimated that their mass was of the order of 1/1000 of the mass of the hydrogen atom. He suggested the name 'corpuscles', although eventually they became known as electrons. Thomson also hypothesized that the electron was a fundamental component of atoms and that its negative charge was canceled by an equal positive charge within the atom. Thomson promoted the so-called plum pudding model where the negatively charged electrons were contained in a sphere of uniform positive charge. In 1900, Becquerel showed that the negatively charged particles from radioactive decay had the same charge to mass ratio as Thomson's electrons and speculated that they were, in fact, the same particle.

Although the plum pudding model of the atom was widely accepted in the early-twentieth century, experimental evidence concerning atomic structure was lacking until scattering experiments were conducted beginning around 1909. Initial experiments were conducted by Rutherford in collaboration with Hans Geiger (1882–1945) and Ernest Marsden (1889–1970), and continued by Rutherford himself. These experiments involved the scattering of α-particles produced by radioactive decay on a gold foil. The α-particles were scattered by the Coulomb interaction with the positive charges in the atom. If the plum pudding model was true, then only scattering by small angles would occur. This is because the positive charge, which was distributed over a large (relatively speaking) volume, would not exert sufficiently large Coulomb forces to produce large angle scattering. However, Rutherford found that one α-particle in about 20 000 was scattered by a large angle (>90°). This could be interpreted as the scattering by an atom in which the positive charge was confined to a very small volume at the center of the atom. Most α-particles would scatter from the Coulomb field at a large distance from the positive nucleus and experience only scattering by a small angle. However, a small fraction of the α-particles would approach the nucleus at a small enough distance to be scattered by a large angle by the greater Coulomb field. Based on these results, Rutherford estimated that the positive atomic nucleus had a diameter of less than 10 fm (fm $= 10^{-15}$ m), which is reasonably consistent with modern measurements.

While Rutherford's experiments confirmed that the atom consisted of a very small positively charged nucleus surrounded by a much larger cloud of electrons, the exact nature of the positively charged nucleus was unknown. Further information concerning the nature of the atomic nucleus came in 1917 when Rutherford studied the reaction of α-particles on ^{14}N nuclei. This was, in fact, the first experimentally observed nuclear reaction. Rutherford first believed that the α-particle merely scattered from the nitrogen nucleus and kicked-out a positively charged hydrogen nucleus. The hydrogen nucleus was given the name 'proton', although it was suggested that it should be called 'prouton' in honor of William Prout (1785–1850; see Browne 1923). In 1925, Rutherford properly identified the reaction that he had observed as

$$\alpha + {}^{14}\mathrm{N} \rightarrow {}^{17}\mathrm{O} + \mathrm{p}$$

where the hydrogen nucleus was now designated as 'p' (for proton) and the reaction involved the absorption of the α-particle by the nitrogen nucleus and the release of the proton.

During the early-1920s, it was commonly assumed that the positive charge of the nucleus was provided by a number of protons that determined the atomic number (Z) of the nucleus and that the additional measured mass of the nucleus was provided by proton–electron pairs. However, by the later part of the 1920s it had become obvious that this model was not viable for at least two reasons, i.e., angular momentum considerations and energy considerations. In the first case, the development of Bose–Einstein statistics by Satyendra Nath Bose (1894–1974) and Albert Einstein (1879–1955) in 1924, and Fermi–Dirac statistics by Enrico Fermi (1901–54) and Paul Dirac (1902–84) in 1926, showed that a proton–electron pair would have integer spin, and this was not compatible with the observed properties of certain atoms. Second, for the electron to be bound inside a nucleus, its binding energy would have to be greater than its kinetic energy. The Coulomb energy, E_C, for an electron in a nucleus of charge Z and radius R is given by,

$$E_C = \frac{Ze^2}{4\pi\varepsilon_0 R}$$

where e is the electronic charge and ε_0 is the permittivity of free space. In this expression it is convenient to use the value of the constant, $\frac{e^2}{4\pi\varepsilon_0} = 1.44$ MeV·fm. Calculated values of E_C range from about 1 MeV for light nuclei to about 18 MeV for heavy nuclei. The total (relativistic) energy of a particle is the sum of the rest mass energy and the kinetic energy as given by

$$E^2 = {m_0}^2c^4 + p^2c^2$$

where m_0 is the rest mass and p is the momentum. When the kinetic energy is much greater than the rest mass energy (as is the case in the present example) then we write the kinetic energy as

$$E_K = pc.$$

In 1924, Louis de Broglie (1892–1987) related a particle's wavelength, λ, to its momentum as

$$\lambda = \frac{h}{p} \tag{4.1}$$

from equation (4.1) the kinetic energy may be written as

$$E_K = \frac{hc}{\lambda}.$$

In this relation it is convenient to use the value of $hc = 1240$ MeV·fm. To confine the electron in the nucleus, we require $\lambda = R$. Using the empirical relation for the nuclear radius (R in fm), in terms of the mass number, A,

$$R = 1.2 \times A^{1/3}$$

gives a range of electron kinetic energies of 160 MeV for a heavy nucleus (e.g. ^{235}U) to 820 MeV for a light nucleus (e.g. ^{2}H). It is clear from this simple analysis that an electron cannot be bound within a nucleus.

The component of the nuclear mass which had been attributed to proton–electron pairs was correctly identified in the early-1930s with the discovery of the neutron. Perhaps the earliest laboratory production of neutrons came in 1930 when Walther Bothe (1891–1957) (along with his student Herbert Becker) used α-particles from the radioactive decay of ^{210}Po,

$$^{210}\mathrm{Po} \rightarrow {}^{206}\mathrm{Pb} + \alpha \tag{4.2}$$

to irradiate a sample containing ^{9}Be. This produced neutrons by the reaction,

$$\alpha + {}^{9}\mathrm{Be} \rightarrow {}^{12}\mathrm{C} + \mathrm{n}. \tag{4.3}$$

Bothe observed the emission of a neutral particle from the reaction in equation (4.3) but since the neutron was not known at the time, he identified it as a gamma-ray. Irène Joliot-Curie (1897–1956) and Jean Frédéric Joliot-Curie (1900–58) used the neutral particles from the reaction in equation (4.3) to irradiate paraffin, which contains a high density of protons (as nuclei of ^{1}H atoms). The neutrons scattered elastically with the protons as described in section 4.5 and transferred much of their kinetic energy to the protons. The Joliot-Curies identified the high energy protons emitted in this experiment and Ettore Majorana (1906–?) correctly interpreted these results as an indication of a massive neutral particle.

James Chadwick (1891–1974) continued work along the lines of that reported by the Joliot-Curies with the goal of determining the mass of the neutral particle (see Chadwick 1932a, 1932b). In some experiments he replaced the beryllium target in equation (4.3) with a boron target and produced neutrons by the reaction

$$\alpha + {}^{11}\mathrm{B} \rightarrow {}^{14}\mathrm{N} + \mathrm{n}. \tag{4.4}$$

Chadwick provided an analysis of the dynamics of this process by looking at the conservation of energy. Equating the total energy on the two sides of the reaction in equation (4.4) provides the expression,

$$\frac{1}{2}m_{\alpha}v_{\alpha}{}^{2} + m_{\alpha}c^{2} + m_{\mathrm{B}}c^{2} = \frac{1}{2}m_{\mathrm{N}}v_{\mathrm{N}}{}^{2} + m_{\mathrm{N}}c^{2} + \frac{1}{2}m_{\mathrm{n}}v_{\mathrm{n}}{}^{2} + \mathrm{m}_{\mathrm{n}}c^{2}$$

in terms of the masses and velocities of the α-particle, boron nucleus, nitrogen nucleus, and the neutron. This expression can be rewritten to solve for the neutron mass as

$$m_{\mathrm{n}} = \frac{\frac{1}{2}m_{\alpha}v_{\alpha}{}^{2} + m_{\alpha}c^{2} + m_{\mathrm{B}}c^{2} - \frac{1}{2}m_{\mathrm{N}}v_{\mathrm{N}}{}^{2} - m_{\mathrm{N}}c^{2}}{\left(c^{2} + \frac{v_{\mathrm{n}}{}^{2}}{2}\right)}.$$

All of the terms on the right-hand side of this equation could be determined experimentally or were known constants, except for the velocity of the nitrogen nucleus and the velocity of the neutron after the reaction. As in the Joliot-Curie experiment, Chadwick directed the neutrons to a target with a high density of protons. He assumed that the neutron mass was similar to the proton mass, so that a direct elastic collision between the moving neutron and the stationary proton would result in a transfer of the total neutron energy (and hence velocity) to the proton (see section 3.5). The velocity of the nitrogen nucleus could then be determined from the conservation of momentum. Based on his studies, Chadwick reported a neutron mass of 938±1.8 MeV c^{-2}, compared to the currently accepted value $m_{\mathrm{n}} = 939.57$ MeV c^{-2}. With the discovery of the neutron, the composition of the atomic nucleus became known.

4.3 Early studies of nuclear fission

Shortly after the discovery of the neutron, Fermi undertook experiments to create transuranic elements by irradiating uranium nuclei with neutrons (see Fermi 1933). It is interesting that Fermi replaced the standard (at that time) polonium–beryllium neutron source using the process in equations (4.2) and (4.3) with a stronger radon-beryllium neutron source where the α-particles were produced by the decay,

$$^{222}\mathrm{Rn} \rightarrow {}^{218}\mathrm{Po} + \alpha$$

and neutrons were subsequently produced through equation (4.3). The results of these experiments were a component of the work for which Fermi was awarded the 1938 Nobel Prize in Physics. However, Fermi's claims of the creation of transuranic elements were not universally accepted. In fact, in 1934 Ida Noddack (1896–1978) suggested that Fermi's experiments of neutron irradiation of uranium nuclei caused the uranium to fission into two lighter nuclei (Noddack 1934) and this suggestion, to a large extent, turned out to be correct.

The uncertainty in Fermi's neutron irradiation experiments caused considerable activity in the field in the mid-1930s. In 1934, Irène and Jean Frédéric Joliot-Curie studied the irradiation of a variety of nuclides and reported the production of artificial radioactive nuclides by processes such as

$$\alpha + {}^{27}\mathrm{Al} \rightarrow {}^{30}\mathrm{P} + \mathrm{n}.$$

They were awarded the 1935 Nobel Prize in Chemistry for this work. Thus, in the late-1930s the transmutation of elements by their bombardment with various particles through reactions such as the (n, γ) process, often followed by α-decay or β-decay, was well established. The first clear evidence of induced nuclear fission came in 1938 when Otto Hahn (1879–1968) along with Fritz Strassmann (1902–80) irradiated uranium with neutrons and observed barium in the by-products (see Hahn and Strassmann 1939). Long-time theoretical physics collaborators on the project, Lise Meitner (1878–1968) and Otto Robert Frisch (1904–79), provided the theoretical basis for induced nuclear fission in uranium and calculated the energy release for the process (Frisch 1939, Meitner and Frisch 1939). Otto Hahn was awarded the 1944 Nobel Prize in Chemistry (individually) for the discovery of nuclear fission. It is commonly felt that Frisch and (particularly) Meitner were unjustly omitted as recipients of this award. In recent years there has been detailed analysis of the reasons for the 1944 Nobel Chemistry Prize and Meitner's omission (Sime 1996, Crawford *et al* 1997).

Two crucial pieces of information concerning nuclear fission came shortly after the reports by Hahn, Strassmann, Meitner, and Frisch. First, these experiments were repeated by Fermi and co-workers (Anderson *et al* 1939) and their analysis of the results strongly suggested that it was the ^{235}U component of the uranium that was undergoing neutron induced fission. This observation indicated the distinction between fissile and non-fissile materials. Second, Frédéric Joliot–Currie and co-workers (Von Halban *et al* 1939) observed that each induced fission emitted more than one neutron. They initially reported 3.5 neutrons per fission but later corrected the value to 2.6 neutrons per fission, which is close to the currently accepted value for ^{235}U. This observation provided clear evidence that a fission chain reaction could be initiated. Fermi and Leo Szilard (1898–1964) almost immediately considered the possibility of the construction of a nuclear reactor. Further evidence for this possibility was provided in 1940 when Georgy Flyorov (1913–90) and Konstantin Petrzhak (1907–98) reported the observation of the spontaneous fission of uranium. These studies were conducted 60 m underground (in a train station) in order to shield them from cosmic rays, which could produce secondary neutrons and cause induced fission. This spontaneous fission process could then initiate a chain reaction. In fact, uranium deposits have been discovered in Gabon which show evidence of a natural self-sustained fission reaction that occurred about 1.7 billion years ago when the natural abundance of ^{235}U was much higher than it is at present.

In 1941, the options for producing fission chain reactions increased when Glenn Seaborg (1912–99) and co-workers discovered that fissile ^{239}Pu could be bred by the neutron irradiation of non-fissile ^{238}U by the reaction

$$n + {}^{238}U \rightarrow {}^{239}U + \gamma \tag{4.5}$$

followed by the β-decays,

$${}^{239}U \rightarrow {}^{239}Np + e^- + \bar{\nu} \tag{4.6}$$

and

$${}^{239}Np \rightarrow {}^{239}Pu + e^- + \bar{\nu} \tag{4.7}$$

4.4 The Manhattan project

The knowledge of nuclear fission reactions that had been acquired up to beginning of the 1940s played a crucial role in the development of the nuclear fission weapon by the United States and its British and Canadian allies during World War II. This project, known as the Manhattan Project, was conducted under the direction of J Robert Oppenheimer (1904–67). The history of the Manhattan Project is described in detail by Rhodes (1986) and the technical details, as they were known in 1943, are described in a series of lectures by Serber (1992). In the present section, a brief overview of the project, dealing particularly with the technical developments and difficulties, is given.

As discussed briefly in the previous chapter, fissile nuclei can readily undergo induced fission. This process produces excess neutrons that can induce further fissions and lead to a chain reaction. The development and construction of a nuclear fission weapon includes a number of challenges, perhaps, principal among these is the production of suitable quantities of fissile material. Two possible fissile nuclides were considered for the Manhattan Project, naturally occurring ^{235}U and artificial ^{239}Pu, produced by breeding from ^{238}U according to the reactions in equations (4.5) to (4.7).

In the case of ^{235}U, it is important to separate this isotope, which occurs with 0.72% natural abundance, from the remaining ~99.3% non-fissile ^{238}U. Today, the most widely used method of producing enriched ^{235}U is through the use of a gas centrifuge, as previously discussed in section 3.6. Uranium, prepared as gaseous uranium hexafluoride (UF_6), could be separated according to its isotopic mass. However, in the 1940s, suitable centrifuges were not sufficiently reliable to produce the necessary quantity of enriched ^{235}U. Alternate techniques, all of which were used during the Manhattan Project to produce uranium enriched to around 85% ^{235}U, were electromagnetic separation, gaseous diffusion, and thermal diffusion.

Electromagnetic separation uses a calutron, which is a type of mass spectrometer. The calutron ionizes the uranium atoms and accelerates them using electric fields. The uranium ions are then passed through a region of magnetic field and different isotopes (with different masses but the same charge) are deflected along different trajectories.

The gaseous diffusion technique is based on Graham's law of diffusion, named after Scottish chemist Thomas Graham (1805–69), and states that the diffusion rate of a gas is inversely proportional to the square root of its molecular mass. Gaseous UF_6 is enclosed in a box with a semipermeable membrane. $^{235}UF_6$ diffuses through the membrane faster than $^{238}UF_6$, where the ratio of diffusion rates is,

$$\frac{\mathrm{rate}(^{235}\mathrm{UF}_6)}{\mathrm{rate}(^{238}\mathrm{UF}_6)} = \left[\frac{m(^{238}\mathrm{UF}_6}{m(^{235}\mathrm{UF}_6)}\right]^{1/2} = \left[\frac{352}{349}\right]^{1/2} = 1.0043 \tag{4.8}$$

After multiple stages of enrichment, suitably pure $^{235}UF_6$ can be obtained.

Thermal separation utilizes principles based on Chapman–Enskog theory which states that mixed gases in a thermal gradient will partially separate with the heavier one tending to concentrate at the cold end of the gradient and the lighter one at the hot end of the gradient. By this method, $^{235}UF_6$ could be separated from $^{238}UF_6$.

The preparation of enriched ^{239}Pu (typically around 95%) involves two steps: first, the breeding of ^{239}Pu from ^{238}U; and second, the separation of the plutonium from the uranium, fission by-products, and any other impurities. For the purpose of breeding ^{239}Pu according to the reactions in equations (4.5)–(4.7), the world's first controlled fission reactor was constructed under the direction of Enrico Fermi. This was located at the University of Chicago and is generally referred to as Chicago Pile-1 (see figure 4.1). Chicago Pile-1 consisted of 330 tonnes of graphite blocks as the moderator and 4.9 tonnes of uranium metal and 41 tonnes of uranium oxide as fuel. It had no radiation shielding or cooling system because its power output was less than one watt.

The basic feasibility of ^{239}Pu production was demonstrated by Chicago Pile-1 and this knowledge was used to construct a larger test reactor, the *X*-10 Graphite Reactor at Oak Ridge National Laboratory in Tennessee, and finally a production reactor at the Hanford Site in Washington state.

Separation of the plutonium (which had a concentration of about 250 ppm) from uranium and other elements was not straightforward. The process, known as the bismuth phosphate process, utilized the fact that plutonium existed in two valence states, 4+ and 6+, and that the chemical properties of the two states were quite different. The plutonium containing sample was dissolved in nitric acid and, through the addition of an appropriate carrier (in this case bismuth phosphate), the

Figure 4.1. Chicago Pile-1, the world's first nuclear reactor at the University of Chicago in 1942. U.S. National Archives and Records Administration (434-RF-62(1)). Image stated to be in the public domain.

plutonium and carrier could be precipitated from the solution, thereby separating it from the uranium and other elements. The plutonium and carrier were then redissolved and the plutonium was oxidized by the addition of (for example) potassium permanganate. The carrier could then be precipitated leaving the plutonium in solution.

Two possible designs of the fission bomb were considered. In the gun-type design (figure 4.2, top panel) two subcritical masses of fissile material were combined using a conventional explosive to produce a mass of fissile material greater than the critical mass (see chapter 3), which would then rapidly undergo an uncontrolled chain reaction. The implosion-type design (figure 4.2, bottom panel) used a conventional explosive to compress a subcritical mass to increase its density and achieve super-criticality. The gun-type design was used for the ^{235}U device, and the implosion-type design was used for the ^{239}Pu device.

While it was generally believed that the gun-type ^{235}U device would function successfully, there was somewhat more concern over the operation of the implosion-type device. Therefore, a test version of the ^{239}Pu fission bomb (see figure 4.3) was constructed and tested at White Sands, Alamogordo, New Mexico. The test, known

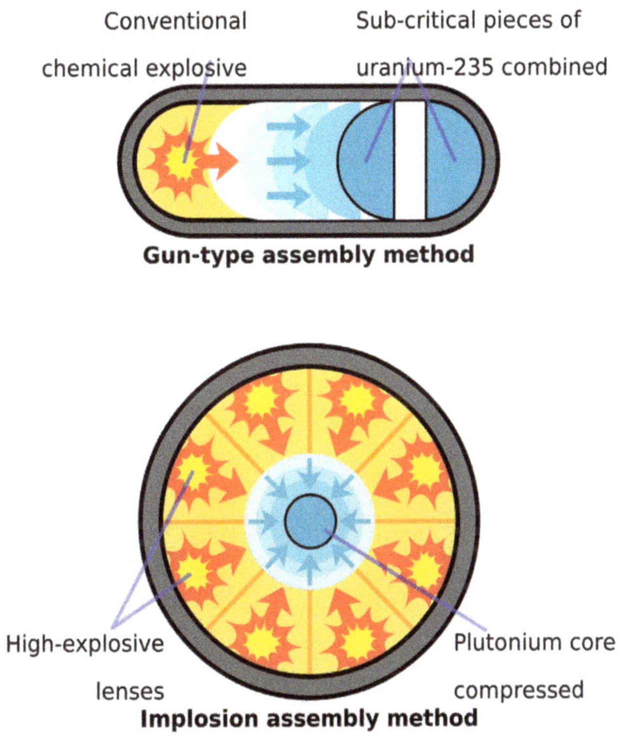

Figure 4.2. Two possible designs for a fission bomb, (top panel) gun-type design and (bottom panel) implosion-type design. This schematic representation of the two methods with which to assemble a fission bomb image has been obtained by the author from the Wikimedia website where it was made available by Fastfission (2006). Image stated to be in the public domain. It is included within this chapter on that basis. It is attributed to Fastfission.

Figure 4.3. Trinity device with Norris Bradbury, group leader. This Trinity Test image has been obtained by the author from the Wikimedia website where it was made available by the Federal Government of the United States (1945). Image stated to be in the public domain. It is included within this chapter on that basis. It is attributed to the Federal government of the United States.

as Trinity, was carried out on 16 July 1945 and an image of the resulting nuclear explosion is shown in figure 4.4.

On 6 August 1945, the United States detonated a gun-type ^{235}U fission bomb over the Japanese city of Hiroshima. On 9 August 1945, the United States, detonated an implosion-type ^{239}Pu fission bomb over the Japanese city of Nagasaki. These two devices are shown in figure 4.5.

4.5 Clementine

The Los Alamos Fast Plutonium Reactor (known by the code name *Clementine*) was an early post-war reactor that was located at Los Alamos National Laboratory in New Mexico. Patenaudea and Freibert (2023) have provided an overview of the design, construction, and use of this reactor. It was unique compared to other reactors that were constructed during this period because it was a fast neutron reactor (although not a breeder reactor) that utilized ^{239}Pu as a fuel and liquid mercury as a coolant. Figure 4.6 shows a photograph of the reactor. The dimensions of the outside of the reactor were approximately 3.4 m × 4.6 m by 2.5 m high.

The reactor core was located at the bottom of a vertical steel cylinder that was 1.17 m long and had a 15.2 cm inside diameter. The fuel assembly was a cylinder 14 cm long by 15 cm in diameter, as shown in figure 4.7. The fuel assembly contained 55 ^{239}Pu fuel rods, as illustrated, along with two control rods comprised of natural

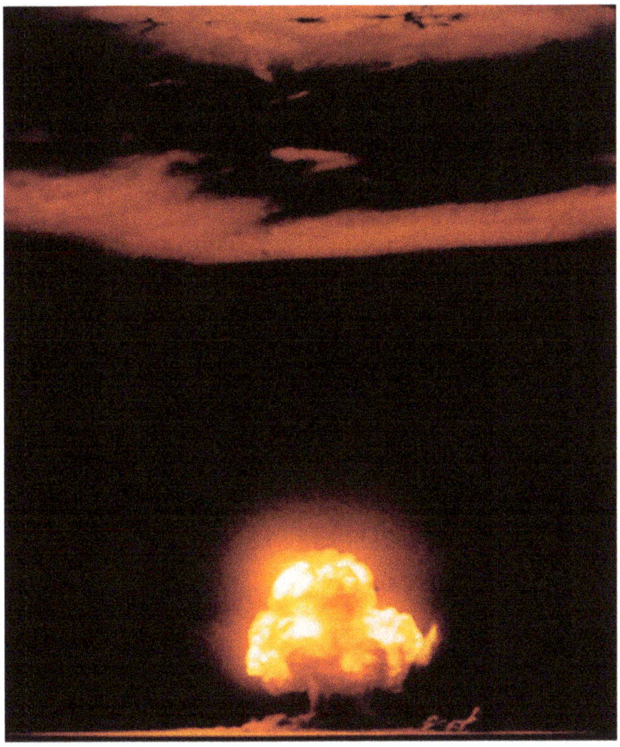

Figure 4.4. Trinity shot, the first nuclear test explosion. This famous color photograph of the "Trinity" shot, the first nuclear test explosion image has been obtained by the author from the Wikimedia website where it was made available by Jack W. Aeby (1945). Image stated to be in the public domain. It is included within this chapter on that basis. It is attributed to Jack W. Aeby.

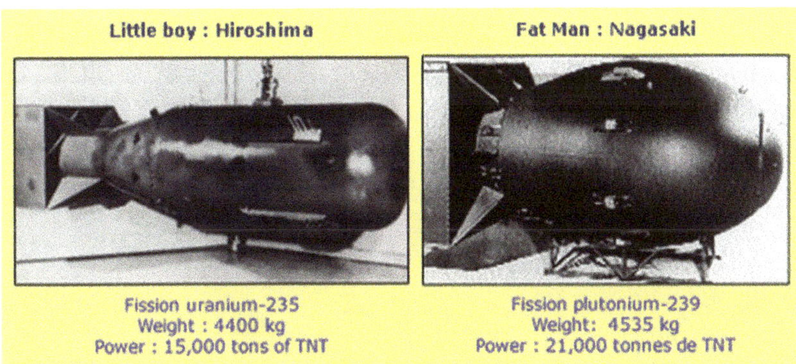

Figure 4.5. (Left-hand panel) The gun-type ^{235}U fission bomb, nicknamed 'Little Boy' and (right-hand panel) the implosion-type ^{239}Pu fission, nicknamed 'Fat Man'. ShanekPPS (2016). CC BY-SA 4.0.

uranium and enriched ^{10}B. Since the reactor utilized fast neutrons, there was no moderator. Each ^{239}Pu fuel rod was 14 cm long by 1.64 cm in diameter and was clad in 0.05 cm of steel. The core was surrounded by a blanket of natural uranium.

Figure 4.6. Los Alamos Fast Plutonium Reactor (also known as Clementine) located at Los Alamos National Laboratory. Reproduced from Jurney *et al* (1954), courtesy of HathiTrust. Image stated to be in the public domain.

Figure 4.7. Core of the Los Alamos Fast Plutonium Reactor. Reproduced from Jurney *et al* (1954), courtesy of HathiTrust. Image stated to be in the public domain.

This could be raised or lowered around the core. Its purpose was to reflect neutrons back to the core to increase output (not to breed new fuel). The reactor was shielded by layers of steel, lead, and concrete.

Construction on Clementine began in 1945 and the reactor achieved criticality in 1946. It first reached its full power of 25 kW (thermal) in 1949 and was operated until 1952. The primary goal of the Clementine project was to study the interaction of neutrons with a variety of materials that could contribute to the nuclear weapons program, as well as to the construction of commercial power reactors. Holes through the shielding allowed fast neutrons to be directed to various experiments.

Clementine provided important information about the neutron cross sections of more than 40 elements. It was also the first reactor to demonstrate reactor control through the use of delayed neutrons (see section 3.7). Finally, it provided information for an assessment of the feasibility of constructing a commercial fast breeder reactor.

4.6 Early nuclear power development

Following the development of the fission bomb during World War II, it was obvious that nuclear reactions were a considerable source of energy that could also be utilized for peaceful purposes. At the end of the war those countries that were directly involved in the Manhattan Project, that is, the United States, the United Kingdom and Canada, had developed expertise in nuclear fission. As well, during the war, the Soviet Union had initiated its own research program to study nuclear fission for weapons and reactor development. Thus, after the war these four countries pursued the development of nuclear fission for peaceful (or at least non-weapons) purposes, such as generating electricity and ship propulsion, although the continued development of nuclear weapons was also the subject of many research efforts. Although a variety of technologies were considered in the early years of reactor development, each country ultimately focused much of its effort on a particular approach to nuclear reactor design. The United States focused on light water reactors, the United Kingdom focused on gas-cooled graphite-moderated reactors, Canada focused on heavy water reactors and the Soviet Union focused on water-cooled graphite-moderated reactors. Many of the reactors constructed during the early years of commercial nuclear power reactor development, were designed as dual-purpose reactors. That is, they were designed to produce electricity and also serve as research reactors or as a means of producing weapons-grade fissile material. In the present section, we overview some of the earliest reactors that were constructed for the purpose of electricity generation or marine propulsion.

4.6.1 Obninsk Nuclear Power Plant

The first nuclear fission reactor that produced usable electricity was the Obninsk Nuclear Power Plant, as shown in figure 4.8. This facility was located at the Institute of Physics and Power Engineering (IPPE) in Obninsk, Kaluga Oblast, Soviet Union, about 110 km southwest of Moscow. Construction began on 1 January 1951, and the power plant was officially commissioned on 26 June 1954. Electricity generation was discontinued in 1959 and the reactor was subsequently used only for research and isotope production until it was decommissioned on 29 April 2002.

Figure 4.8. The world's first nuclear power plant was built at the Institute of Physics and Power Engineering (IPPE) in Obninsk, Russia. Reproduced from IAEA Imagebank (2013). CC BY-SA 2.0.

The Obninsk nuclear reactor was a water-cooled graphite-moderated reactor known as AM-1 or Atom Mirny-1 (Russian for 'Peaceful Atom'). It was the predecessor to the current Russian RBMK reactor (Reaktor Bolshoy Moshchnosti Kanalnyy, Russian for 'high-power channel-type reactor'), which is discussed further in sections 5.6 and 6.5. The AM-1 produced about 5 MW_e (electrical) output for about 30 MW_{th} (thermal) output and used 5% enriched uranium for fuel. Recent RBMK reactors typically use 1.8% enriched uranium.

4.6.2 USS Nautilus

The first use of a nuclear fission reactor for marine propulsion was in the USS *Nautilus* submarine (SSN 571). Development of the USS *Nautilus* occurred at about the same time as the construction of the Obninsk reactor. The first prototype marine nuclear reactor is shown in figure 4.9. This reactor was known as the S1W—S for submarine, 1 for first generation, and W for Westinghouse, the contractor that constructed it. It was a pressurized light water reactor (see section 5.4) that used enriched uranium as a fuel. The USS Nautilus used a second generation of this reactor, the S2W, which produced 10 MW_{th}, and was propelled by geared steam turbines driven by thermal energy from the reactor. The general layout of the propulsion and electrical system of a pressurized light water reactor submarine is illustrated in figure 4.10.

Construction of the USS *Nautilus* began on 14 June 1952. It was launched on 21 January 1954 in the Thames River in Groton, CT (see figure 4.11), and officially commissioned on 30 September 1954. The USS *Nautilus* was in service until 3 March

Figure 4.9. The land-based S1W prototype reactor for the world's first nuclear submarine. This land-based S1W prototype nuclear engine for the world's first nuclear submarine image has been obtained by the author from the Wikimedia website where it was made available by Idaho National Laboratory (2010). Image stated to be in the public domain. It is included within this chapter on that basis. It is attributed to Idaho National Laboratory.

1980 and is currently on display as part of the Submarine Force Library and Museum in Groton, CT.

4.6.3 Calder Hall

After the end of World War II, the United Kingdom's Ministry of Supply established a nuclear site at Sellafield near Seascale, Cumbria, England. Initially, the facility consisted of two gas-cooled graphite-moderated reactors, designated the Windscale Piles, that were designed for the production of weapons-grade plutonium. Pile number 1 became operational in October 1950 and Pile number 2 became operational in June 1951. They were shut down after the Windscale fire in 1957 (see section 6.3 for further details of the Windscale accident).

The world's first full-scale commercial power reactor facility, known as Calder Hall Nuclear Power Station, was constructed at the Sellafield site and was connected

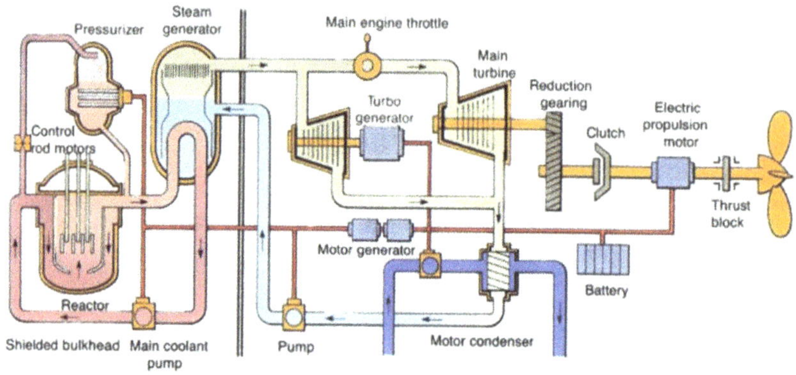

Figure 4.10. Generic diagram of the propulsion and electrical systems of a nuclear submarine utilizing a pressurized light water reactor. Reproduced from T6985wsx (2016). CC BY-SA 4.0.

Figure 4.11. The 98 m long nuclear-powered submarine USS *Nautilus* in the Thames River (Groton, CT) after her launch on 21 January 1954. This file photo taken 21 January 1954 of the nuclear-powered submarine USS Nautilus (SSN 571) in the Thames River shortly after a christening ceremony has been obtained by the author from the Wikimedia website where it was made available by the U.S. Navy (2012). Image stated to be in the public domain. It is included within this chapter on that basis. It is attributed to the U.S. Navy.

to the grid on 27 August 1956. The Calder Hall facility, as shown in figure 4.12, consists of four 60 MW_e Magnox reactors. The Magnox reactor is a gas-cooled graphite-moderated reactor operating on natural uranium fuel. Carbon dioxide is used as the cooling gas, as is typically the case for graphite-moderated reactors, because graphite readily oxidizes in air at elevated temperatures. Natural uranium, rather than enriched uranium, is a suitable fuel for these reactors due to the

Figure 4.12. Calder Hall nuclear power station. Reproduced from UK government agencies (ND). OGL 3.0

effectiveness of graphite as a moderator compared to light water. Carbon dioxide cooling gas is pumped through the reactor core, where it is heated and then flows though a heat exchanger, where the heat is transferred to water to produce steam to operate turbine/generators. The Magnox reactor is a dual-purpose reactor that is designed to produce weapons-grade plutonium and to generate electricity for commercial distribution. Details of the operation of gas-cooled graphite-moderated fission reactions are given in section 5.5.

During the early years of its operation, Calder Hall was primarily considered to be a plutonium production facility, although in later years it was used exclusively for electricity production. The facility was shut down on 31 March 2003 and de-fueling of the reactors was completed on 3 September 2019.

4.6.4 Shippingport Atomic Power Station

The Shippingport Atomic Power Station (see figure 4.13) on the Ohio River in Beaver County, Pennsylvania was located about 40 km from Pittsburgh. It was the first nuclear reactor to commercially produce electricity in the United States. It was also the first full-scale nuclear generating station that did not produce materials for a nuclear weapons program. Construction on the station began on 6 September 1954 and the facility was commissioned on 26 May 1958. It operated until 1982. The reactor vessel is shown during construction in figure 4.14.

Although the Shippingport reactor was a pressurized light water reactor, its design was quite unlike typical light water power reactors and was significantly modified during its lifetime. In a sense, Shippingport could be thought of as an

Figure 4.13. The Shippingport Atomic Power Station. Library of Congress, Prints & Photographs Division, HAER PA-81, 4-SHIP,1. Image stated to be in the public domain.

Figure 4.14. Reactor vessel of the Shippingport Atomic Power Station. Library of Congress, Prints & Photographs Division, HAER PA, 4-SHIP, 1–87. Image stated to be in the public domain.

experimental reactor, although it did provide electricity to the grid. The first core was originally constructed for a nuclear aircraft carrier. When the aircraft carrier construction was canceled, the reactor became available for the Shippingport facility. The core consisted of a 93% enriched ^{235}U inner core (called a seed) surrounded by a natural uranium outer core (referred to as a blanket). A second core, which replaced the first core in 1965, was of a similar design but had a larger seed and produced greater output power. The third, and final, core was installed in

1977 and retained the seed/blanket design. However, it used ^{233}U for the seed and ^{232}Th for the blanket and breeds fissile ^{233}U fuel from non-fissile ^{232}Th, as previously discussed in section 3.8. However, Shippingport was a thermal breeder reactor, in contrast to the typical fast breeder reactor design that is in use at present. The net breeding ratio for this core was 1.01.

Overall, Shippingport was not commercially successful. Over its 25-year existence, it operated at an average capacity factor of around 10%. This can be compared to current commercial light water reactors that operate at a capacity factor of around 93%. With a typical output of 60 MW_e, it was much smaller than typical commercial light water reactors, which are in the range of 1000 MW_e. However, its unique (and changing) design was a useful testbed for the development of later commercial power reactors.

References

Aeby J W 1945 Famous color photograph of the 'Trinity' shot, the first nuclear test explosion. https://commons.wikimedia.org/wiki/File:Trinity_shot_color.jpg

Anderson H L, Booth E T, Dunning J R, Fermi E, Glasoe G N and Slack F G 1939 The fission of uranium *Phys. Rev.* **55** 511–2

Browne A W 1923 A suggested modification of 'proton' to 'prouton' as a memorial to William Prout *Nature* **112** 793

Chadwick J 1932a Possible existence of a neutron *Nature* **129** 312

Chadwick J 1932b The existence of a neutron *Proc. Royal Soc. Lond.* **136** 692–708

Crawford E, Sime R L and Walker M 1997 A nobel tale of postwar injustice *Phys. Today* **50** 26–32

Fermi E 1933 Possible production of elements of atomic number higher than 92 *Nature* **134** 898–9

Fastfission 2006 Schematic representation of the two methods with which to assemble a fission bomb https://en.wikipedia.org/wiki/File:Fission_bomb_assembly_methods.svg

Federal Government of the United States 1945 Trinity Test. Norris Bradbury, group leader for bomb assembly, stands next to the partially assembled Gadget atop the test tower https://commons.wikimedia.org/wiki/File:HD.4G.053_(10540204545).jpg

Frisch O R 1939 Physical evidence for the division of heavy nuclei under neutron bombardment *Nature* **143** 276

Hahn O and Strassmann F 1939 Über den nachweis *u*nd das verhalten der bei der bestrahlung des urans mittels neutronen entstehenden erdalkalimetalle *Sci. Nat.* **27** 11–5

IAEA Imagebank 2013 The world's first nuclear power plant was built at the Institute of Physics and Power Engineering (IPPE) at Obninsk, Russia www.flickr.com/photos/iaea_imagebank/8383320538/

Idaho National Laboratory 2010 Nautilus core https://commons.wikimedia.org/wiki/File:Nautilus_core.jpg

Jurney E T *et al* 1954 *The Los Alamos Fast Plutonium Reactor* (Los Alamos, NM: Los Alamos Scientific Laboratory of the University of California) available at https://babel.hathitrust.org/cgi/pt?id=mdp.39015086416370&seq=1

Meitner L and Frisch O R 1939 Disintegration of uranium by nutrons: a new type of nuclear reaction *Nature* **143** 239–40

Noddack I 1934 Über das element 93 *Angew. Chem.* **47** 653–5

Patenaudea H K and Freibert F J 2023 Oh, my darling clementine: a detailed history and data repository of the los alamos plutonium fast reactor *Nucl. Tech.* **209** 963–1007

Rhodes R 1986 *The Making of the Atomic Bomb* (New York: Simon & Schuster)

Serber R 1992 *The Los Alamos Primer—The First Lectures on How to Build an Atomic Bomb* (Berkeley: University of California Press)

Shanek P P S 2016 The Little Boy and Fat Man bombs dropped on Japan https://commons.wikimedia.org/wiki/File:Little_Boy%2BFat_Man.jpeg

Sime R L 1996 *Lise Meitner: A Life in Physics* (Berkeley: University of California Press)

T6985wsx 2016 Propulsion and electrical https://commons.wikimedia.org/wiki/File:Diagram_Submarine_Reactor_%26_Steam_Plant.jpg

UK Government Agencies ND Calder Hall nuclear power station, after opening.jpg https://commons.wikimedia.org/wiki/File:Calder_Hall_nuclear_power_station,_after_opening.jpg

U.S. Library of Congress 1968a View of power station from west—Shippingport Atomic Power Station, Image 3 from Survey HAER PA-81 www.loc.gov/resource/hhh.pa1658.photos/?sp=3

U.S. Library of Congress 1968b Reactor vessel of the Shippingport Atomic Power Station, Image 87 from Survey HAER PA-81 www.loc.gov/resource/hhh.pa1658.photos/?sp=87

U.S. National Archives 1946 The first nuclear reactor was erected in 1942 in the West Stands section of Stagg Field at the University of Chicago (Image 434-RF-62(1)) https://catalog.archives.gov/id/558600

U.S. Navy 2012 The nuclear-powered submarine USS Nautilus https://commons.wikimedia.org/wiki/File:US_Navy_120120-N-ZZ999-002_In_this_file_photo_taken_Jan._21,_1954,_the_nuclear-powered_submarine_USS_Nautilus_(SSN_571)_is_in_the_Thames_River_shor.jpg

Von Halban H, Joliot F and Kowarski L 1939 Number of neutrons liberated in the nuclear fission of Uranium *Nature* **143** 680

Chapter 5

Current fission reactor designs and the use of nuclear energy

This chapter will describe the nuclear power reactor designs that are currently in use. These reactors include light water designs, developed primarily in the United States, which utilize both boiling water and pressurized water as moderator and coolant. The gas-cooled graphite-moderated reactors that were developed in the United Kingdom and water-cooled graphite-moderated reactors that were developed in the Soviet Union, as well as heavy water-moderated reactors, such as the CANDU reactor developed in Canada, are also included. The growth of the nuclear power industry from the late-1950s to present is discussed, with particular emphasis on the effects of nuclear accidents on the development of new power reactors. The longevity of the use of nuclear power is discussed in the context of uranium and thorium resources, along with the possibility of fuel reprocessing.

5.1 Introduction

The development of nuclear reactors for military applications led the way to the use of this technology for the construction of reactors for the commercial generation of electricity. Some of the earliest reactors for this purpose were discussed in chapter 4. Since then, a number of reactor designs have been developed for the power industry. This chapter reviews the timeline of this development and provides technical details of the various reactor designs that are in use today. This chapter also overviews the use of nuclear power since its development in the 1950s and provides some insight into its possible future directions.

5.2 Nuclear reactor generations

Commercial power reactors constructed up until to the early-1970s are referred to as Generation I reactors. The Wylfa Nuclear Power Station, which contained two Magnox reactors, was located on the island of Anglesey in Wales and became

doi:10.1088/978-0-7503-6069-2ch5

operational in 1971. It was the last functioning Generation I reactor facility when it was shut down in 2015.

The Generation I reactors led the way to the development of more advanced Generation II reactors that were constructed in the 1970s and early-1980s. They were designed to be safer than Generation I reactors and were the first to be designed and constructed with economic viability as a major consideration. These reactors include the water-cooled graphite-moderated reactors at Chernobyl and the boiling water reactors at Fukushima. They had a variety of different core designs, as illustrated in figure 5.1. Most Generation II reactors had an anticipated lifespan of around 40 years, but many have had their operation extended substantially beyond this estimate. The majority of operating commercial reactors in the world today are Generation II.

Generation III nuclear reactors were an evolutionary development from Generation II reactors. They include features that improve safety and economic viability. Some features that have been incorporated into Generation III reactor designs are improved fuel technology and increased thermal efficiency, as well as a greater degree of design standardization. The first Generation III reactors were constructed in the mid-1990s. Generation III+ reactors are a continuation of the trends that were initiated in Generation III reactors. An important safety feature that is common in Generation III+ reactors is a core catcher. This is a device that

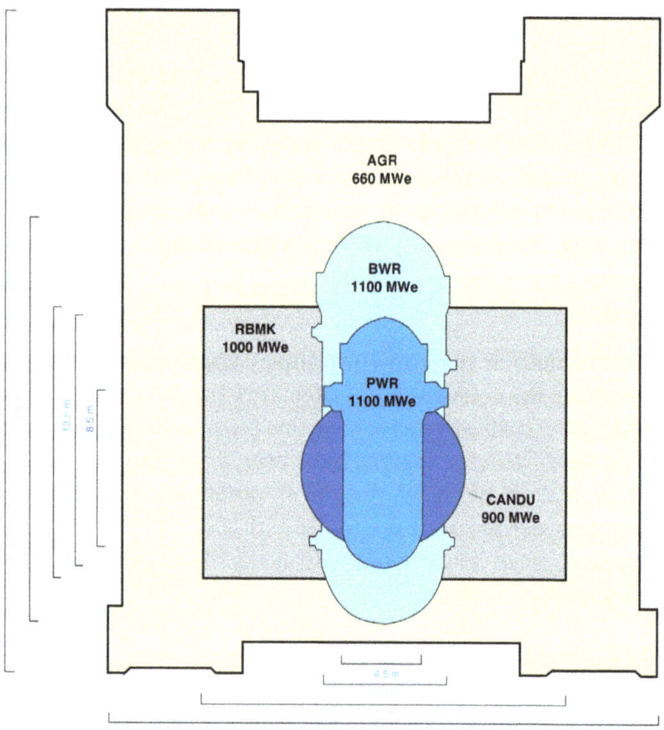

Figure 5.1. Comparison of the sizes of reactor vessels for of some common Generation II nuclear reactors. Reproduced from Fleitz (2016). CC BY-SA 4.0.

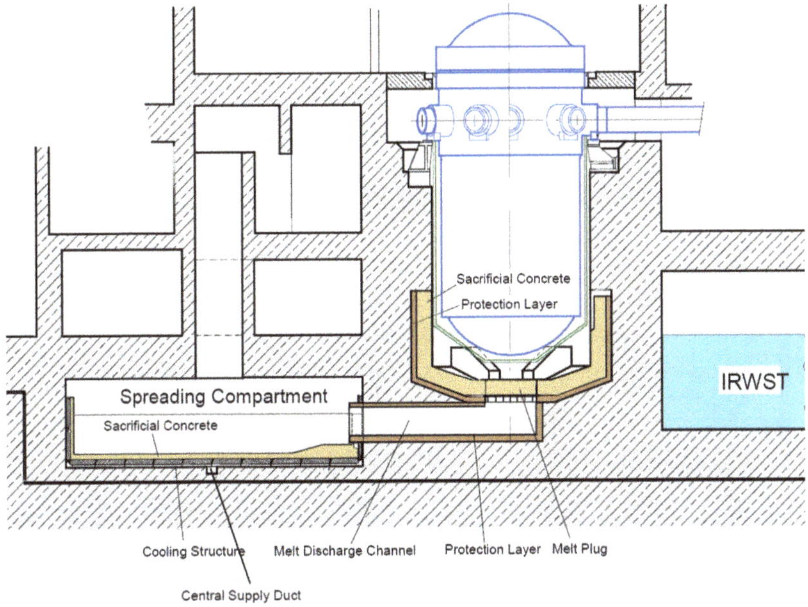

Figure 5.2. Design of the core catcher used in the EPR. This schemata of EPR's core catcher has been obtained by the author from the Wikimedia website where it was made available by Keine Schöpfungshöhe reusing the file from Areva NP (2006). Image stated to be in the public domain. It is included within this chapter on that basis. It is attributed to Areva NP and Keine Schöpfungshöhe.

will catch and confine materials from the core in the event of a core meltdown. An example of this technology is included in the design of the European Pressurized Reactor (EPR), a Generation III+ pressurized light water reactor, as shown in figure 5.2. The first power station utilizing this reactor design was the Taishan Nuclear Power Plant in Guangdong Province, China which became operational in 2018.

Since the mid-1990s, only a dozen or so Generation III/III+ reactors have been constructed. This is partly due to the decline (since the mid-1980s) in interest in nuclear power plant construction. Although there has been increased activity in the nuclear industry in recent years (see section 5.10), new constructions tend to favor Generation II designs.

Generation IV reactors are a revolutionary step from Generation III/III+ reactors. These new concepts are discussed in detail in chapters 7 through 12 of the present book.

5.3 Boiling water reactors

The boiling water reactor is a light water-moderated thermal neutron reactor, as illustrated in figure 5.3. Water is used as the moderator, as described in chapter 3, and also for cooling. Because of the moderating properties of water, fuel is typically enriched to about 3% in ^{235}U, which is the case for all light water reactors discussed below. The cooling water is held at a pressure of about 7.6 MPa (about 75 atm) and boils at around 285 °C. The water that is heated by the core produces steam, which exits the pressure vessel near the top and is used to drive turbines connected to

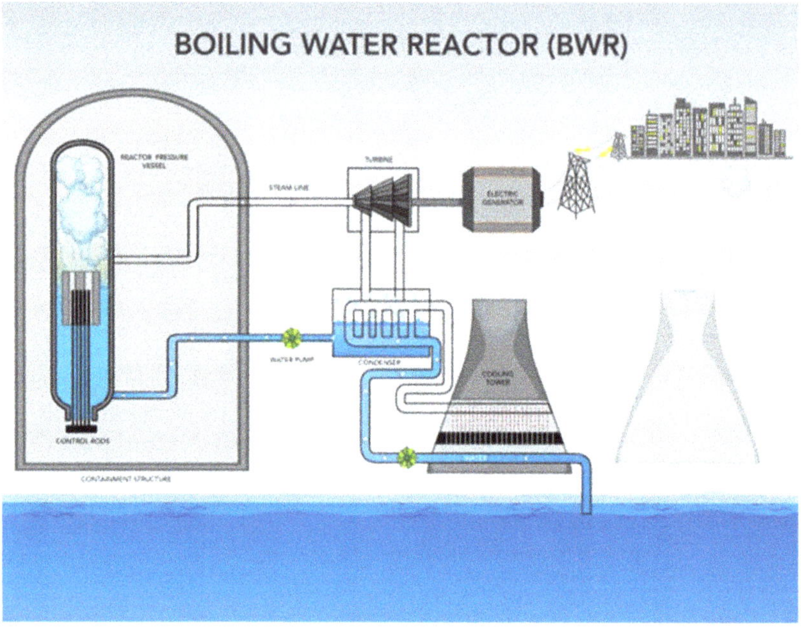

Figure 5.3. Diagram of a boiling water reactor. Reproduced from U.S. Department of Energy (2023). Image stated to be in the public domain.

Figure 5.4. Cooling towers at the Dukovany Nuclear Power Station near Dukovany in the Czech Republic. Reproduced from Frettie (2009). CC BY-SA 3.0.

generators. The steam is then cooled in a condenser and returned to the pressure vessel. Cooling is accomplished by means of cooling water from a suitable reservoir (i.e., river, lake, or ocean) or by means of cooling towers, as illustrated in figure 5.4.

Figure 5.5 shows the details of the reactor pressure vessel for a typical boiling water reactor. Control rods are inserted from the bottom of the reactor pressure vessel and water is circulated by means of pumps.

The mounting of the reactor pressure vessel in the containment structure for an advanced boiling water reactor (the Hitachi-GE UK ABWR) is shown in figure 5.6. In the event of a failure of the reactor pressure vessel, excess pressure can be

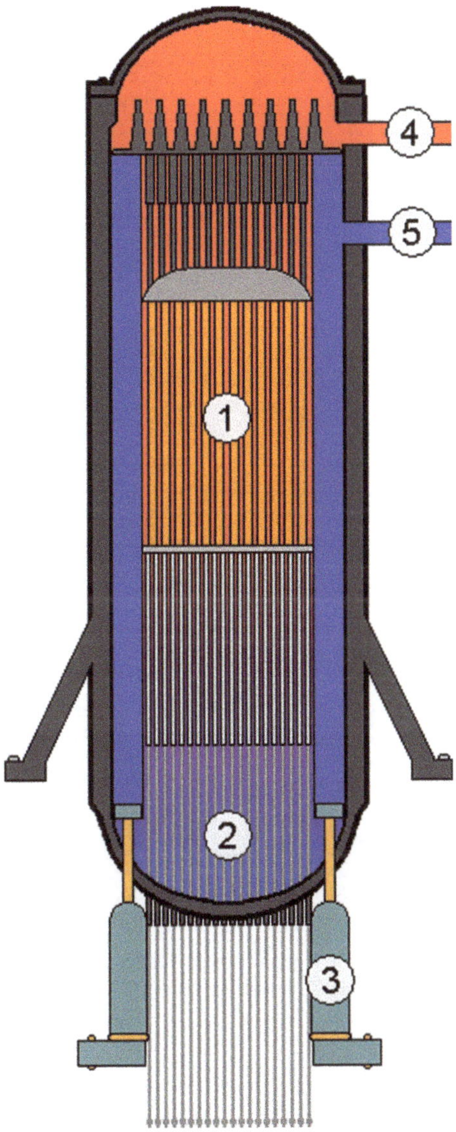

Figure 5.5. Boiling water reactor pressure vessel and core: (1) reactor core, (2) control rods, (3) circulating pumps, (4) steam outlet, and (5) water inlet. Reproduced from ChNPP (2008). CC BY-SA 3.0.

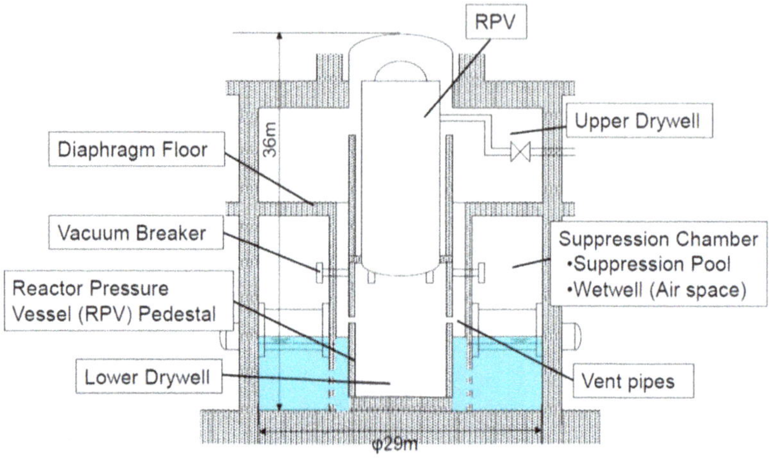

Figure 5.6. Reactor pressure vessel (RPV) and containment structure of an ABWR. Reproduced from Office for Nuclear Regulation (2017). OGL 3.0

Figure 5.7. Reactor number 3 at Forsmark Nuclear Power Plant in Forsmark, Sweden. This is an ABB Atom BWR 75 boiling water reactor that produces 1190 MW_e and which became operational in 1985. Sea water from the Bothnian Sea is used for cooling. Reproduced from robin-root (2006). CC BY-SA 2.0.

minimized by venting through the air chambers around the reactor and steam can be rapidly condensed by routing it to the suppression pool (the blue region in the figure). A typical nuclear power plant utilizing a boiling water reactor is shown in figure 5.7.

5.4 Pressurized water reactors

A diagram of a pressurized light water reactor is shown in figure 5.8. This reactor uses a closed primary water loop for the moderator and reactor cooling. The primary cooling water enters the reactor pressure vessel, is heated by the reactor core, and then exits the vessel near the top of the core. It is held under a pressure of about 15.5 MPa (153 atm) and achieves a maximum temperature of around 315 °C without boiling. The primary cooling water transfers heat to the secondary loop through a heat exchanger (i.e., steam generator in the figure). Steam produced in the secondary loop is used to drive a turbine and generator. Steam from the turbine is condensed in a condenser (as in the case of the boiling water reactor) and is returned to the heat exchanger.

Figure 5.9 shows the details of a reactor pressure vessel from a pressurized water reactor. The diagram shows the pathways for the water that enters the reactor and passes through the core to the outlet. Control rods are typically inserted from the top of the reactor, as illustrated in the figure.

Pressurized water reactors have the advantage over boiling water reactors that the water which passes through the core is in a closed loop and is not used to operate the turbine. This means that any possible radioactive contamination from the core that gets into the cooling water is contained within the closed system. Pressurized water reactors are, by far, the most commonly utilized fission reactor technology for present power reactors (see section 5.9). The technology can also be readily adapted for marine reactors that are used in submarines and aircraft carriers. Figure 5.10 shows a typical example of a nuclear power plant that incorporates pressurized water reactors.

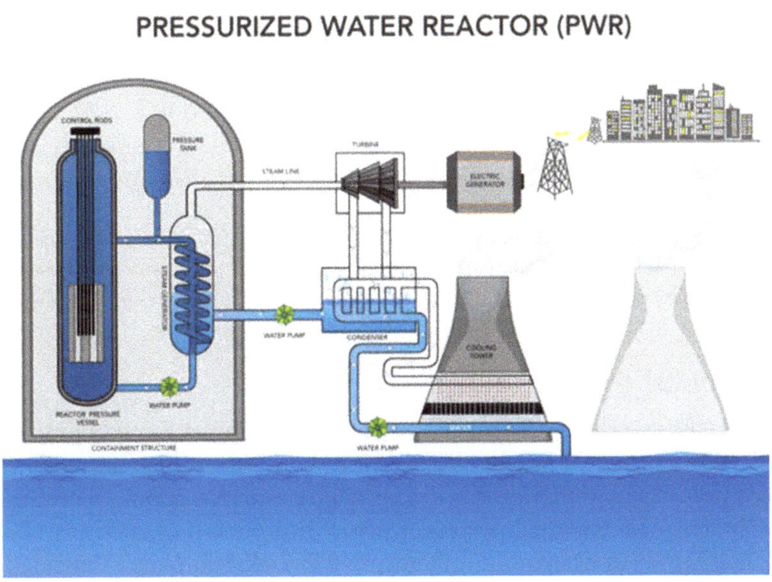

Figure 5.8. Diagram of a pressurized water reactor. Reproduced from U.S. Department of Energy (2023). Image stated to be in the public domain.

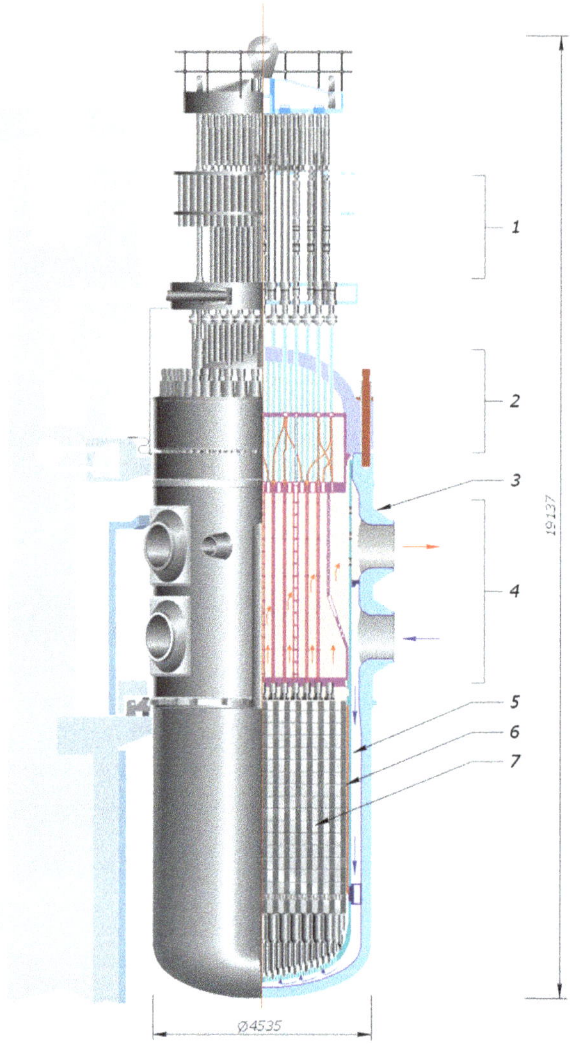

Figure 5.9. Illustration of a pressurized water reactor pressure vessel: (1) control rods, (2) reactor cover, (3) reactor chassis, (4) inlet and outlet nozzles, (5) reactor vessel, (6) active reactor zone, and (7) fuel rods. Reproduced from Panther (2005). CC BY-SA 3.0.

5.5 Gas-cooled graphite-moderated reactors

The original Chicago Pile-1 constructed under the direction of Enrico Fermi was a gas-cooled graphite-moderated reactor. After World War II, the further development of this reactor technology was primarily undertaken in the United Kingdom, where the Magnox reactor was designed as a dual-purpose reactor. It produced electricity and also bred ^{239}Pu for nuclear weapons use. Like other early gas-cooled graphite-moderated reactors that were designed as dual-purpose facilities, the Magnox reactor could be refueled while operating (see section 6.5 on the Chernobyl accident). Figure 5.11 shows a diagram

Figure 5.10. Units 1 and 2 of the Kori Nuclear Power Plant located near Busan, Korea. This facility contains eight pressurized light water reactors with a total generating capacity of 7489 MW_e. Two additional units (2680 MW_e) are under construction. Reproduced from International Atomic Energy Agency (2013). CC BY-SA 2.0.

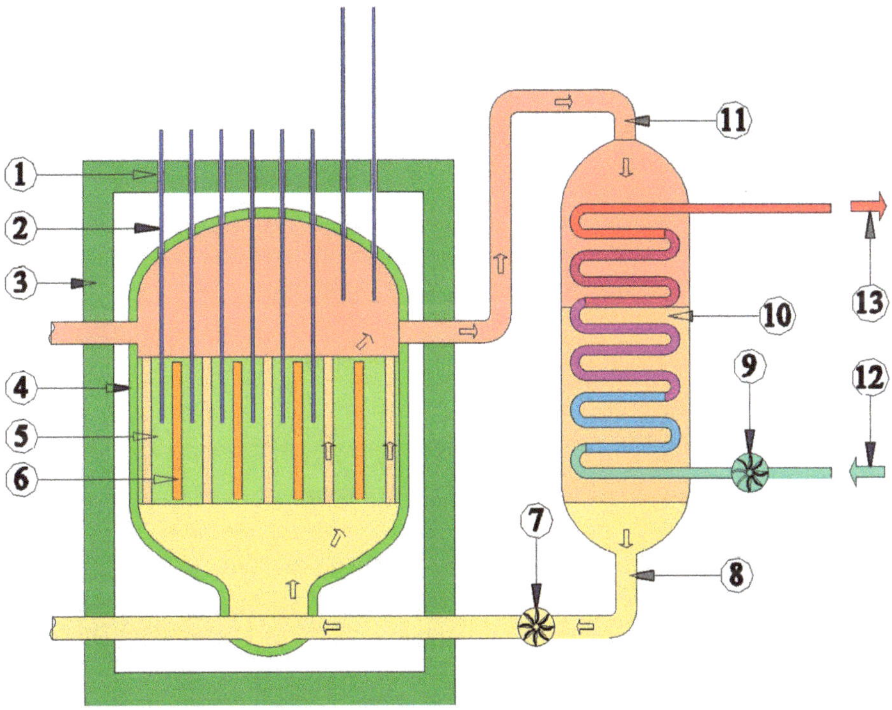

Figure 5.11. Diagram of a Magnox gas-cooled graphite-moderated fission reactor: (1) charge tubes, (2) control rods, (3) concrete shielding, (4) pressure vessel, (5) graphite moderator, (6) fuel rods, (7) gas circulator, (8) return carbon dioxide, (9) water circulator, (10) heat exchanger, (11) hot carbon dioxide, (12) return water, and (13) steam. This Srpskohrvatski image has been obtained by the author from the Wikimedia website where it was made available by Mmarre (2012). Image stated to be in the public domain. It is included within this chapter on that basis. It is attributed to Mmarre.

of a Magnox reactor. Graphite surrounding the fuel rods acts as the neutron moderator, while flowing gas (carbon dioxide) acts as the coolant and carries heat from the core to a heat exchanger where the heat is transferred to water to produce steam to operate the turbines. Because of the moderating properties of graphite, natural uranium is used as the fuel. The name 'Magnox' is derived from magnesium non-oxidizing, i.e., the name of the magnesium-aluminum alloy used to clad the fuel rods (see figure 5.12) which were about 1.2 m in length. Calder Hall (as discussed in section 4.6) was the first Magnox reactor.

The gas-cooled graphite-moderated reactors that are currently in use in the United Kingdom are referred to as advanced gas-cooled reactors (AGR) and utilize a design that is derived from the early Magnox reactor. However, the advanced gas-cooled reactor is designed solely for electricity production. A diagram of a contemporary advanced gas-cooled graphite-moderated reactor is shown in figure 5.13. In this design, the heat exchanger is contained within the containment vessel. Carbon dioxide has

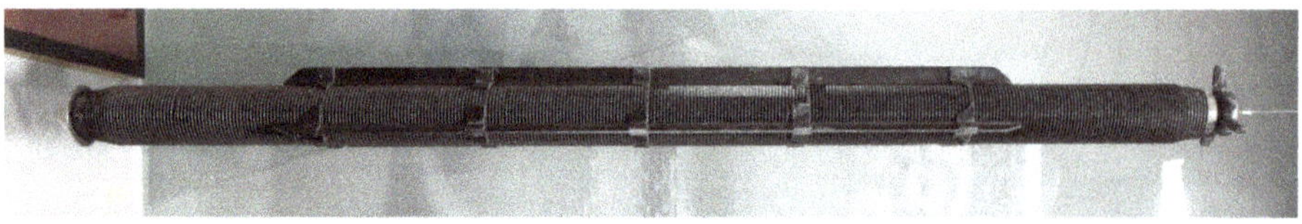

Figure 5.12. Fuel rod from a Magnox reactor. The top of the rod is to the right in the image. Reproduced from Geni (2019). CC BY-SA 4.0.

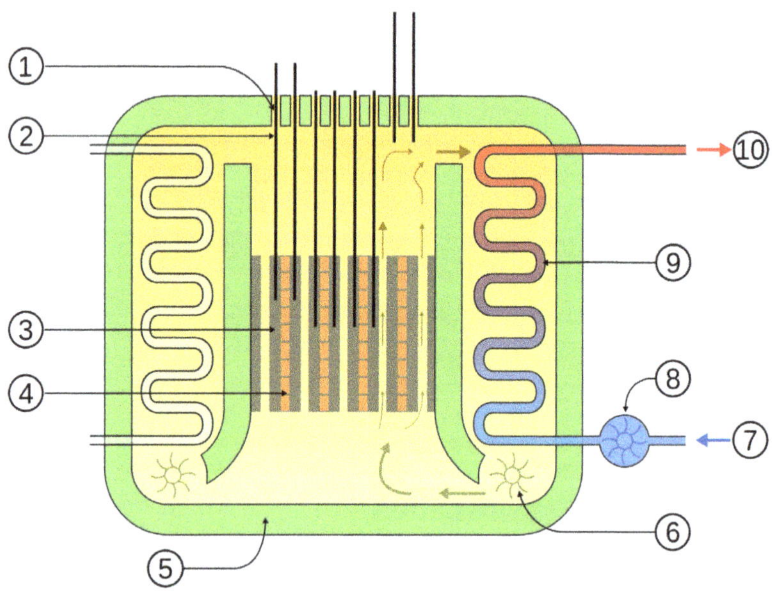

Figure 5.13. Diagram of an advanced gas-cooled graphite-moderated nuclear reactor: (1) charge tubes, (2) control rods, (3) graphite moderator, (4) fuel assemblies, (5) concrete pressure vessel and radiation shielding, (6) gas circulator, (7) water, (8) water circulator, (9) heat exchanger, and (10) steam. Reproduced from MesserWoland (2007). CC BY-SA 3.0.

Figure 5.14. Torness Nuclear Power Station at Torness Point near Dunbar in East Lothian, Scotland. This station consists of two gas-cooled graphite-moderated reactors with a total capacity of 1290 MW_e. Reproduced from Young (2016).

been retained as the cooling gas. In order to improve thermal efficiency for steam generation, the operating temperature of the reactor has been increased, relative to the Magnox design. As a result of the increased operating temperature, stainless steel has been used as cladding on the fuel rods. The increased neutron absorption in the cladding due to the neutron cross section of iron requires the use of fuel enriched to 2.2% to 3.5% ^{235}U. A current nuclear power plant utilizing advanced gas-cooled graphite-moderated reactors is illustrated in figure 5.14.

5.6 Water-cooled graphite-moderated reactors

Water-cooled graphite-moderated reactors were primarily developed in the Soviet Union after World War II. Like the Magnox reactor, they were designed as dual-purpose reactors for the production of both electricity and ^{239}Pu for the nuclear weapons program. The RBMK reactor was the result of this development. The name RBMK is from the Russian 'reaktor bolshoy moshchnosti kanalnyy', meaning 'high-power channel-type reactor'. Because of the neutron cross section of hydrogen, the use of light water as the coolant requires the use of enriched ^{235}U as fuel. Typical RBMK enrichment levels are around 1.8% ^{235}U.

Figure 5.15 shows a schematic of a typical RBMK reactor where the cooling channels for water through the reactor are illustrated. Figure 5.16 shows the top of an RBMK reactor. The array of reactor tubes can be seen in the image, where the square protrusions correspond to control rod locations in the core. The arrangement of control rods of various types (some automatic, some manual, some emergency, etc.) with respect to the fuel assemblies is illustrated in figure 5.17.

Details of the construction of an RBMK reactor fuel rod are illustrated in figure 5.18. The enriched uranium fuel pellets, shown in red, are enclosed in the fuel tubes. The fuel tubes are arranged in bundles and inserted in the channels in the graphite bricks, shown in gray.

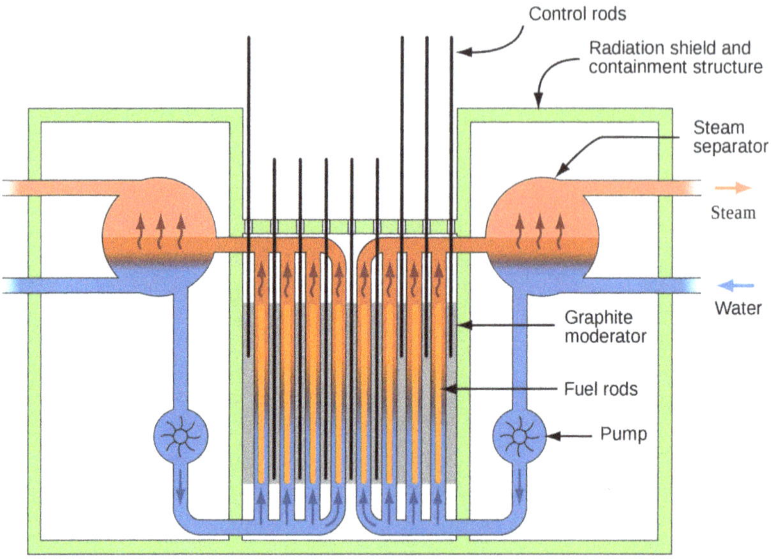

Figure 5.15. Diagram of a water-cooled graphite-moderated nuclear reactor. Reproduced from Emoscopes (2008). CC BY-SA 3.0.

Figure 5.16. Tops of the reactor rubes in the RBMK reactor at the Ignalina Nuclear Power Plant in Visaginas Municipality, Lithuania. This reactor facility was commissioned in 1983 and was shut down in 2009. This image of the reactor tube tops of a RBMK reactor at Ignalina has been obtained by the author from the Wikimedia website where it was made available by Argonne National Laboratory (2008). Image stated to be in the public domain. It is included within this chapter on that basis. It is attributed to the Argonne National Laboratory.

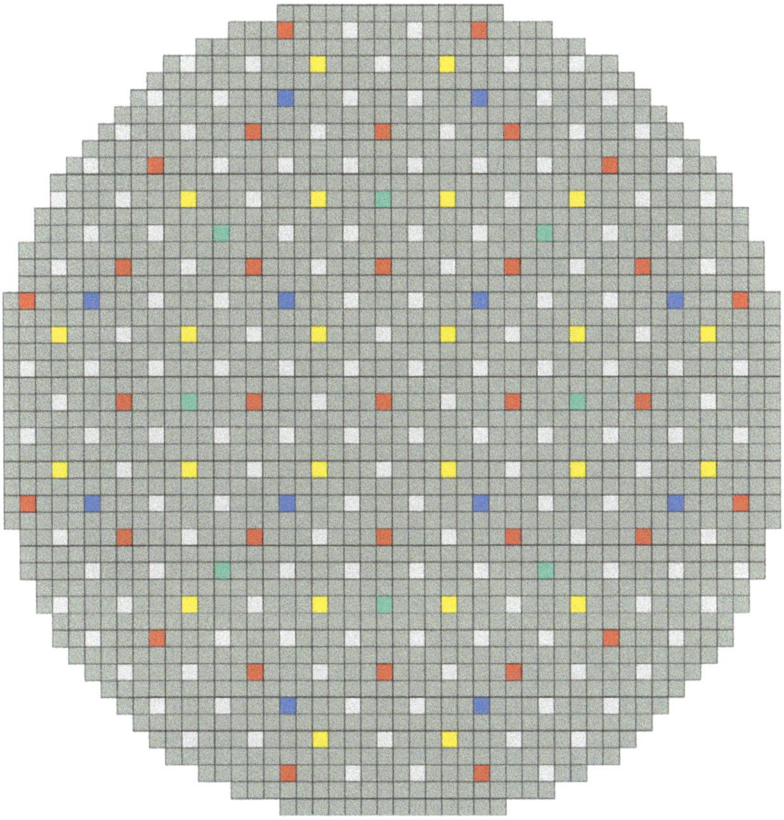

Figure 5.17. Arrangement of fuel rods and control rods in a second generation RBMK reactor. In total there are 1884 reactor tubes: 1661 are fuel rods (gray), 223 are control rods. The 33 emergency control rods are indicated in red. This image of a Unified RBMK-Core for 1000 MW-reactors of the second generation (OPB-82) after modernization has been obtained by the author from the Wikimedia website where it was made available by ChNPP (2013). Image stated to be in the public domain. It is included within this chapter on that basis. It is attributed to ChNPP.

Figure 5.19 shows the Smolensk Nuclear Power Plant in the Desnogorsk province in Russia. This facility consists of three second generation water-cooled graphite-moderated nuclear reactors of the RBMK type. A fourth unit was under construction but was canceled after the Chernobyl accident.

5.7 Heavy water reactors

Nuclear reactors utilizing heavy water (D_2O) as a coolant and moderator were primarily developed in Canada. The result of this development was the Canadian deuterium uranium (CANDU) reactor. It is necessary to seal the heavy water that is used as the coolant and moderator in a primary loop and to transfer thermal energy from the primary loop to a secondary loop to produce steam to operate the turbine. This essentially eliminates the loss of expensive D_2O. The CANDU design has been

Figure 5.18. Cut-away illustration of an RBMK fuel element and the graphite moderator bricks. Reproduced from Tadpolefarm (2019). CC BY-SA 4.0.

Figure 5.19. Smolensk Nuclear Power Plant in the Smolensk region of Desnogorsk province in Russia. This facility consists of three water-cooled graphite-moderated nuclear reactors (of the RBMK type) with a total output of 3000 MW_e. The reactors began operation between 1983 and 1990. The power plant is still operational, as of 2023. Reproduced from Fedchenko (2008). CC BY-SA 3.0.

quite successful and today there are 31 operating reactors in Canada, Europe, South America, and Asia. In addition, India has 17 operating heavy water reactors that have been modeled after the CANDU design.

Figure 5.20 shows a schematic of a CANDU reactor. The use of heavy water as the moderator allows for the use of natural uranium as the fuel. This results from the low neutron absorption cross section of deuterium compared to hydrogen. The neutrons that are absorbed by the deuterium nuclei produce radioactive tritium, as discussed in section 3.6. The tritium that is produced is separated from the heavy water in order to minimize the risk of radioactive contamination in the event of a leak in the primary cooling loop. This tritium can be sold commercially for use in (for example) powerless emergency lighting systems. The basic design of the CANDU reactor core is that of a pressure tube reactor (more details about pressure tube reactors are given in chapter 9). In this design, as illustrated in figure 5.20, pressurized heavy water in the primary loop flows through tubes containing the fuel assemblies in order to transfer heat to the turbine system. 'Cool' heavy water also fills the space between the pressure tubes and acts as the neutron moderator. Heat from the primary (heavy water loop) is transferred to light water in a heat exchanger to produce steam and to operate the turbine/generators. Control rods (also called adjuster rods in the CANDU reactor) are inserted into the reactor vessel from the top.

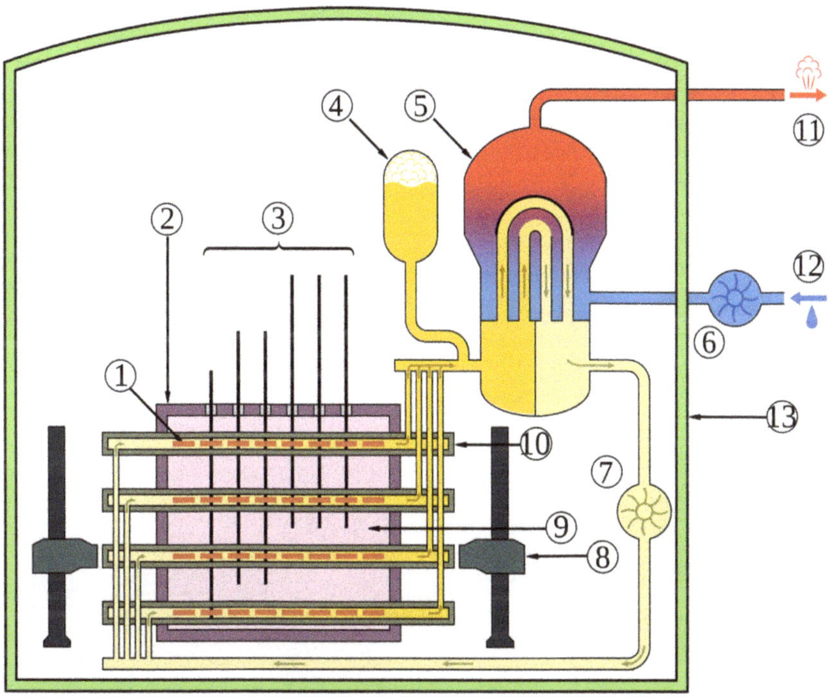

Figure 5.20. Diagram of a CANDU pressurized heavy water-cooled nuclear reactor: (1) nuclear fuel rod, (2) calandria, (3) control rods, (4) pressurizer, (5) steam generator, (6) light water condensate pump (secondary cooling loop), (7) heavy water pump (primary cooling loop), (8) nuclear fuel loading machine, (9) heavy water (moderator), (10) pressure tubes, (11) steam, (12) water condensate, and (13) reactor containment building. This schematic diagram of the pressurized heavy water cooled version of a CANDU reactor has been obtained by the author from the Wikimedia website where it was made available by Inductiveload (2007) under a CC BY-SA 2.5 licence. It is included within this chapter on that basis. It is attributed to Inductiveload.

Because the energy density of natural uranium is less than that of uranium enriched in ^{235}U, the core of a CANDU reactor is typically larger than that of a similar power reactor that utilizes enriched fuel (see figure 5.1). The fuel bundles, which are about 50 cm in length and 10 cm in diameter, are illustrated in figure 5.21. Figure 5.22 shows the Bruce Nuclear Generating Station in Kincardine, Bruce County, Ontario. This station consists of eight CANDU pressurized heavy

Figure 5.21. CANDU Fuel Bundle. This CANDU fuel bundle image has been obtained by the author from the Wikimedia website where it was made available by Atomic Energy of Canada Limited (2007). Image stated to be in the public domain. It is included within this chapter on that basis. It is attributed to Atomic Energy of Canada Limited.

Figure 5.22. Bruce Nuclear Generating Station (Unit B) located near Kincardine, Ontario, Canada on the eastern shore of Lake Huron, which provides the cooling water for the turbines. The photograph shows four of the eight CANDU reactors at the plant. Reproduced from Szmurlo (2006). CC BY-SA 3.0.

water-cooled nuclear reactors with a total capacity of 6610 MW_e. It is the largest heavy water reactor power plant in the world and the fourth largest nuclear power plant of any kind.

5.8 Liquid metal-cooled fast breeder reactors

Clementine (see section 4.5) was the world's first liquid metal-cooled fast reactor. Although it was not designed to breed fissile fuel, its operation provided insight into the design of such a reactor. While Clementine used liquid mercury as a coolant, it was determined that mercury had poor heat transfer characteristics and the toxicity of mercury was also a significant detriment. Subsequent liquid metal-cooled reactors utilized several different coolants, including sodium, sodium-potassium eutectic (NaK), lead, lead-bismuth eutectic (PbBi) and tin. Several important criteria must be considered in the choice of a suitable liquid coolant for a fast reactor. These criteria include,

- Temperature range of the liquid phase
- Corrosivity for reactor cooling system components
- Reactivity with materials that might be present in the cooling system (either intentionally or as a result of a leak)
- Thermal properties
- Neutron cross section

Experimental liquid metal-cooled fast reactors have been constructed using NaK and sodium as coolants. Notable experimental or prototype fast breeder reactors include the Dounreay Fast Reactor in Scotland, which used NaK as a coolant and operated from 1959 to 1977, Phénix in France, which used sodium as a coolant and operated from 1975 to 2010 and Monju Nuclear Power Plant in Japan, which used sodium as a coolant and operated from 1995 to 2017. There are currently two operating commercial sodium-cooled fast breeder reactors in Russia: BN-600 and BN-800. China and India both have commercial sodium-cooled fast breeder reactors that are under construction. Sodium- and lead-cooled fast breeder reactors are considerations for Generation IV reactors. Further details of previous experimental and prototype sodium-cooled fast breeder reactors, as well as possible future designs, are given in chapter 11. This discussion includes an overview of India's plans to integrate sodium-cooled fast breeder reactors into their overall nuclear energy program.

5.9 Growth of nuclear power

The nuclear power industry grew rapidly in the 1960s after the development of early Generation II power reactors and particularly in the 1970s when public awareness of the limitations of fossil fuel reserves became prevalent. Since the mid- to late-1980s (i.e., after Chernobyl) nuclear development has been less rapid. Since the early-2000s, expansion of nuclear energy facilities has been minimal. The total annual power generated by nuclear energy worldwide since 1965 is illustrated in figure 5.23. The share of nuclear as a function of total primary energy use follows different

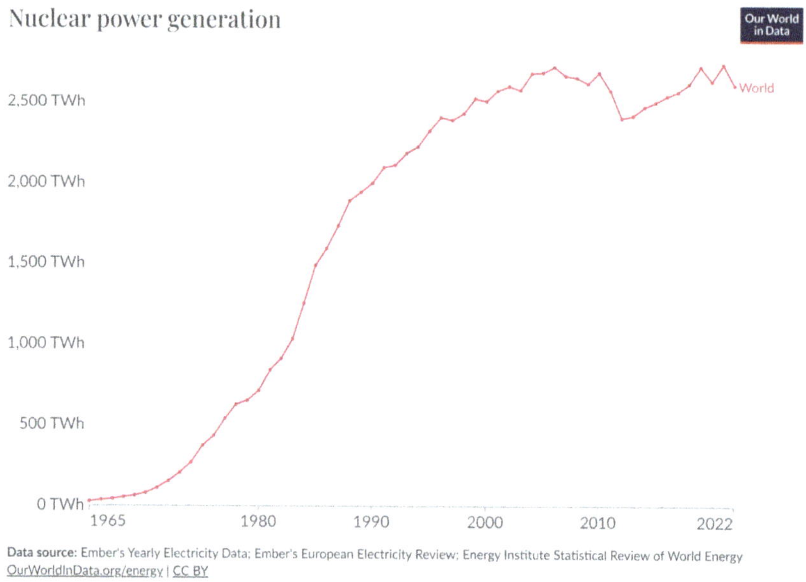

Figure 5.23. Annual world nuclear power generation since 1965. Reproduced from Ritchie and Rosado (2020). CC BY 4.0.

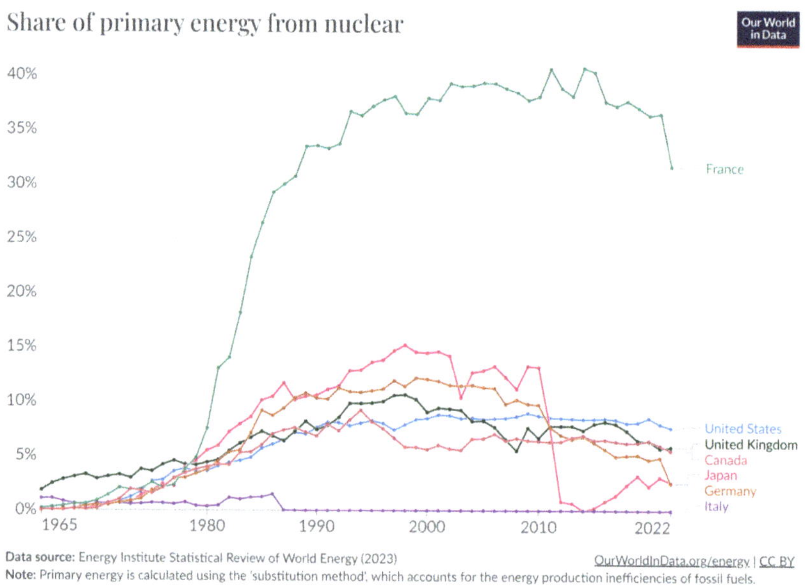

Figure 5.24. Share of primary energy from nuclear for different countries as a function of time since 1965. Reproduced from Ritchie and Rosado (2020). CC BY 4.0.

trends in different countries. These trends are illustrated in figure 5.24. While some countries, such as France, have made a major commitment to nuclear power, other countries have been less ambitious about nuclear power development. The United

States and the United Kingdom have used nuclear energy for about 5% to 10% of their primary energy fairly consistently since the 1980s. Other countries have, for various reasons, scaled down their nuclear energy production. For example, Italy and Germany have, for social and political reasons, shut down their nuclear energy programs. Japan drastically reduced its use of nuclear energy after the Fukushima accident. The map in figure 5.25 shows that most countries that rely significantly on nuclear energy are in North America, Europe, and some parts of Asia. At present, nuclear electricity accounts for about 11% of worldwide electricity production and about 4% of worldwide primary energy use.

The interest in the use of nuclear power as a function of time is more accurately illustrated by the graph in figure 5.26, where newly installed nuclear capacity, as well as permanently shut down nuclear capacity, is shown as a function of year. The decline following the Chernobyl accident is evident. However, this decline continued for more than two decades following Chernobyl and is the result of many factors. In recent years there has been some increase in nuclear reactor construction. The situation in the United States, as shown in figure 5.27, is similar except there has been little, if any, revival of activity in the nuclear power industry since the decline in the late-1980s. The graph shows, for comparison, the newly installed capacity for other energy sources. Clearly, up until the mid-1980s new capacity was dominated by coal and since 1990 or so new capacity has been largely natural gas. Since the mid-2000s, wind, and more recently solar, have also been areas of active development. Only two new reactors have come online in the United States since 1997 (in 2016 and 2023) and a third is scheduled to begin operations in 2024.

The current number of different types of nuclear power reactors in the world is summarized in table 5.1. Pressurized light water reactors are, by far, the most numerous worldwide. They are also the most widespread and account for virtually all recently commissioned reactors and nearly all recent new construction starts.

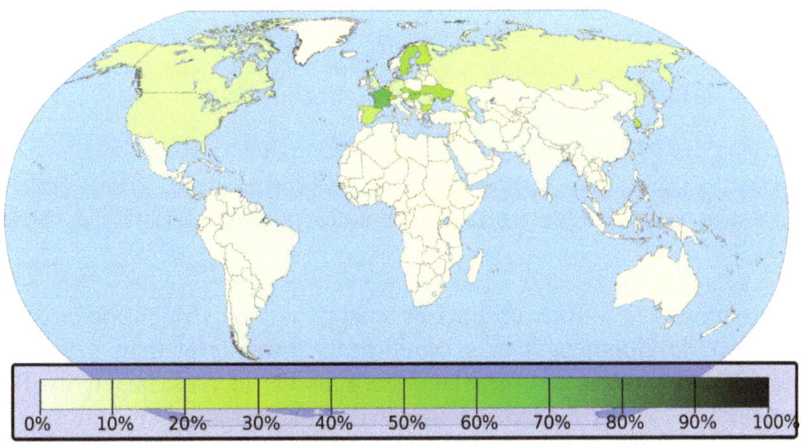

Figure 5.25. Percentage of electricity produced by nuclear energy for all countries. Reproduced from NuclearVacuum (2009). CC BY-SA 3.0.

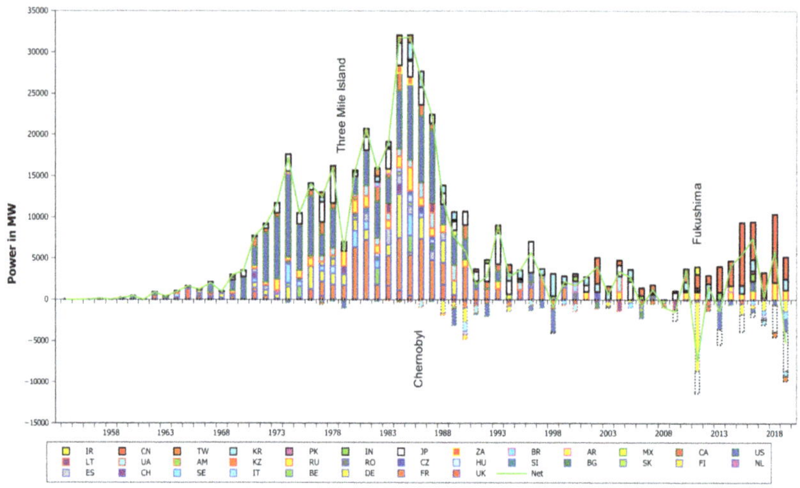

Figure 5.26. Newly installed world nuclear capacity as a function of year since 1953. Data from different countries is shown in the color-coded graph. Decommissioned (or permanently shut down) capacity is shown as a negative contribution. Years of major nuclear accidents are indicated. Reproduced from by Torsch (2020). CC BY-SA 3.0.

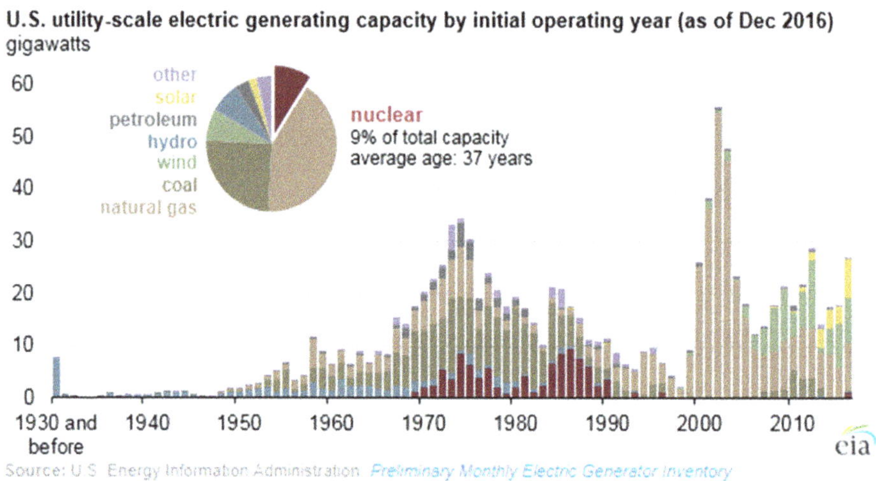

Figure 5.27. Newly installed electricity generating facilities in the United States as a function of year for various technologies. Reproduced from U.S. Energy Information Administration (2017). Image stated to be in the public domain.

Understanding the future of nuclear energy is not straightforward. Overall nuclear power development is at a significantly lower level than it was prior to 1990. However, in recent years there has been a modest increase in activity. Based on the assessment in chapter 1, there is evidence that the increased utilization of nuclear energy may be a significant or even essential component of a future energy plan that could successfully deal with climate change. Projected nuclear electricity generation

Table 5.1. Number of different types of nuclear fission reactors that are operational (as of 2023). See chapter 7 for further information on high-temperature gas-cooled reactors. Data adapted from International Atomic Energy Agency (2023).

Type	Number	Total capacity (GW_e)
Pressurized water reactors (light water)	303	289.3
Boiling water reactors (light water)	41	43.1
Heavy water reactors	46	24.1
Gas-cooled graphite reactors	8	4.7
Water-cooled graphite reactors	11	7.4
Fast breeder reactor	2	1.4
High-temperature gas-cooled reactors	1	0.2
World total	412	370.2

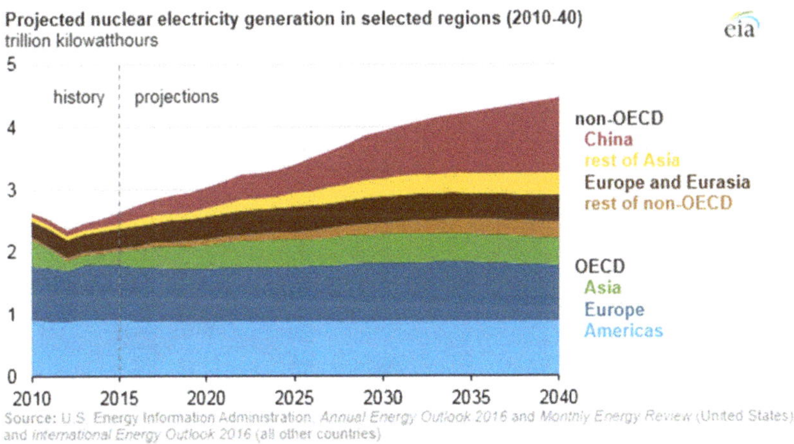

Figure 5.28. Nuclear electricity generation as projected by the U.S. Energy Information Administration. The Organisation for Economic Co-operation and Development (OECD) is an intergovernmental organization consisting of 38 member countries. Reproduced from U.S. Energy Information Administration (2016). Image stated to be in the public domain.

until 2040 from the U.S. Energy Information Administration is shown in figure 5.28. The projection shows minimal, if any, growth in nuclear electricity generation for OECD countries, which includes the United States, Canada, the United Kingdom, much of the European Union, Japan, and Korea. The greatest increase in nuclear electricity is projected for developing countries, particularly China, where the greatest increase in overall energy use is expected (see chapter 1). The actual situation going forward depends on a large number of factors, including the development of viable Generation IV reactors.

5.10 Nuclear energy resources

In the past, nuclear reactors have primarily utilized the energy content of the fissile ^{235}U component of natural uranium. Breeder reactors (section 5.8) and fuel reprocessing (section 5.11) also allow for the use of at least some of the energy from fertile ^{238}U. The development of thorium-fueled breeder reactors would also allow for the utilization of the energy content of fertile ^{232}Th. In this section, we consider the availability and production of these different fissile and fertile resources.

5.10.1 Uranium

Annual world uranium production since the late-1940s is shown in figure 5.29. The graph also shows total annual uranium need (for civilian power reactors and military use). About 90% of the uranium that is used is for commercial power reactors and about 10% for military purposes. Future projections to 2035 are shown. It is interesting to note that from the late-1980s to the present, uranium need has substantially exceeded production. This discrepancy was accounted for by using stockpiled weapons-grade uranium and by fuel reprocessing (as discussed in the next section).

The situation illustrated in figure 5.29, is even more obvious in uranium production data from the United States. This is shown in figure 5.30, where it is seen that uranium production is less than 1% of what it was from the mid-1950s through the mid-1980s. A breakdown of uranium production (for 2021) by country

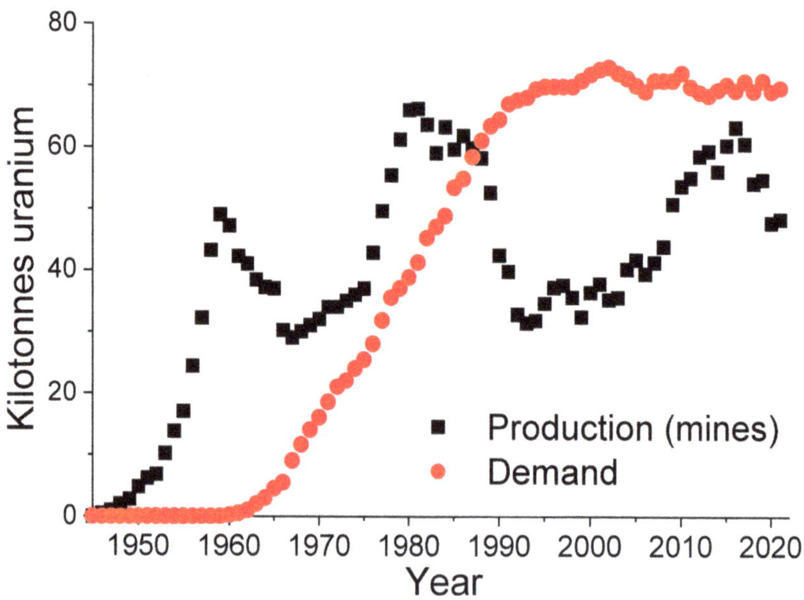

Figure 5.29. Graph of world uranium production and global need. Reproduced from Materialscientist (2023). CC BY-SA 3.0.

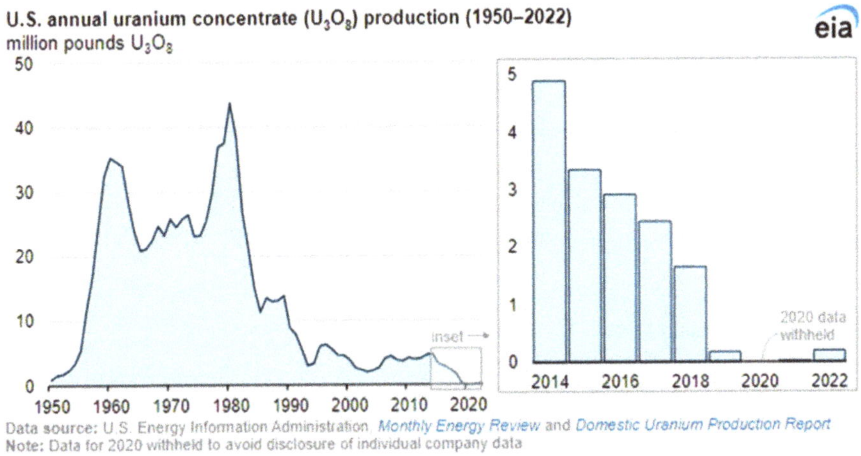

Figure 5.30. Uranium production in the United States from 1950 to present. Reproduced from U.S. Energy Information Administration (2023). Image stated to be in the public domain.

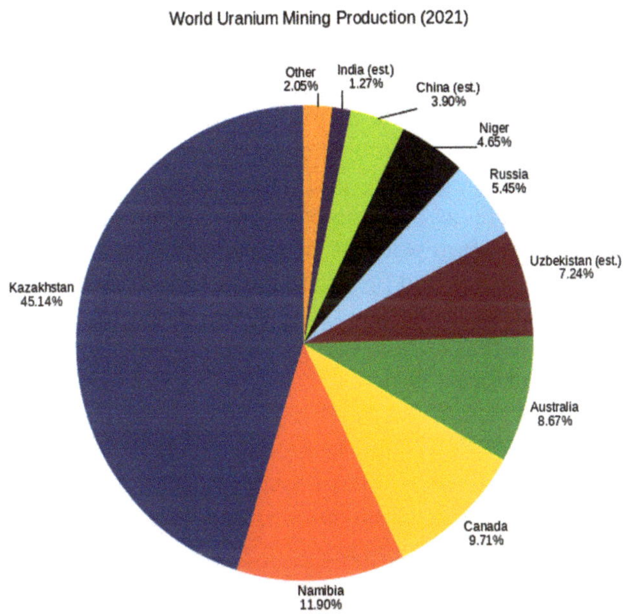

Figure 5.31. Percentages of world uranium production from 2021 from leading countries. Reproduced from Liberum scientia (2022). CC BY-SA 4.0.

is shown in figure 5.31. The major uranium producers are seen to be Kazakhstan, Namibia, Canada, Australia, and Uzbekistan.

The future of uranium production and use will depend on the availability of resources, as well as the demand. The former can be understood on the basis of geological exploration. The latter depends on technical, political, and social factors governing the use of nuclear power.

Table 5.2. Countries with the largest known uranium reserves (at less than 260 USD kg^{-1}) as of 2016. Data adapted from Nuclear Energy Agency (2016).

Country	Uranium reserves (thousand tonnes)	Percent world total
Australia	2049	25.4
Kazakhstan	969	12.0
Canada	873	10.8
Russia	662	8.2
Namibia	504	6.2
Rest of world	3013	37.4
Total	8070	100.0

An assessment of the viable uranium reserves for commercial reactor use depends on the cost per kilogram to extract and process the resource. For example, uranium that occurs in deposits at too low a concentration will be too expensive to be commercially useful. Table 5.2 gives the known reserves that can be extracted and processed at less than 260 USD kg^{-1}. The total known world uranium reserves are about 8 million tonnes. At less than 130 USD kg^{-1}, the known resources are about 6 million tonnes. At the current uranium demand of around 70 thousand tonnes per year, as shown in figure 5.30, the known resources would last for somewhere in the range of 85 to 110 years, depending on the maximum acceptable uranium price. It is also possible that new resources will be located or that future mining and production costs may change, and these factors will affect the longevity of resources.

It is important to note that nuclear energy currently provides about 11% of world electricity. If nuclear capacity increases, then demand for uranium will increase proportionally. If there is a widespread implementation of breeder reactors that utilize the ^{238}U component of uranium to produce fissile ^{239}Pu, then the longevity of resources would increase significantly. Estimates suggest that the energy produced per unit of natural uranium could be increased by about a factor of 60 through the use of breeder reactors.

5.10.2 Thorium

Thorium is extracted from monazite, which is a naturally occurring phosphate with the composition (Ce, La, Th)PO_4. On average, the thorium concentration in monazite is around 2.5%, although some deposits can contain up to about 20% thorium. Thorium has a number of interesting physical and chemical properties that are unrelated to it potential application as a reactor fuel. In the past, these properties have been utilized in a number of industries. While there are still some non-nuclear uses for thorium, much of this use has been replaced by other elements because of concerns related to thorium's radioactivity. Some of its current uses include,

- As an additive to tungsten to improve its mechanical properties
- As a ceramic in the form of thorium dioxide, alone or as an additive to other ceramics, because of its high melting temperature
- To increase the refractive index of glass used in optical components and as an antireflective optical coating

Since thorium is not currently used commercially as a reactor fuel, the overall demand is fairly low. Total world monazite production from which thorium is extracted is around 2700 tonnes per year. This mining activity is primarily for the purpose of producing rare earth metals and thorium is, more or less, a by-product of such activities.

Because of the low demand for thorium, exploration into its occurrence in nature has not been extensive. As a result, estimates of the total amount of commercially viable thorium are subject to substantial uncertainties. The Nuclear Energy Agency (2016) has estimated world thorium reserves to be around 6.355 million tonnes. India, Australia, the United States and Brazil are known to have significant thorium reserves (see section 11.4).

A comparison of the estimated reserves as given above for thorium and uranium suggests that the availability of thorium may be similar or less than that of uranium. Based on the half-lives of ^{232}Th (1.4×10^{10} y) and ^{238}U (4.47×10^{9} y), it would be expected that thorium would be more abundant on earth than uranium. Analyzes have suggested an overall Th/U ratio for the Earth of between 3.6 and 4.2 (see Allègre *et al* 1986). This discrepancy may be explained by the fact that uranium exploration is much more mature than thorium exploration and it is quite possible that significantly more viable thorium resources could be discovered if there is commercial incentive to do so.

The basics of thorium breeder reactors were considered in section 3.8. Further details of the thorium fuel cycle, the use of thorium in thermal neutron reactors are presented in chapter 8, and the use of thorium in fast breeder reactors is covered in chapter 11.

5.11 Fuel reprocessing

During the operation of operation of a nuclear reactor, fissile material in the core undergoes induced fission and produces energy. This process changes the composition of the fuel in two ways: first, the fissile material is used; and second, fission by-products accumulate. The proper operation of the reactor requires that the fuel has the correct amount of fissile material. At some point, fresh fuel will need to be loaded into the reactor to replace the fuel that is depleted in fissile material. Another important factor is the effect of the fission by-products in the fuel on the operation of the reactor. From the graph of the fission yield shown in figure 3.3, it is seen that there are light by-products with A around 95 and heavy by-products with A around 140. The heavy by-products are an important concern because many of these can be lanthanides (i.e., rare earths) and these typically have very large neutron absorption (i.e., (n, γ)) cross sections. The presence of these nuclei in the fuel affects the concentration of neutrons in the core and will have adverse effects on the operation of the reactor. Such materials are referred to as neutron poisons.

When spent fuel is removed from the reactor, it must be dealt with in an appropriate manner. Basically, there are two possibilities: disposal and reprocessing. Waste disposal is considered in detail in section 6.8. In the present section, we provide an overview of fuel reprocessing. We begin with a consideration of fuel reprocessing for a non-breeder uranium reactor.

Figure 5.32 shows a diagram of a typical uranium fuel cycle. Reactor fuel is prepared from natural uranium by appropriate enrichment (in ^{235}U) followed by chemical processing and preparation of appropriate shape fuel elements (see section 3.6.1). The details of these processes will depend on the design of the reactor. After use in the reactor, spent fuel can be reprocessed by removing fission by-products and separating the remaining actinides into non-fissile uranium (^{238}U) and fissile materials (unused ^{235}U and ^{239}Pu produced by the (n, γ) reaction from ^{238}U, see section 3.8). Reactor facilities that do not reprocess fuel but merely dispose of spent fuel are said to have an open fuel cycle or once-through fuel cycle. Reactor facilities that reprocess the spent fuel, are said to have a closed fuel cycle. Reprocessed fissile material can be used to fabricate new reactor fuel. In the case that mixed uranium and plutonium is used, mixed oxide (MOX) fuel is produced. In some processes, reprocessed uranium and plutonium are separated as shown in figure 5.33. In all cases, waste material that cannot be used for further fuel fabrication must be disposed of by appropriate means. Although much of this material is radioactive and

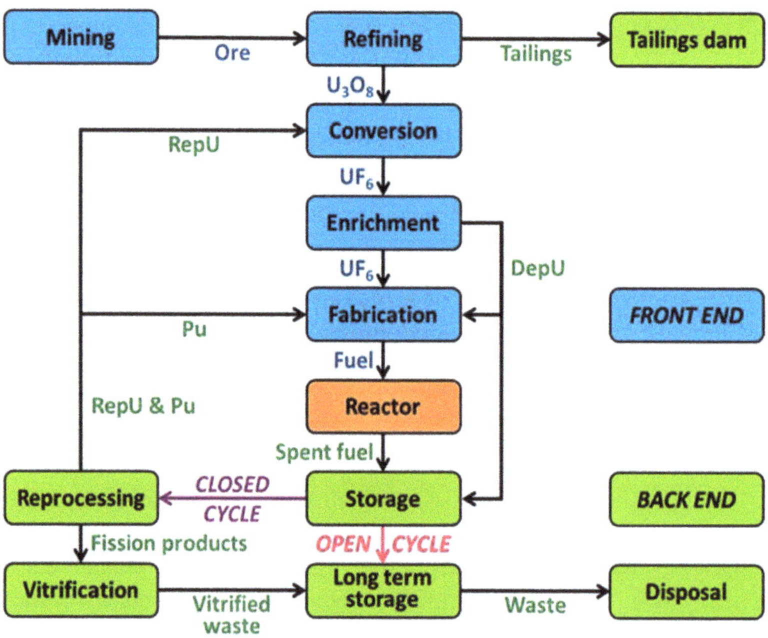

Figure 5.32. The nuclear fuel cycle for uranium. Reproduced from Taylor *et al* (2022). CC BY 4.0.

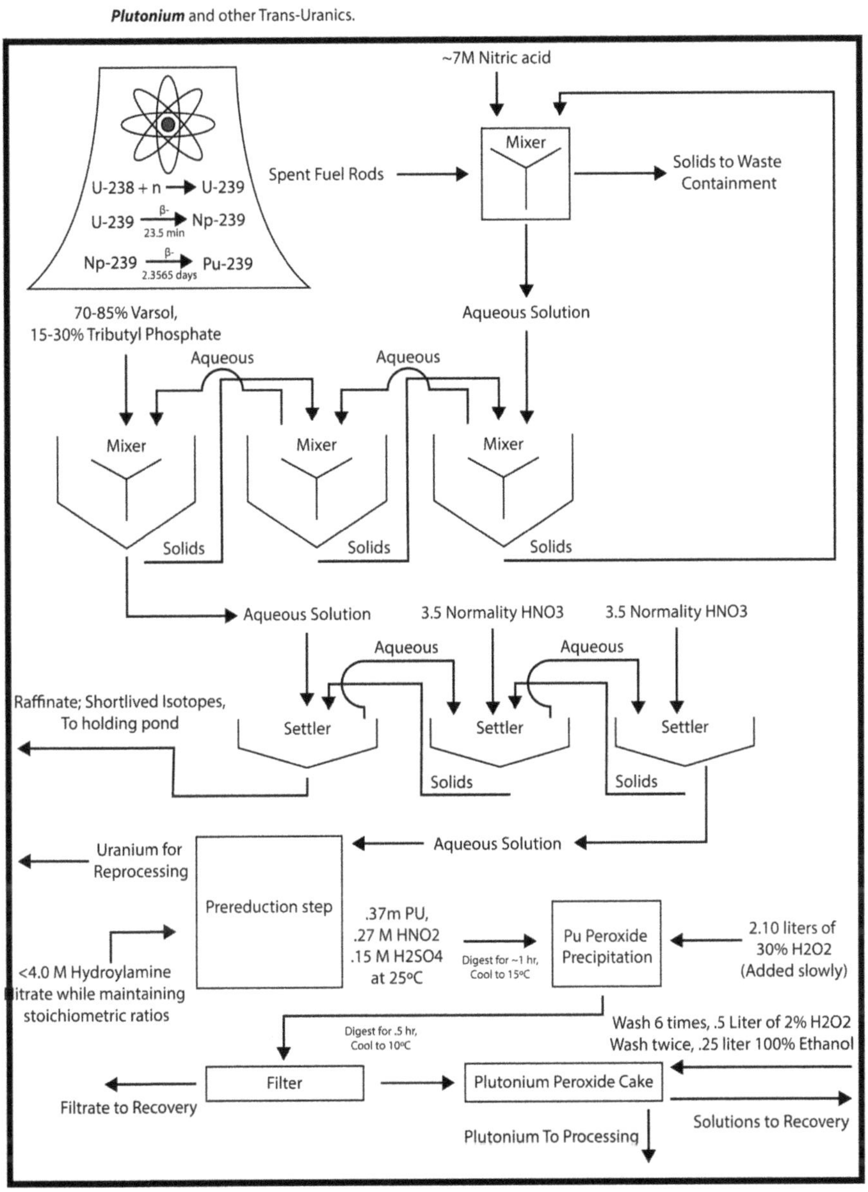

Figure 5.33. Nuclear fuel reprocessing to extract fissile uranium and plutonium. Reproduced from Koonzybear (2014). CC BY-SA 3.0.

must be treated accordingly, the waste disposal procedures are less stringent than those for unprocessed spent fuel. See further details in section 6.8.

Reprocessing reactor fuel has both advantages and disadvantages. Among the advantages are the extended longevity of uranium resources. The use of reprocessing is essential in the context of a breeder reactor program because it allows for the recovery of fissile material (i.e., ^{239}Pu) produced from non-fissile ^{238}U. Reprocessing is also essential for the use of ^{232}Th as a reactor fuel. ^{233}U bred from natural thorium can be extracted for reactor use, see chapters 8, 9 and 11. Finally, the radioactive waste that is left after reprocessing is more easily dealt with. It is important to note that advanced fuel cycles, as utilized in many Generation IV reactors, vastly reduce the quantity and activity of radioactive waste.

Among the disadvantages of fuel reprocessing is that fissile material extracted from spent fuel, e.g., ^{239}Pu can easily be converted into weapons-grade material for bomb construction. Thus, heightened security must be implemented at reprocessing facilities. In addition, reprocessing is expensive. Studies have shown that under present market conditions the cost of processing new uranium and direct disposal of spent fuel is less than that of reprocessing.

At present, China, France, India, Pakistan, and Russia reprocess at least some of their spent reactor fuel. In the past, Belgium, Germany, Italy, Israel, Japan, the United Kingdom, and the United States have at some time operated nuclear fuel reprocessing plants. In a number of cases, these plants have been operated in conjunction with fast breeder reactor programs.

References

Allègre C J, Dupré B and Lewin E 1986 Thorium/uranium ratio of the earth *Chem. Geol.* **56** 219–27

Areva NP 2006 Schemata of EPR's core-catcher https://commons.wikimedia.org/wiki/File:Schemata_core_catcher_EPR.jpg

Argonne National Laboratory 2008 The reactor tube tops of a RBMK reactor at Ignalina https://commons.wikimedia.org/wiki/File:RBMK_reactor_from_Ignalina_ArM.jpg

Atomic Energy of Canada Limited 2007 CANDU fuel bundle https://commons.wikimedia.org/wiki/File:CANDU_fuel_bundles.jpg

ChNPP 2008 Core from the Advanced Boiling Water Reactor (ABWR) https://commons.wikimedia.org/wiki/File:ABWR.PNG

ChNPP 2013 Unified RBMK-Core for 1000 MW-reactors of the second generation https://commons.wikimedia.org/wiki/File:RBMK-1000_OPB-82_Kartogramm.svg

Emoscopes 2008 RBMK reactor schematic https://commons.wikimedia.org/wiki/File:RBMK_reactor_schematic.svg

Fedchenko K 2008 Smolensk Nuclear Power Plant https://commons.wikimedia.org/wiki/File:Smolensk_Nuclear_Power_Plant.jpg

Fleitz A 2016 Generation II nuclear reactor vessel sizes of similar power output https://commons.wikimedia.org/wiki/File:Gen_II_nuclear_reactor_vessels_sizes.png

Frettie 2009 Cooling towers of Dukovany Nuclear Power Station https://commons.wikimedia.org/wiki/File:Cooling_towers_of_Dukovany_Nuclear_Power_Station_near_Dukovany,_T%C5%99eb%C3%AD%C4%8D_District.jpg

Geni 2019 Photo of a Magnox fuel rod https://commons.wikimedia.org/wiki/File:Magnox_fuel_rod_science_museum_2019.JPG

International Atomic Energy Agency 2013 Unit 1–2 of Korea Shin-Kori Nuclear Power Plant www.flickr.com/photos/iaea_imagebank/8505820635/

International Atomic Energy Agency 2023 Power reactor information system https://pris.iaea.org/PRIS/WorldStatistics/OperationalReactorsByType.aspx

Inductiveload 2007 Schematic diagram of the pressurised heavy water cooled version of a CANDU (CANada Deuterium-Uranium) nuclear reactor https://commons.wikimedia.org/wiki/File:CANDU_Reactor_Schematic.svg

Koonzybear 2014 Plutonium processing https://commons.wikimedia.org/wiki/File:Uranium_Reprocessing.jpg

Liberum scientia 2022 World uranium mining production 2021 https://commons.wikimedia.org/wiki/File:World_Uranium_Mining_Production_2021.png

Materialscientist 2023 World uranium production (mines) and demand https://commons.wikimedia.org/wiki/File:U_production-demand.png

MesserWoland 2007 Schematic diagram of an advanced gas-cooled reactor https://commons.wikimedia.org/wiki/File:AGR_reactor_schematic.svg

Mmarre 2012 Magnox reactor https://commons.wikimedia.org/wiki/File:Magnox_reaktor.jpg

Nuclear Energy Agency 2016 *Uranium 2016: Resources, Production and Demand* NEA No. 7301 (Paris: OECD Publishing) www.oecd-nea.org/jcms/pl_15004

NuclearVacuum 2009 Nuclear power percentage https://commons.wikimedia.org/wiki/File:Nuclear_power_percentage.svg

Office for Nuclear Regulation 2017 *Step 4 Assessment of Severe Accidents for the UK Advanced Boiling Water Reactor* Assessment Report: ONR-NR-AR-17-015 www.onr.org.uk/new-reactors/uk-abwr/reports/step4/onr-nr-ar-17-015.pdf

Panther 2005 Wwer-1000-scheme https://commons.wikimedia.org/wiki/File:Wwer-1000-scheme.png

Ritchie H and Rosado P 2020 Nuclear energy https://ourworldindata.org/nuclear-energy

Robin-Root 2006 Reactor number 3 at Forsmark Nuclear Power Plant https://commons.wikimedia.org/wiki/File:Forsmark3.jpg

Szmurlo C 2006 Aerial photo of the Bruce Nuclear Generating Station https://commons.wikimedia.org/wiki/File:Bruce-Nuclear-Szmurlo.jpg

Tadpolefarm 2019 Cut-away view illustrating the fuel pellets, fuel tubes, spacer rings inside the graphite bricks https://commons.wikimedia.org/wiki/File:RBMK_fuel_set_cut-away_view.jpg

Taylor R, Bodel W, Stamford L and Butler G 2022 A review of environmental and economic implications of closing the nuclear fuel cycle—part 1: wastes and environmental impacts *Energies* **15** 1433

Torsch 2020 Nuclear energy by year https://commons.wikimedia.org/wiki/File:Nuclear_Energy_by_Year.svg

U.S. Department of Energy 2023 Nuclear 101: how does a nuclear reactor work? www.energy.gov/ne/articles/nuclear-101-how-does-nuclear-reactor-work

U.S. Energy Information Administration 2016 China expected to account for more than half of world growth in nuclear power through 2040 www.eia.gov/todayinenergy/detail.php?id=28132

U.S. Energy Information Administration 2017 Most U.S. nuclear power plants were built between 1970 and 1990 www.eia.gov/todayinenergy/detail.php?id=30972

U.S. Energy Information Administration 2023 U.S. uranium production up in 2022 after reaching record lows in 2021 www.eia.gov/todayinenergy/detail.php?id=60160

Young T 2016 Torness Nuclear Power Station, Scotland, United Kingdom https://commons.wikimedia.org/wiki/File:Torness_Nuclear_Power_Station_-_April_2016.jpg

IOP Publishing

Generation IV Nuclear Reactors
Design, operation and prospects for future energy production
Richard A Dunlap

Chapter 6

Risks associated with nuclear energy

Chapter 6 discusses the risks associated with nuclear power, including thermal pollution, occupational risks, nuclear accidents, nuclear security, and nuclear waste disposal. The International Nuclear and Radiological Event Scale (INES) is described as a method of categorizing nuclear reactor incidents and accidents. Four major nuclear accidents, i.e., Windscale, Three Mile Island, Chernobyl, and Fukushima, are discussed in detail and an analysis of the events leading to these accidents is presented. This chapter also discusses nuclear security as it relates to the security of nuclear materials, as well as possible attacks on nuclear facilities. Finally, the long-term storage of radioactive waste from nuclear reactors is considered.

6.1 Introduction

Unlike the burning of fossil fuels, the production of electricity by means of nuclear fission energy does not release carbon dioxide to the atmosphere, and therefore does not contribute directly to global climate change. However, the use of nuclear energy has its own risks. These are

- Thermal pollution
- Occupational risks
- Nuclear accidents
- Nuclear security
- Nuclear waste disposal

We look briefly at the first two concerns.

Because nuclear fission reactions produce heat, this heat must be converted into electricity by means of a heat engine. Since heat engines are typically no more than about 40% efficient, the majority of the heat that is produced by nuclear reactions is expelled to the environment, either to a source of water (i.e., lake, river, ocean) or to the atmosphere through cooling towers (see figure 5.4). Thermal pollution produced by nuclear generating stations has local environmental effects. Since the same

doi:10.1088/978-0-7503-6069-2ch6

approach to converting thermal energy to electricity is used for fossil fuel fired generating stations, the replacement of fossil fuels with nuclear energy is fairly neutral in this respect.

Occupational risks include risks to miners and personnel involved in transporting fuel. Since the energy density of uranium is several orders of magnitude greater than the energy density of coal, much less material needs to be mined and transported, leading to a lower overall occupational risk for nuclear power.

The remainder of this chapter deals with the final three points above.

6.2 Nuclear reactor safety concerns

The possibility of an accident at a nuclear reactor is a major concern for the safety of these facilities. In fact, a number of accidents at nuclear reactors have occurred over the years. These accidents vary in severity, with minor events occurring fairly frequently and more significant event occurring less often. In order to quantify the severity of nuclear accidents, the International Atomic Energy Agency has established the INES system to categorize adverse nuclear events.

The INES system categorizes nuclear events on a scale from 0 to 7, as illustrated in figure 6.1. The scale is logarithmic, much as the scale for measuring the severity of earthquakes. The lowest category (zero) is referred to as an operating deviation. The next three categories are referred to as incidents, while the highest four categories are referred to as accidents. The assignment of a particular event to the appropriate category involves an assessment of three factors: off-site effects, on-site effects, and degradation of defense in depth measures. The final assessment is a measure of the degree to which standard layered or stepwise countermeasures are not effectively implemented. The classification of a nuclear event is determined by the worst of the assessments in these three categories and is intended to assist in the deployment of disaster aid.

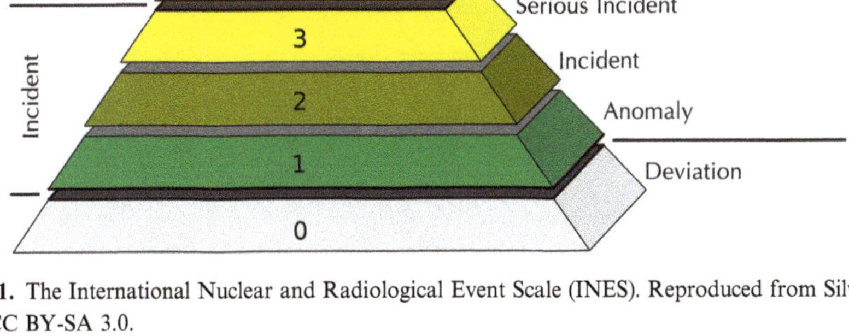

Figure 6.1. The International Nuclear and Radiological Event Scale (INES). Reproduced from Silver Spoon (2011). CC BY-SA 3.0.

All levels described by the INES system are related to adverse radiological events, although not necessarily to reactor accidents, as we would envision them. For example, the Kyshtym disaster occurred in 1957 at the Mayak Chemical Combine (MCC) in Chelyabinsk Oblast, Soviet Union. This was a radioactive contamination accident that resulted from an explosion at a plutonium production and nuclear reprocessing facility that was part of the Soviet nuclear weapons program. This was a level 6 accident on the INES. The Goiânia accident, which occurred at a hospital in Goiânia, Goiás, Brazil, is another non-reactor example of a radiological accident. In this event, an unsecured ^{137}Cs radiotherapy source was stolen and led to four deaths among irradiated individuals. This event is classified as an INES level 5 accident. Only two events, Chernobyl and Fukushima, are classified as INES level 7 accidents.

In the following four sections, we look at some examples of serious nuclear accidents that occurred at reactor facilities. These are presented in chronological order. The reasons for the accidents and how they might affect future nuclear reactor operations are discussed. A complete history of nuclear accidents is presented in Mahaffey (2014).

6.3 The Windscale fire

The Windscale Piles (as reactors were called at the time) were two air-cooled graphite-moderated reactors designed for the production of plutonium for the United Kingdom's nuclear weapons program. These reactors were discussed briefly in section 4.6.3. The reactors became operational in 1950–51 and are shown in figure 6.2. The two towers (called chimneys) are 120 m tall and are for reactor cooling by convective air flow.

Figure 6.2. Windscale plutonium factory. Reproduced from Eaton (1985). CC BY-SA 2.0.

When graphite (or other materials) is bombarded with neutrons, radiation damage occurs. This results in lattice dislocations in the crystalline graphite structure. If this damage becomes too severe, the structure can spontaneously relax, releasing a significant amount of heat (referred to as Wigner energy). This rapid energy release can cause the graphite to ignite. The method used at the Windscale facility to avoid a possible reactor fire was to periodically anneal the graphite core by heating it in a controlled way to a temperature above 250 °C. At this temperature, the dislocations anneal out and return the graphite to its well-ordered crystalline state. This procedure was undertaken several times at the Windscale reactors after they became operational.

On 7 October 1957, it was discovered that Windscale reactor 1 was running hotter than normal. It was decided to do a core anneal to release the Wigner energy. When the temperature in the reactor became uneven, a second Wigner release was attempted on 8 October. At this point, the entire core began to overheat. On 10 October, the flow of cooling air was increased in an attempt to cool the core. However, at this point the graphite had begun to burn, and the additional air only exacerbated the situation.

Attempts were made to extinguish the fire using carbon dioxide, but these were unsuccessful. Water was also used to put out the fire. This is a dangerous approach because the high temperatures can lead to a reaction between the water and the metal cladding on the fuel rods. The metal will oxidize, stripping the water molecules of their oxygen and freeing hydrogen gas. The subsequent build-up of hydrogen gas can lead to an explosion. Fortunately, this did not occur in the case of the Windscale accident, although see sections 6.5 and 6.6 concerning this situation in the Chernobyl and Fukushima accidents. However, the attempt to extinguish the fire with water was unsuccessful. Finally, the fire was extinguished by cutting off the supply of air to the reactor.

The Windscale fire has a rating of five on the INES scale. A significant amount of radioactive material was released to the atmosphere. The estimated release of radionuclides is given in table 6.1. ^{131}I and ^{210}Po are of particular concern from a health perspective because ^{131}I can cause thyroid cancer and ^{210}Po can cause lung cancer. Studies of the environmental radioactive contamination from the Windscale fire have predicted that this event was responsible for 100 fatal and 90 non-fatal thyroid cancers and 70 fatal and 10 non-fatal lung cancers (Wakeford 2007).

Table 6.1. Estimated release of radioactive material from the Windscale fire. Data adapted from Crick and Linsley (1984).

Nuclide	Half-life	Released (TBq)
^{133}Xe	5.2 d	12 000
^{131}I	8.0 d	740
^{210}Po	138 d	8.8
^{137}Cs	30 y	22

The level of radioactive contamination at the reactor site has limited clean-up efforts and it is estimated that clean-up and decommissioning of Windscale reactor 1 will not be completed until sometime after 2040, or more than 80 years after the accident.

6.4 The Three Mile Island accident

The Three Mile Island Nuclear Generating Station is a nuclear power plant near Harrisburg, Pennsylvania. It consists of two pressurized light-water reactors: Three Mile Island Unit 1 (TMI-1) was commissioned in 1974 and had a generating capacity of 819 MW_e, while Three Mile Island Unit 2 (TMI-2) was commissioned in 1978 and had a generating capacity of 906 MW_e.

On 28 March 1979, the core of TMI-2 suffered a partial meltdown. This accident was classified as INES level 5. The events leading up to the accident and the reasons for the meltdown can be understood in the context of the design of TMI-2.

Figure 6.3 shows a schematic of the design of the TMI-2 reactor facility. The figure indicates the locations of the various reactor components. The typical design of a light pressurized water reactor includes three water/steam loops, as discussed in section 5.4. Water in the primary loop is circulated under pressure through the core. Typically, a pressure of around 16 MPa is used, which prevents the water from boiling at a normal operating temperature of about 315 °C. The water in the primary loop transfers heat to the secondary loop by means of a heat exchanger (steam generator) and the steam produced is used to operate the turbine/generator to generate electricity. Steam from the turbine is condensed in the condenser and returns to the steam generator. The condenser is cooled by the third (cooling) loop, which transfers heat to the cooling tower (in the case of TMI-2) or to a heat reservoir such as a lake or river for some other pressurized water reactors.

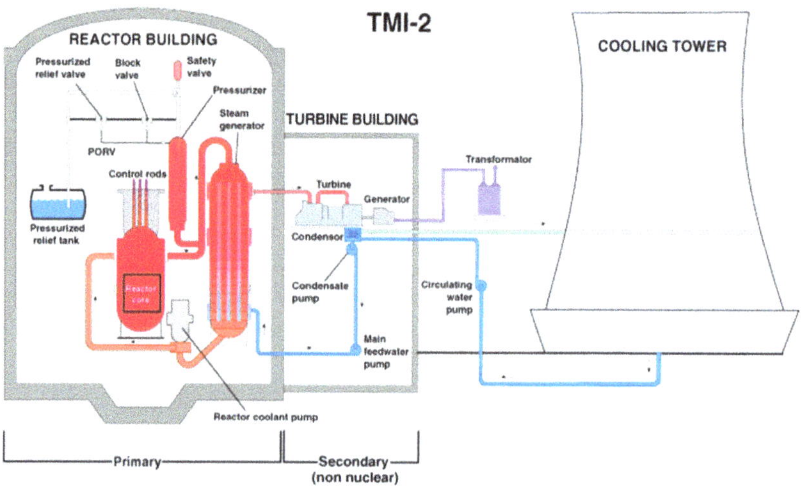

Figure 6.3. Schematic of the Three Mile Island reactor (TMI-2). Reproduced from U.S. Nuclear Regulatory Commission (2022). Image stated to be in the public domain.

Around 4 am on 28 March 1979, there was a failure of the main feedwater pump in the secondary loop. This stopped the flow of cooling water through the steam generator and eliminated the means of removing heat from the primary loop and the reactor core. This caused the turbine/generator and the reactor itself to automatically shut down. After a shut down, the fission products in the core which decay by β^- decay continue to produce heat (the so-called decay heat) for some time and cooling water is necessary to dissipate this heat. The decay heat decreases with time, as shown in figure 6.4, but in the Three Mile Island reactor core caused the temperature and pressure in the primary loop to initially increase as a result of the loss of coolant. When the pressure reached a certain level, the pressurized relief valve opened to divert pressure to the relief tank. When the pressure decreased to an acceptable level, the pressurized relief valve should have closed automatically. However, it did not close but remained open, releasing water in the form of steam from the primary loop. This caused the coolant level in the reactor vessel to drop, which ultimately uncovered the core and led to a partial core meltdown. This is referred to as a loss of coolant accident (LOCA). The molten core material contaminated the primary coolant water and some of this radioactive water was released to the environment through the open pressurized relief valve. This radioactive material contained, primarily, ^{85}Kr and ^{133}Xe (see below).

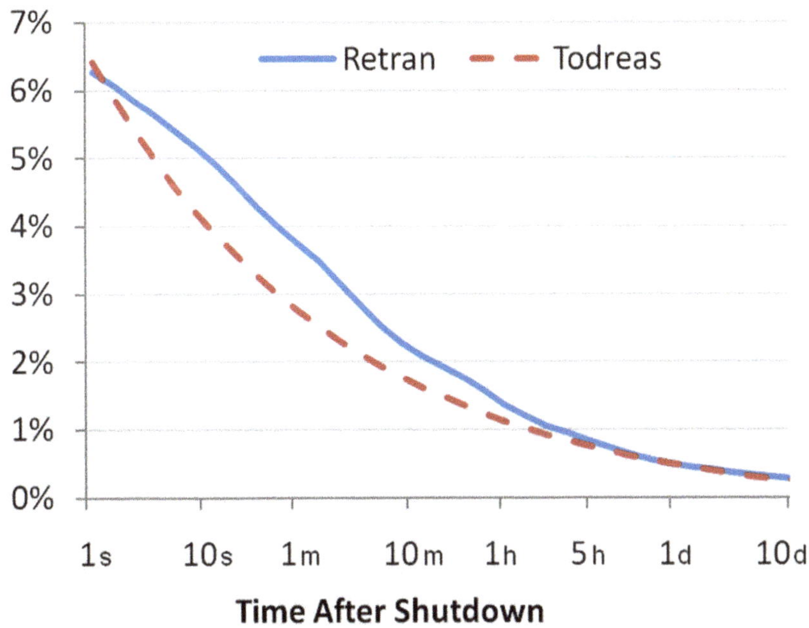

Figure 6.4. Decay heat for a scrammed reactor as a percentage of reactor full power. An emergency reactor shut down is referred to as a scram. Model results are from Retran (Exitech ND) and Todreas and Kazimi (1989). This graph of decay heat as fraction of full power from a SCRAMed operating power reactor has been obtained by the author from the Wikimedia website where it was made available by Theanphibian (2011) under a CC0 licence. It is included within this chapter on that basis. It is attributed to Theanphibian.

TMI-2 was permanently closed after the accident and the reactor core has been removed for disposal (see figure 6.5). Final clean-up and decommissioning are expected to be completed in 2052. Three Mile Island Unit 1 (TMI-1) was shut down for refueling at the time of the accident in TMI-2. It was restarted in 1985 and operated until it was closed in 2019 for financial reasons.

Studies have determined that both reactor design and operator error contributed to the accident (see e.g., Walker 2004). Warning systems provided incomplete or misleading information about the status of reactor components. In addition, operators were not suitably trained to deal with emergency situations.

An analysis of the accident (see Rogovin 1980) concluded that radiation exposure experienced by approximately two million people living near the plant from radioactive krypton and xenon gas that was released was comparable to less than one half the exposure from a medical chest *x*-ray. ^{85}Kr decays by β^- decay to ^{85}Rb,

$$^{85}\mathrm{Kr} \rightarrow {}^{85}\mathrm{Rb} + \mathrm{e}^- + \overline{\nu}_\mathrm{e} + \gamma$$

with γ-emission primarily at 149.5 keV. ^{133}Xe decays by β^- decay to ^{133}Cs,

$$^{133}\mathrm{Xe} \rightarrow {}^{133}\mathrm{Cs} + \mathrm{e}^- + \overline{\nu}_\mathrm{e} + \gamma.$$

with γ-emission at 81 keV. Figure 6.6 shows a map of the relative γ-ray dose in the area around the reactor. It is clear that the largest doses were received by residents to the north of the reactor, although these are still considered to be small from a health standpoint. Initial conclusions were that the Three Mile Island accident posed no health risk to people living in the area.

However, recent reanalysis (Datesman 2020) has considered exposure to ^{133}Xe gas in more detail. The health effects of ^{133}Xe are somewhat more complex than

Figure 6.5. Aerial view of the Three Mile Island Nuclear Power Plant in 2012. The two cooling towers on the left are part of TMI-2, which was shut down, and the two cooling towers on the right are part of TMI-1, which was operating at the time. The two reactor containment buildings with domed tops are seen near the center of the image. Reproduced from Johnson (2012). CC BY-SA 2.0.

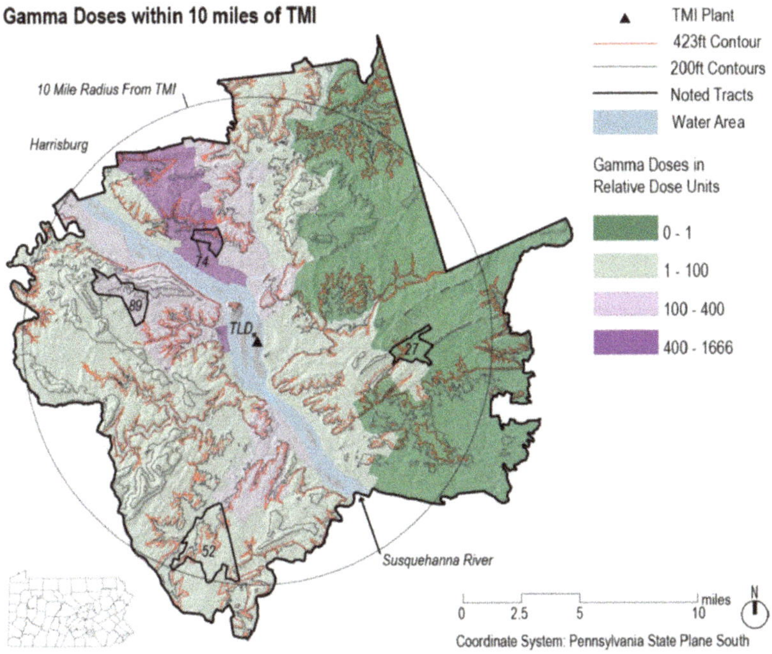

Figure 6.6. Gamma-ray dose estimates in the region around Three Mile Island Unit 2 after the accident of 28 March 1979. Map created with QGIS 2.14 package (https://qgis.org). Reproduced from Datesman (2020). CC BY 4.0.

merely looking at γ-dosimetry. For external exposure, the 81 keV γ-ray is of considerably more concern than the electron emitted in the β^- decay with an energy of 346 keV. ^{133}Xe that is inhaled will be in equilibrium concentration inside the lungs with ^{133}Xe in the surrounding air. Internal exposure to the 346 keV electrons from the ^{133}Xe β^- decay can be a considerable concern from a health standpoint. It must also be considered that hemoglobin has a significant affinity for xenon. Thus, radioactive xenon that is inhaled can be readily distributed throughout the body. Within the timescale on which the ^{133}Xe decays ($\sim$5.3 days), adverse effects could be significant. Subsequent studies of cancer rates in Eastern Pennsylvania (e.g., Han *et al* 2011) have shown excess cancer and other health effects in residents in the area surrounding Three Mile Island.

6.5 The Chernobyl accident

The Chernobyl Nuclear Power Plant was located near the city of Pripyat in what is now Ukraine. It consisted of four RBMK-1000 water-cooled graphite-moderated reactors. The four units became operational in 1977 (reactor 1), 1978 (reactor 2), 1981 (reactor 3), and 1983 (reactor 4). The total capacity was 3515 MW_e. Reactors 1 and 2 were Generation I reactors, and reactors 3 and 4 were Generation II reactors. The Generation II reactors included a more robust containment structure. The reactors were cooled by water from an artificial pond that was fed by the Pripyat

River. Two additional reactors (5 and 6) were planned and construction on reactor 5 was about 70% complete at the time of the accident.

The accident in Chernobyl reactor 4 occurred as a result of an intrinsically unsafe reactor design combined with operator errors. There were a number of design features, some of which are described below, that played a significant role in the accident.

1. The RBMK reactors were designed as dual-purpose reactors to provide electricity and also to produce ^{239}Pu for the Soviet nuclear weapons program. In order to accomplish these goals, the reactor containment structure was designed with a considerable amount of space above the reactor core so that some of the fuel rods could be removed (to extract the plutonium) without shutting down the reactor. This design compromised the integrity of the containment structure, although this was improved somewhat in the Generation II RBMKs (such as the one that experienced the accident at Chernobyl).
2. The RBMK design is intrinsically unstable at low power levels and in such cases it is important that appropriate operator actions are taken. This is largely due to the effects of ^{135}Xe, a fission by-product, which is a strong neutron poison.
3. The control rods in the RBMK reactor are made from neutron absorbing boron carbide. However, the end of the rod which enters the reactor core first is made from graphite. If the control rods that are fully removed from the reactor are inserted to shut the reactor down, the graphite control rod ends that initially displace water in the reactor core will actually lead to a momentary increase in power.
4. The RBMK reactors, at least at that time, had a positive void coefficient (unlike light-water-moderated reactors, e.g., pressurized water reactors). This meant that if water in the core boiled and steam replaced water, then the lower neutron absorption cross section of steam compared to water would cause the reactor power to increase.
5. The manner in which the reactor obtains electricity to operate computers and cooling water pumps was flawed. The reactor normally obtained its electricity from the grid. However, in the event of a grid outage, the reactor would switch to back-up diesel generators. The delay, 60 to 75 s, between a power outage and generator power taking over was considered excessive. It was hypothesized that in the event of a grid outage and reactor shut down, the residual rotational energy of the turbines could be used to run the generators in order to keep cooling pumps running for about 45 s or so while the generators were up starting up. In fact, it was a test of this possibility, the so-called turbine run-down test, that began the sequence of events that led to the accident.

The test of the emergency electrical system as described above, was scheduled for 25 April 1986 as part of a routine reactor maintenance shut down. The test was to consist of the following steps:

- Reactor power would be reduced to between 700 to 1000 MW_{th} out of the rated capacity of 3200 MW_{th}.
- Four (of the eight) main circulating pumps would be configured to run on the turbine/generator output, while the other four pumps would be left to run on grid power.
- When the correct conditions were attained, the steam to the turbines would be shut off and the reactor would be shut down.
- The generator output and pump performance would be monitored until the back-up generators were up to full power.

Reduction of reactor power began shortly after midnight on 25 April. However, due to a variety of delays, the test, which was scheduled to begin at 2 pm on 25 April, actually began very late in the evening, when a different shift of operators was in the control room. Early on 26 April the reactor power had dropped to 720 MW_{th}, consistent with the requirements of the test. However, as noted above, the influence of neutron poisons became pronounced at low operating powers and the reactor output continued to drop until it reached only 30 MW_{th}. In order to get the reactor under control, the operators manually removed nearly all of the reactor's control rods. This increased the reactor power to around 200 MW_{th}. At that point, it was decided to proceed with the test, even though the reactor power was substantially lower than that which was specified. As the turbines ran down after the steam supply to them was shut off, the pumping capacity of the circulating pumps decreased, leading to an increase in heat in the core. The actual course of events following this are not known precisely. However, it appears that in order to control the reactor, a scram was initiated. This procedure inserts all the control rods and as noted above, would lead to an initial increase in reactor power. Given the instability of the operating conditions, this produced steam voids in the water in the core, which led to further heating. It is speculated that this caused some of the fuel rods to fracture, thereby preventing the control rods from being inserted further. Thermal output of the reactor very quickly increased to at least 30 000 MW_{th}.

Two explosions followed and these destroyed the reactor, thereby distributing radioactive material over a wide area and destroying much of the containment building. It is generally believed that the first explosion was a steam explosion resulting from the build-up of pressure in the reactor vessel. The reasons for the second explosion, which followed a few seconds later, are somewhat uncertain, although it has been speculated that it was a hydrogen explosion. Hydrogen is liberated when very high temperature stream reacts with the zirconium used in the fuel rod cladding. The reaction is

$$Zr + 2H_2O \rightarrow ZrO_2 + 2H_2. \tag{6.1}$$

Hydrogen can also be produced by the reaction of steam with high temperature graphite by the reaction

$$C + H_2O \rightarrow H_2 + CO.$$

Figure 6.7. Chernobyl reactor 4 after the accident. Reproduced from International Atomic Energy Agency (2011). CC BY-SA 2.0.

The hydrogen could have ignited when exposed to air at high temperature after the reactor vessel was destroyed.

Figure 6.7 shows an image of Chernobyl reactor 4 after the accident. In order to contain the radioactive contamination that remained in reactor 4, a concrete structure, called the sarcophagus, was constructed around the remains of the reactor building. This was completed in December 1986 and is shown in figure 6.8.

The sarcophagus had an expected lifetime of around 30 years, which was reached in the mid-2010s. In order to resolve this issue, a new structure, the New Safe Confinement (NSC), was constructed (see figure 6.9) and installed over the existing sarcophagus (see figure 6.10). The NSC will allow for the sarcophagus and the remaining reactor components to be dismantled and transported away from the site

Figure 6.8. Reactor 4 of the Chernobyl Nuclear Power Plant in 2005 showing the sarcophagus that was constructed over the reactor building. Reproduced from International Atomic Energy Agency (2013a). CC BY-SA 2.0.

Figure 6.9. Chernobyl Nuclear Power Plant in June 2013 showing the construction of the NSC. Buildings (from left to right) are the New Safe Containment, reactor 4 with the sarcophagus, reactor 3, reactor 2, and reactor 1. Reproduced from Ingmar Runge (2013). CC BY-SA 3.0.

without releasing radioactive contamination to the environment. It is expected that the final clean-up will be completed by 2065.

Three years after the Chernobyl accident, construction work on reactors 5 and 6 was terminated. However, the remaining three operational Chernobyl reactors were restarted in late-1986 and 1987, and continued to operate for some time after the accident. After a fire in October 1991, reactor 2 was permanently shut down. Reactor 1 was decommissioned in 1996 and reactor 2 was finally shut down in 2000.

Figure 6.10. The NSC in its final position over reactor 4 at Chernobyl Nuclear Power Plant in 2017. Reproduced from Tim Porter (2017). CC BY-SA 4.0.

The explosions that destroyed Chernobyl reactor 4 distributed radioactive material over a wide area. This radioactive contamination consisted of nuclear fuel, as well as fission by-products. ^{137}Cs, which decays by β^- decay with a half-life of 30 y, is, perhaps, the most concerning fission by-product. Figure 6.11 shows a map of the distribution of ^{137}Cs activity around Chernobyl. It is clear that areas of high activity occur at distances of up to about 200 km from the reactor and these occur in areas that are presently Ukraine, Belarus, and Russia.

The exclusion zone was initially established with a radius of 30 km around the reactor site to designate an area where the level of radioactive contamination was unacceptable for human habitation. This region was subsequently extended considerably to include an area of about 2600 km^2 and was officially referred to as the zone of alienation. Estimates of when this zone will be suitable for human habitation range from around 300 years to tens of thousands of years.

The health risks associated with the Chernobyl accident are somewhat difficult to assess. Two reactor workers were killed as a result of the explosion. Twenty-eight additional reactor workers or emergency responders died within the following three weeks as a result of severe radiation exposure. Estimates of the long-term effects of radiation exposure as a result of the Chernobyl accident vary widely. It is also difficult to definitively correlate deaths on a timescale of years to decades with radiation exposure from the accident. Many estimates of increased cancer deaths suggest thousands to tens of thousands of additional deaths caused by Chernobyl related radiation exposure. However, since 1986 only a few tens of cancer deaths

Figure 6.11. Map of radiation levels in the area around Chernobyl. Contamination levels are shown in the figure. Reproduced from Central Intelligence Agency (1996), courtesy of HathiTrust. Image stated to be in the public domain.

have been able to be clearly related to the accident. It is not only difficult to assess the long-term physical health consequences of an event such as the Chernobyl accident but it is also difficult to assess the psychological consequences leading to depression, alcoholism, and suicide. Recent analyzes (Ritchie 2017) place the total number of deaths from the Chernobyl accident in the range of 300 to 5000, with the majority being due to the delayed effects of thyroid cancer.

6.6 The Fukushima accident

The Fukushima Daiishi Nuclear Power Plant, shown in figure 6.12, is located in the towns of Ōkuma and Futaba in Fukushima Prefecture, Japan. The plant consists of six boiling water reactors with capacities between 460 and 1100 MW_e. Reactors 1 through 5 had Mark I containment, while reactor 6 had an improved Mark II containment structure. Seawater provided cooling for the reactors. The units were commissioned between 1971 and 1979.

Figure 6.12. Fukushima Daiichi Nuclear Power Station photographed in 2007. The six reactors are enclosed in the cube-like buildings. From the lower left of center, the reactors are numbered 4, 3, 2, 1, 5, and 6, where there is a space between reactors 1 and 5. Reproduced from International Atomic Energy Agency (2013b). CC BY-SA 2.0.

A basic diagram of a Mark I boiling water reactor is shown in figure 6.13. An important feature of this design is the spent fuel pool (SFP) where fuel that has been removed from the reactor is kept prior to final disposal. This fuel must be cooled because of the decay heat from residual radioactivity from fission fragments and actinides. As noted previously, this is also true of fuel in reactors that have been shut down (i.e., control rods fully inserted). Since the 1990s, there had been safety concerns about the design of the Fukushima reactors. These concerns included the reliability of the back-up generators in the event of a grid power outage and the adequacy of the seawall in the event of a tsunami. No design changes were made in response to these concerns. While the Fukushima accident is sometimes referred to as a 'beyond design basis event', it is clear that more careful consideration of potential adverse events could have reduced the severity of the accident.

On 11 March 2011, the 9.0 magnitude Tōhoku earthquake occurred about 72 km off the East coast of Japan. This was the fourth most powerful earthquake recorded since seismographic records began in 1900. At the time of the earthquake, Fukushima reactors 1, 2, and 3 were operating. Reactors 4, 5, and 6 were shut down for scheduled maintenance. The fuel from the core of reactor 4 had been transferred to the spent fuel pool, but reactors 5 and 6 still had fuel present in their cores. In addition to cooling the cores of the operating reactors, cooling water was

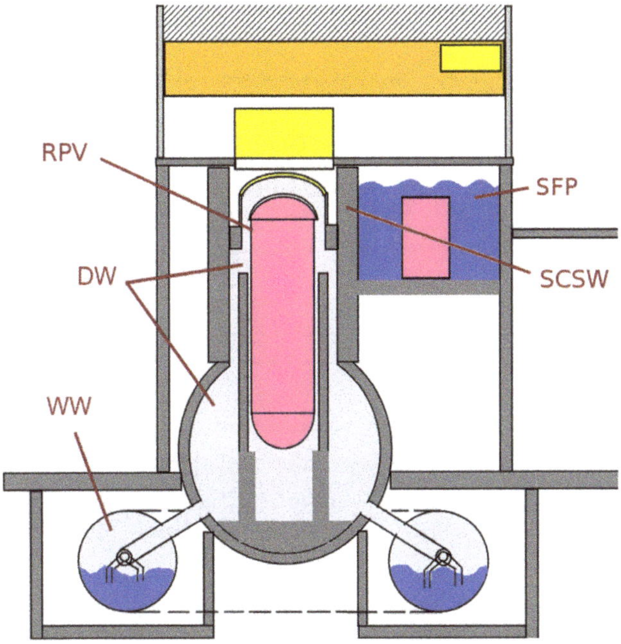

Figure 6.13. Basic diagram of a boiling water reactor with Mark I containment. RPV: reactor pressure vessel, DW: dry well, WW: wet well, SFP: spent fuel pool, and SCSW: secondary concrete shield wall. This rough sketch of a typical boiling water reactor (BWR) mark I concrete containment with steel torus including downcomers, as used in the BWR/1, BWR/2, BWR/3 and some BWR/4 model reactors has been obtained by the author from the Wikimedia website where it was made available by 84user (2011). Image stated to be in the public domain. It is included within this chapter on that basis. It is attributed to 84user.

being supplied to non-operating reactors to cool the fuel in the spent fuel pool of reactor 4 and the cores of the shut down reactors 5 and 6 to eliminate decay heat.

When the earthquake occurred, reactors 1, 2, and 3 were automatically shut down. Since a grid outage was expected, cooling water to all reactors was maintained, as noted above, by pumps operated by the emergency diesel generators. The pumps also had additional DC power back-up from batteries. The tsunami caused by the undersea earthquake hit the reactor site about 50 min later. As shown in figure 6.14, the tsunami overtopped the seawall and reached the height of the reactor buildings, about 15 m above normal sea level.

Flooding of the reactor buildings by the tsunami resulted in a loss of diesel generator power for reactors 1 through 5 and a loss of DC battery power for reactors 1, 2, and 4. The emergency diesel generators for reactor 6, which had not been damaged, were configured to meet the needs of both reactors 5 and 6 (which were separated from the other four reactors, as shown in figure 6.12). No serious damage to either of these two reactors occurred.

As a result of inadequate core cooling, reactors 1, 2, and 3 experienced core meltdowns. Hydrogen liberated from the reaction of water and zirconium-containing fuel cladding according to equation (6.1) led to explosions that severely damaged the

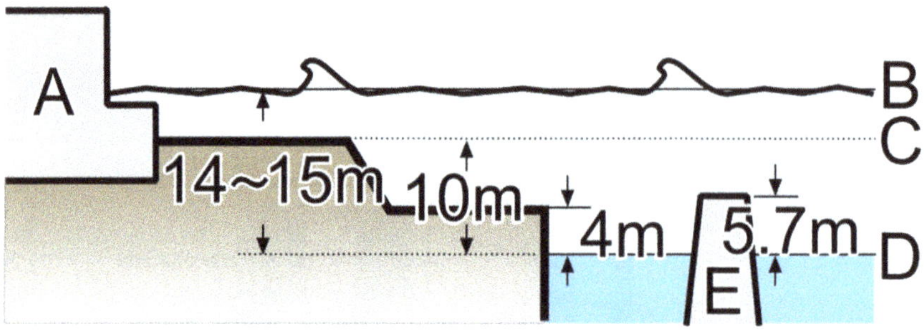

Figure 6.14. Tsunami height relative to reactor structures at Fukushima. A: reactor building, B: tsunami height, C: ground level, D: normal sea level, and E: seawall. Reproduced from Shigeru23 (2011). CC BY-SA 3.0.

Figure 6.15. Fukushima reactors 4, 3, 2, and 1 (left to right) five days after the tsunami. Compare with figure 6.12. Reproduced from Digital Globe (2011). CC BY-SA 3.0.

containment buildings for reactors 1 and 3. Although reactor 4 had no fuel in its core at the time of the accident, it was also damaged by an explosion that is thought to have been caused by hydrogen from reactor 3 leaking through common plumbing. Figure 6.15 shows a photograph of reactors 1 through 4 of the Fukushima Daiichi Power Plant five days after the tsunami. The substantial damage to the buildings housing reactors 1, 3, and 4 caused by explosions is clear in the image.

The damage to the reactors in Fukushima spread considerable radioactive material in the environment. Estimates place the total environmental contamination at about 10% of that produced by Chernobyl. The health effects are, again, difficult to estimate. Two reactor workers died as a direct result of the tsunami. One subsequent lung cancer death (in 2015) has clearly been linked to the accident. Estimates of future fatalities and psychological effects vary widely, although the Japanese government has placed the total number of deaths at around 2300

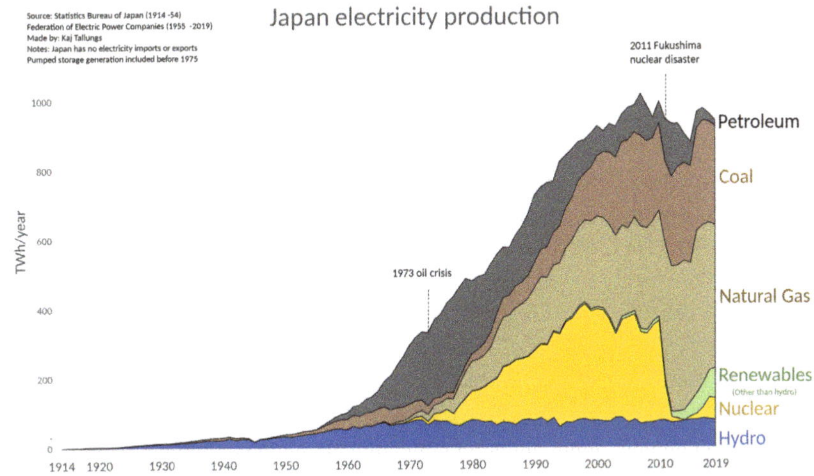

Figure 6.16. Sources of electricity production in Japan. Reproduced from Tallungs (2021). CC BY-SA 4.0.

(see Ritchie 2017 and references therein). The vast majority of these were indirect deaths that resulted from the evacuation and relocation of residents from the area. The clean-up and decommissioning are expected to take 30–40 years.

After the Fukushima accident, all nuclear reactors in Japan were shut down for a re-evaluation of their safety in the event of another earthquake/tsunami. As of 2023, only 6 of the 17 nuclear power plants in Japan have reopened. In most cases, even those which have resumed operations still have one or more reactors that have not been restarted. The effect of the Fukushima accident on nuclear power production in Japan is obvious from an analysis of the sources of electricity, as illustrated in figure 6.16. While there has been some growth in the use of renewable sources since 2011, a large fraction of the lost nuclear capacity has been made up with fossil fuels.

6.7 Nuclear security concerns

The security of nuclear materials deals not only with issues related to the nuclear power industry but also any radioactive materials (see for example see comments in section 6.2 regarding the Goiânia accident). In the present section, we consider nuclear security only as it relates to nuclear power reactors. Nuclear security can be roughly divided into two categories: first, actions that are intended to cause damage to the reactor facility and, in some cases, to the environment; and second, the acquisition of nuclear materials by unauthorized persons for the purpose of nuclear proliferation. We can look in more detail at some examples of some of these issues.

Attacks on nuclear reactor facilities have occurred in the past and there are a variety of reasons why such attacks occur. Some reasons for these incidents include terrorist attacks, attacks for political reasons, and acts of war. Several attacks of reactor facilities by militant groups have occurred. These include,

- An attack of the Atucha Nuclear Power Plant in Argentina by leftist guerrillas in 1973.

- Attacks on the Lemóniz Nuclear Power Plant in Spain by Basque separatists between 1977 and 1979.
- An attack on the Superphénix fast reactor in France by the Red Army Faction in 1982.

In each of these cases, the attacks occurred during the construction of the power plant, and therefore did not compromise the security of any radioactive material. However, after the 9/11 terrorist attacks in the United States, the possibility that aircraft could be used to attack an operating nuclear power plant has been considered as a real threat. The concrete and steel containment structure around the reactor core is the first line of defense against such an attack and the design of recent reactors considers this possibility.

The possibility of an incident occurring at a nuclear facility as a result of an act of war by a government was responsible for the 1979 addition to the Geneva Convention prohibiting the attack of a nuclear power plant that could cause the release of radioactive material. There were several attacks by government forces on facilities related to nuclear energy in the Middle East during the 1980s and 1990s but there was no release of radioactive material in any of these cases. Probably the most notable attack on an operating nuclear power reactor by government forces occurred during the Russian invasion of Ukraine. On 4 March 2022, Russian forces attacked and took control of the Zaporozhe Nuclear Power Plant. There is no evidence that the integrity of any of the six reactors was compromised.

In addition to physical attacks on nuclear power plants or fuel processing facilities, cyberattacks on nuclear reactor computer systems are also possible. Although no such cyberattack has been recorded, security measures are in effect at nuclear power stations to avoid this possibility.

The security of radioactive material associated with the nuclear power industry may be compromised when they are in the reactor, when they are in transit, or when they are in storage. The nuclear industry and the governments of countries utilizing nuclear power have protocols in place to secure radioactive materials in these three scenarios. The theft of any significant radioactive material would likely put those involved in danger of severe radiation exposure. The most serious concern might seem to be the unauthorized acquisition of a suitable quantity of fissile material for nuclear proliferation by the construction of an actual nuclear bomb. However, a more realistic situation might be the utilization of radioactive material (fissile or not) to construct a dirty bomb. A dirty bomb, more accurately referred to as a radiological dispersal device, consists of radioactive material mixed with a conventional explosive. While there would be destruction and fatalities as a result of the conventional explosion, the main purpose of the dirty bomb is to spread radioactive contamination to as large an area as possible. The effect of such contamination would limit the use of this area for civilian and/or military purposes, and would introduce a disruption of resources related to the subsequent clean-up. The psychological impact on the population in the area would also be substantial. The area affected by a dirty bomb depends on the amount and type of radioactive material, the amount of conventional explosive, and the local weather conditions.

Typically, distances of the order of hundreds of meters to perhaps a kilometer might be involved. To date, no theft of radioactive material related to the nuclear power industry for unlawful purposes has taken place.

6.8 Nuclear waste disposal

Nuclear waste, i.e., waste material that is radioactive, comes from a variety of sources. Principal among these are the nuclear power industry; nuclear weapons programs; medial research, diagnostics, and treatment; and scientific research. Waste is also produced during fuel reprocessing, as discussed in section 5.11. In the present section we consider waste products from a once through fission reactor fuel cycle and the disposal of radioactive waste, in general.

Waste associated with nuclear fission reactors can come from initial fuel processing (i.e., front end waste) or spent fuel (back end waste). Front end waste is almost entirely left over ^{238}U when processed uranium is enriched in ^{235}U for use in (for example) typical light-water-moderated reactors. ^{238}U has a low specific activity because of its long half-life and is not a major waste disposal concern. In fact, depleted uranium has a number of uses, such as high-density anti-tank shells or for downblending surplus weapons-grade material that is redirected to the nuclear power industry. It is the spent fuel from a fission reactor that creates the major concern for radioactive waste disposal. This fuel consists of several radioactive components, as summarized below.

Fission products and products of their β^- decay chains: These are illustrated in figure 3.3 and account for a total of about 3% of the spent fuel. Many of these are short lived nuclides that decay to longer lived species by β^- decay. The major components of spent fuel that result from fission products that have medium to long half-lives are given in table 6.2.

Uranium: Uranium in spent fuel accounts for about 96% of its mass. It consists primarily of ^{238}U from the original fuel, ^{235}U that has not undergone fission and a small amount of ^{236}U that is produced by the reaction

Table 6.2. Major fission products of ^{235}U and ^{239}Pu in the spent fuel of a thermal neutron reactor.

Nuclide	Fission yield (%)	Half-life (y)
^{85}Kr	0.22	10.8
^{90}Sr	4.5	28.9
^{93}Zr	5.5	1.5×10^6
^{99}Tc	6.1	2.2×10^5
^{107}Pd	1.2	6.5×10^6
^{126}Sn	0.11	2.3×10^5
^{135}Cs	6.9	2.3×10^6
^{137}Cs	6.3	30
^{151}Sm	0.53	90

$$n + {}^{235}U \rightarrow {}^{236}U + \gamma.$$

If the reactor fuel contains thorium, then the spent fuel will also contain residual ^{232}Th and ^{233}U produced by the reaction

$$n + {}^{232}Th \rightarrow {}^{233}U + \gamma.$$

Plutonium: Plutonium (^{239}Pu and ^{240}Pu) is produced by the reactions

$$n + {}^{238}U \rightarrow {}^{239}U + \gamma$$

followed by the β^- decays

$$^{239}U \rightarrow {}^{239}Np + e^- + \overline{\nu}$$

and

$$^{239}Np \rightarrow {}^{239}Pu + e^- + \overline{\nu}.$$

Neutron capture produces ^{240}Pu by the process

$$n + {}^{239}Pu \rightarrow {}^{240}Pu + \gamma.$$

Plutonium accounts for about 1% of the spent fuel mass and contains about 80% ^{239}Pu and 20% ^{240}Pu. The non-fissile ^{240}Pu content is too high for direct use for weapons production.

Minor actinides: Minor amounts of actinides, other than uranium and plutonium, are present as a result of neutron capture and subsequent β^- decays of fuel components. These actinides can include (e.g.) neptunium, americium, and curium.

The activity of typical spent fuel from a fission reactor is shown in figure 6.17. The short-term activity (up to as few years) is dominated by short lived fission products. The medium-term activity (tens to hundreds of years) is dominated by medium lifetime fission products (see table 6.2). The long-term activity is dominated by actinides and their radioactive daughters (and some long-lived fission products). It is clear in the figure that the reduction of activity to levels comparable to that of natural uranium requires about 100 000 years. This timescale is an important consideration in the approach to the long-term storage of spent fuel.

Figure 6.18 shows the total amount of spent fuel that has been produced by the world's reactors since 1990. The graph shows that at present about a quarter of spent fuel is reprocessed, as discussed in section 5.11. The total mass of stored spent fuel is about 330 kt (3.3×10^8 kg). Methods for dealing with stored spent fuel are discussed below.

In a consideration of future fission reactor utilization and the approach to dealing with radioactive waste, it is important to consider the effects of fuel reprocessing, as discussed in section 5.11. Figure 6.19 shows the activity as a function of time for spent fuel for a once through fuel cycle (see also figure 6.17), a twice through fuel cycle, and a closed cycle, where spent fuel is continuously reprocessed. Compared to the once through cycle where it requires on the order of 100 000 years for the activity of spent fuel to drop below that of the original natural resource, waste from the

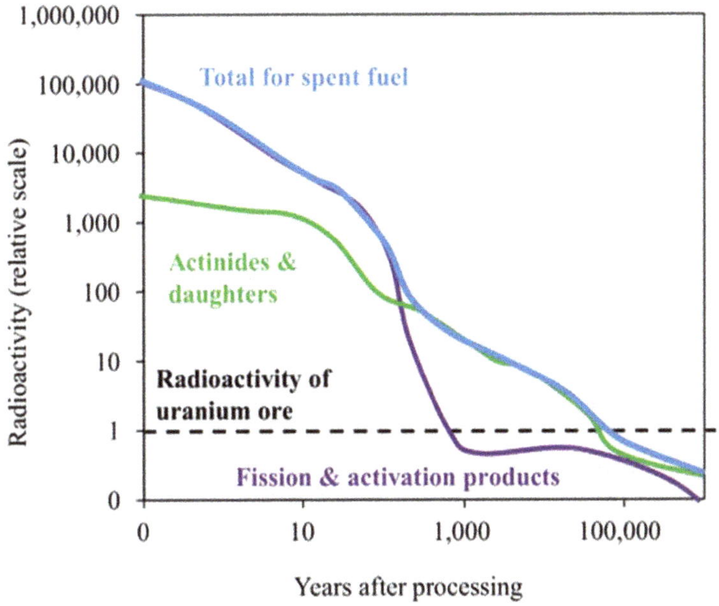

Figure 6.17. Decay of typical spent nuclear fission reactor fuel. Contributions from fission products and actinides are shown. The activity of the original uranium ore is shown for comparison and activities are normalized to a value of 1 for the original uranium. Reproduced from Corkhill and Hyatt (2017).

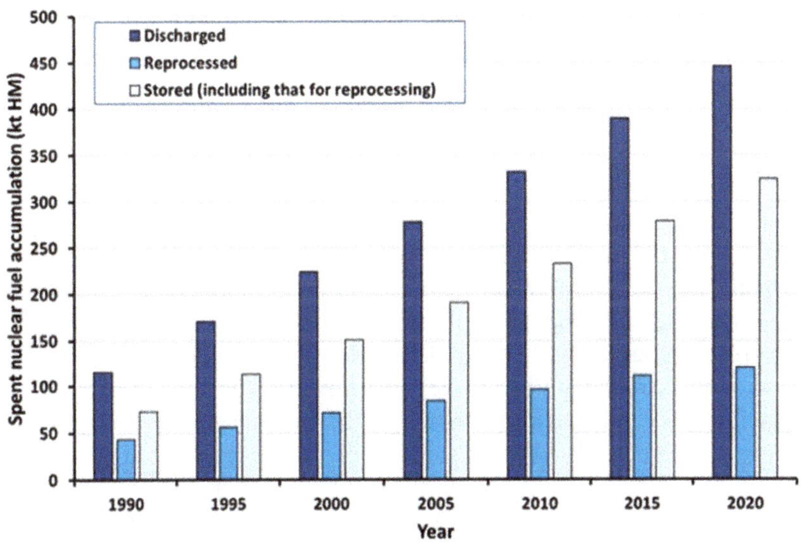

Figure 6.18. World spent reactor fuel from 1990 to 2020, showing the amount produced (discharged) and the proportions of stored and reprocessed fuel. Data are from International Atomic Energy Agency (2016). Reproduced from Taylor *et al* (2022).

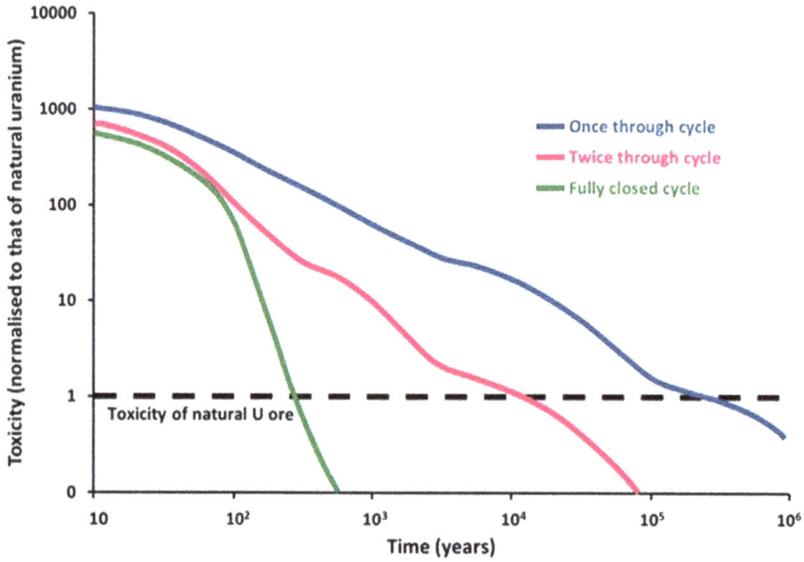

Figure 6.19. Nuclear fuel cycle on activity levels of nuclear waste. Data are from Poinssot *et al* (2015). Reproduced from Taylor *et al* (2022). CC BY 4.0.

closed fuel cycle requires a few hundred years to drop to a reasonably safe level. The importance of this feature will be discussed further in the sections on Generation IV reactor designs.

There are several approaches to the storage of high-level radioactive waste. When spent fuel is first removed from a reactor core, it continues to produce heat, as previously shown in figure 6.4. It must, therefore, be cooled in a spent fuel pool, typically for a few years. Spent fuel pools are usually located at the reactor site and are generally cooled by an active system that transfers excess heat to the environment through a heat exchanger. Figure 6.20 shows a typical example of a spent fuel pool at a nuclear reactor. See also the discussion above (section 6.6) on the Fukushima accident.

When the decay energy is reduced so that the spent fuel no longer requires cooling, it is typically transferred to a dry cask (see figure 6.21). Spent fuel rods are sealed in a steel cylinder in order to prevent leakage of radioactive material to the environment. The cask is surrounded by suitable shielding to minimize radiation exposure for workers and the general public.

As illustrated in figures 6.17 and 6.19, long-term storage of spent fuel (particularly in the case of a once through fuel cycle) is necessary. The long-term stability of the storage facility is of utmost importance. In the case of storage in a human-made facility, it is impossible to predict the stability of human civilization on a timescale of hundreds of thousands of years. In the case of storage in natural formations, the long-term stability (e.g., from earthquakes) is also difficult to predict. However, several approaches to long-term storage of spent nuclear fuel have been considered.

Typically, it is necessary to stabilize the fuel to avoid chemical changes and to avoid interaction of the fuel with its surroundings (i.e., corrosion). This is done by

Figure 6.20. A spent fuel pool at the San Onofre Nuclear Generating Station. Reproduced from U.S. Nuclear Regulatory Commission (2014a). CC BY 2.0.

Figure 6.21. Dry cask storage facility. Reproduced from U.S. Nuclear Regulatory Commission (2014b). CC BY 2.0.

Figure 6.22. Onkalo spent nuclear fuel repository near the Olkiluoto Nuclear Power Plant in Eurajoki, Finland. The cave has been excavated to a depth of about 430 m. Spent reactor fuel storage is expected to begin in the near future. This Onkalo spent nuclear fuel repository, Olkiluoto, Finland image has been obtained by the author from the Wikimedia website where it was made available by Kallerna (2020) under a CC BY-SA 4.0 licence. It is included within this chapter on that basis. It is attributed to Kallerna.

vitrification, which is a process by which the fuel particles are embedded in a suitable glass. The long-term storage of this material may involve storage in appropriate sealed containers in either terrestrial or oceanic locations. Terrestrial possibilities that have been investigated include deep geological locations, such as salt mines, specifically constructed caves (see figure 6.22), or bore holes. Ocean disposal can include deep ocean sediments. It has also been suggested that radioactive waste can be placed in subduction zones, where they will be naturally carried downward by the movement of tectonic plates into the Earth's mantle.

As an alternative to long-term storage of spent nuclear fuel, the radioactivity of the fuel may be eliminated by transmutation. This is done by inducing nuclear reactions in the radioactive nuclei in the fuel using particle beams from an accelerator of intense laser radiation to convert long-lived nuclides into short lived nuclides. It has also been suggested that radioactive waste could be shot into space. However, neither of these approaches is economically viable at present.

Although there is no definitive view on long-term spent fuel storage, it seems, at present, that underground storage in salt mines or artificial caves is the most economical and viable.

6.9 Nuclear safety analysis

The utilization of nuclear fission energy can play a significant role in the minimization of the effects of climate change, as previously discussed in chapter 1. However, the future development of nuclear fission power depends on the degree to which the concerns discussed previously in this chapter can be addressed. The design and construction of Generation IV nuclear reactors, as presented in the remaining chapters of this book, will play an important role in the viability of such development.

While several significant events (e.g., Chernobyl and Fukushima) have influenced public opinion on the safety of nuclear energy (which is discussed further in chapter 13) and have had an influence on government policies concerning nuclear energy, it is important to view the overall safety of nuclear energy in comparison to other existing and potentially future energy sources. Figure 6.23 shows the death rate per unit electricity generated for different sources. Fossil fuels clearly show the greatest death rate per unit energy generated. This relates directly to illness that results from air pollution and also from accidents related to the large quantity of material (particularly coal) that must be mined and transported. The death rate for biomass is dominated by the effects of air pollution.

The death rate for hydroelectric energy is dominated by deaths from a single dam failure. In 1975, the Banqiao Dam in China failed resulting in about 170 000 deaths. Without this one event, the death rate per unit electricity for hydroelectric would be around 0.04 TWh^{-1}, which is comparable to wind and solar.

The nuclear power death rate includes estimated deaths from Chernobyl and Fukushima (see sections 6.5 and 6.6, above). Even with these major nuclear accidents, the death rates for nuclear, solar, and wind (as well as hydroelectric with the exception of one major event) are all very low and of similar magnitude

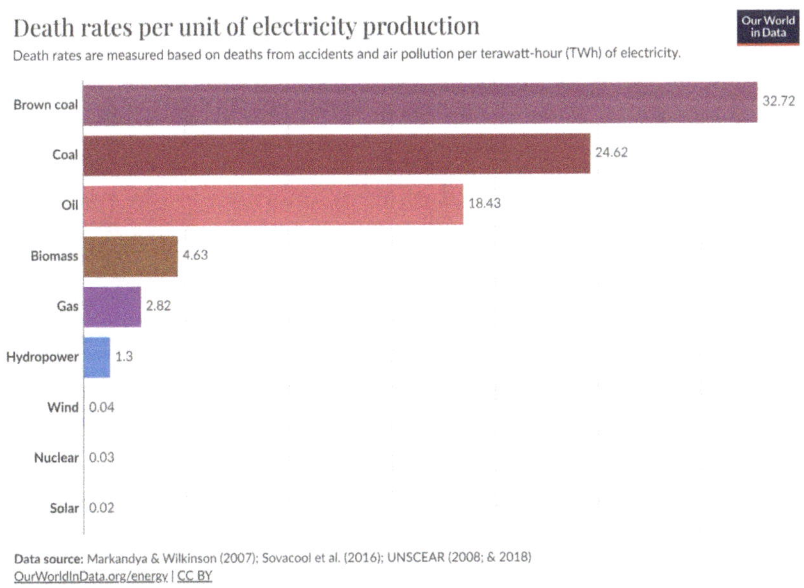

Figure 6.23. Death rates per unit electricity generated for different sources. Deaths from accidents and air pollution are given per TWh of electricity produced. Data adapted from Markandya and Wilkinson (2007), Sovacool *et al* (2016), and United Nations Scientific Committee on the Effects of Atomic Radiation (2008). Reproduced from Ritchie (2020). CC BY 4.0.

compared to those forms of electricity generation that contribute to illness through air pollution.

It is also informative to look at the progress in nuclear safety over the years. Two important measures of nuclear safety are the core damage frequency (CDF) and the large release frequency (LRF), i.e., the frequency of accidents resulting in reactor core damage per reactor (CDF) and the frequency of large radiation release (LRF) per reactor. Table 6.3 shows the estimated CDF and LRF for different generations of fission reactors. The improvement in reactor safety since the earliest commercial reactors in the 1950s is substantial. Current Generation III reactors have an expected large release frequency that corresponds to one such event every 2.7 million years, or so, per operating reactor.

Future reactors are expected to continue to improve on safety and a major feature of many Generation IV reactor designs is the reduced risk of accidents. A study by the U.S. Nuclear Regulatory Commission (2009) provided the results for core damage frequency given in table 6.4. It is clear that advanced reactor designs have improved safety. The results also illustrate the difference between reactors that rely on active safety measures, i.e., measures that require operator intervention and/or external power, and those that rely on passive measures that require neither operator control nor external power. These differences will be discussed in detail in the following chapters on Generation IV reactor design.

This chapter has shown that advanced reactor designs are expected to go far in alleviating concerns related to safety. The utilization of breeder reactors, as

Table 6.3. Average CDF and LRF for different generations of light-water nuclear fission reactors. Data adapted from International Atomic Energy Agency (2004).

Generation	CDF (y^{-1})	LRF (y^{-1})
I	1.8×10^{-3}	7.9×10^{-4}
II	7.4×10^{-5}	2.0×10^{-5}
III	2.8×10^{-6}	3.7×10^{-7}

Table 6.4. Range of values for CDF of some classes of operating and future nuclear power reactors. PWR = pressurized water reactor, BWR = boiling water reactor, and LWR = light-water reactor. Data adapted from U.S. Nuclear Regulatory Commission (2009) and references therein.

Reactor type	CDF (y^{-1})
Operating PWR	2.5×10^{-6}–7.0×10^{-5}
Operating BWR	1.1×10^{-6}–4.0×10^{-5}
Advanced LWR (active)	2.2×10^{-7}–2.2×10^{-6}
Advanced LWR (passive)	1.2×10^{-8}–2.3×10^{-7}

discussed in section 5.8, will provide long-term resource availability and will, at least partially, reduce potentially adverse effects of nuclear waste disposal, as noted in section 6.8.

References

Central Intelligence Agency 1996 *Handbook of International Economic Statistics* https://babel.hathitrust.org/cgi/pt?id=osu.32435055467591&seq=185

Corkhill C and Hyatt N 2017 *Nuclear Waste Management* (Bristol: IOP Publishing)

Crick M J and Linsley G S 1984 An assessment of the radiological impact of the windscale reactor fire, october 1957 *Int. J. Radiat. Biol. Relat. Stud. Phys. Chem. Med.* **46** 479–506

Datesman A M 2020 Radiobiological shot noise explains three mile Island biodosimetry indicating nearly 1,000 mSv exposures *Sci Rep.* **10** 10933

Digital Globe 2011 Fukushima nuclear power plant https://commons.wikimedia.org/wiki/File:Fukushima_I_by_Digital_Globe.jpg

Eaton C 1985 Storm Clouds over Sellafield https://en.wikipedia.org/wiki/File:Storm_Clouds_over_Sellafield_-_geograph.org.uk_-_330062.jpg

Exitech N D *Core Nucleonics Model* https://web.archive.org/web/20120118143841/http://exitech.com/models/core_neutronics.htm

Han Y Y, Youk A O, Sasser H and Talbott E O 2011 Cancer incidence among residents of the three mile Island accident area: 1982–1995 *Environ. Res.* **111** 1230–5

International Atomic Energy Agency 2004 *Status of Advanced Light Water Reactor Designs IAEA TECDOC-1391* (Vienna: International Atomic Energy Agency))

International Atomic Energy Agency 2011 April 26 1986 Signified the boundary between life and death https://www.flickr.com/photos/iaea_imagebank/5613115146/in/photostream/

International Atomic Energy Agency 2013a The ill-fated 4th block of the Chernobyl Nuclear Power Plant www.flickr.com/photos/iaea_imagebank/8388690847/

International Atomic Energy Agency 2013b Fukushima Daichi Nuclear Power Station www.flickr.com/photos/iaea_imagebank/8388174045/

International Atomic Energy Agency 2016 *Nuclear Technology Review* www.iaea.org/sites/default/files/16/08/ntr2016.pdf

Johnson J L 2012 An interesting window view www.flickr.com/photos/37569637@N07/7145287481

Kallerna 2020 Onkalo spent nuclear fuel repository, Olkiluoto, Finland https://commons.wikimedia.org/wiki/File:Onkalo_spent_nuclear_fuel_repository.jpg

Mahaffey J 2014 *Atomic Accidents—A History of Nuclear Meltdowns and Disasters: From the Ozark Mountains to Fukushima* (New York: Pegasus)

Markandya A and Wilkinson P 2007 Electricity generation and health *Lancet* **370** 979–90

Poinssot C, Boullis B and Stéphane B 2015 Role of recycling in advanced nuclear fuel cycles *Reprocessing and Recycling of Spent Nuclear Fuel* ed R Taylor (Sawston: Woodhead Publishing) pp 27–48

Porter T 2017 The New Safe Confinement in final position over reactor 4 at Chernobyl Nuclear Power Plant https://commons.wikimedia.org/wiki/File:NSC-Oct-2017.jpg

Ritchie H 2017 What was the death toll from Chernobyl and Fukushima? https://ourworldindata.org/what-was-the-death-toll-from-chernobyl-and-fukushima

Ritchie H 2020 What are the safest and cleanest sources of energy? https://ourworldindata.org/safest-sources-of-energy

Rogovin M 1980 *Three Mile Island: A Report to the Commissioners and to the Public* **vol I** (Washington, DC: U.S. Nuclear Regulatory Commission)

Runge I 2013 A stitched panorama view of the Chernobyl Nuclear Power Plant site in Ukraine, pictures taken in June 2013 https://commons.wikimedia.org/wiki/File:Chernobyl_NPP_Site_Panorama_with_NSC_Construction_-_June_2013.jpg

Shigeru23 2011 Tsunami height of Fukushima I Power plant https://commons.wikimedia.org/wiki/File:Fukushima_I_Powerplant_(Tsunami_height).png

Silver Spoon 2011 International Nuclear Event Scale https://commons.wikimedia.org/wiki/File:INES_en.svg

Sovacool B K *et al* 2016 Balancing safety with sustainability: assessing the risk of accidents for modern low-carbon energy systems *J. Clean. Prod.* **112** 3952–65

Tallungs K 2021 Electricity production in Japan https://commons.wikimedia.org/wiki/File:Electricity_production_in_Japan.svg

Taylor R, Bodel W, Stamford L and Butler G 2022 A review of environmental and economic implications of closing the nuclear fuel cycle—part one: wastes and environmental impacts *Energies* **15** 1433

Theanphibian 2011 Graph of decay heat as fraction of full power from a SCRAMed operating power reactor https://commons.wikimedia.org/wiki/File:Decay_heat_illustration.PNG

Todreas N E and Kazimi M S 1989 *Nuclear Systems I, Thermal Hydraulic Fundamentals* (Boca Raton, FL: CRC Press)

U.S. Nuclear Regulatory Commission 2009 Risk metrics for operating new reactor www.nrc.gov/docs/ML0909/ML090910608.pdf

U.S. Nuclear Regulatory Commission 2014a A spent fuel pool at the San Onofre Nuclear Generating Station www.flickr.com/photos/nrcgov/16042442105/

U. S. Nuclear Regulatory Commission 2014b Dry cask storage www.flickr.com/photos/nrcgov/14678900905/in/gallery-138473945@N07–72157661891602880/

U.S. Nuclear Regulatory Commission 2022 Backgrounder on the Three Mile Island Accident www.nrc.gov/reading-rm/doc-collections/fact-sheets/3mile-isle.html#tmiview

United Nations Scientific Committee on the Effects of Atomic Radiation 2008 *Sources and Effects of Ionizing Radiation* www.unscear.org/unscear/uploads/documents/unscear-reports/UNSCEAR_2008_Report_Vol.I-CORR.pdf

User:84user 2011 BWR mark I Containment sketch https://commons.wikimedia.org/wiki/File:BWR_Mark_I_Containment_sketch_with_downcomers.png

Wakeford R 2007 The windscale reactor accident—50 years on *J. Radiol. Prot.* **27** 211–5

Walker J S 2004 *Three Mile Island: A Nuclear Crisis in Historical Perspective* (Berkeley, CA: University of California Press)

Chapter 7

Very high-temperature reactors (VHTRs)

Very high-temperature reactors (VHTRs) are covered in this chapter. These are thermal neutron reactors that utilize fuel pellets that are temperature stabilized with a ceramic coating. The reactor may have either a prismatic core, where fuel particles are contained in a graphite moderator matrix in the form of fuel pins that are stacked to produce fuel rods, or in the form of fuel pebbles that are used in a pebble bed. The earliest VHTRs date back to the mid-1960s, where both experimental prismatic core and pebble bed reactors were constructed. This chapter reviews high-temperature reactors that have been constructed since that time. These reactors include the two HTR-PM reactors that have been operational in China since 2021 and which represent the world's first commercial Generation IV reactors. The advantages of VHTRs, including high thermodynamic efficiency and cogeneration opportunities, are discussed in this chapter.

7.1 Introduction

VHTRs, sometimes called high-temperature gas-cooled reactors (HTGRs) or just high-temperature reactors (HTRs), have several potential benefits compared to conventional thermal reactors. Most important, perhaps, are improved thermal efficiency of the turbines and increased potential for cogeneration. Designs of high-temperature reactors, as specified by the Generation IV International Forum, also include inherent safety features that reduce the risk of adverse incidents.

A major concern in the design of a high-temperature reactor is the choice of materials. Many materials that are in common use in traditional thermal neutron reactors are inappropriate for use in high-temperature reactors. An important example is the design of the fuel elements. As noted in chapter 3, conventional thermal neutron reactors typically use a zirconium alloy (Zircaloy) for fuel tubes, which will tolerate steam temperatures up to about 450 °C. For higher temperatures, stainless steel is sometimes use for fuel tubes and these are suitable up to about 650 °C. For VHTRs, a different approach to fuel encapsulation is needed. This is described in the next section.

doi:10.1088/978-0-7503-6069-2ch7

7.2 Fuel elements for very high-temperature reactors

VHTRs typically fall into two categories based on the geometry of their fuel elements: those with prismatic cores and those with pebble bed cores. Both types of fuel elements utilize fissile fuel that is encapsulated in a ceramic that will withstand the high operating temperatures of the reactor. Tristructural-isotropic (TRISO) fuel consists of 0.5 mm sized pellets (see figure 7.1) where a fissile uranium containing (or possibly thorium containing) kernel is coated with four protective layers, as illustrated in figure 7.2. For a uranium fueled reactor the fissile kernel could be uranium dioxide (UO_2), uranium carbide (UC), or uranium oxycarbide (UCO).

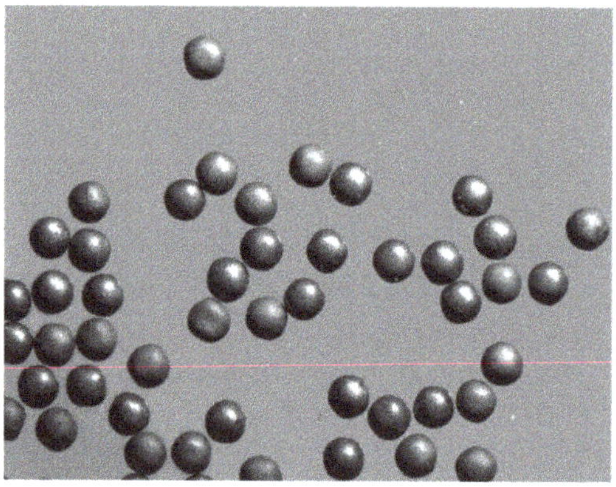

Figure 7.1. Individual TRISO nuclear fuel pellets. Idaho National Laboratory (2013a) CC BY 2.0.

Figure 7.2. Internal structure of a typical TRISO fuel pellet showing the fissile kernel (yellow) coated with a layer of porous carbon, a layer of pyrolytic carbon, a layer of silicon dioxide (yellow), and an outer layer pyrolytic carbon. Idaho National Laboratory (2013b) CC BY 2.0.

Further details of the composition of the fissile fuel for high-temperature reactors are given below. The uranium containing kernel is coated with a layer of porous carbon, followed by a layer of pyrolytic carbon and a layer of silicon dioxide (SiO_2). The outer layer consists of a second layer of pyrolytic carbon. The fissile fuel in this form is very stable and, even in the event of unexpected increases in reactor operator temperature, is protected from dispersal into the environment.

An additional safety feature that has been proposed for VHTRs that utilize TRISO fuel is the incorporation of QUADRISO (quadruple isotropic) fuel particles mixed with the TRISO fuel particles in the reactor core (see Talamo 2008). The QUARDISO particle, as shown in figure 7.3, contains a layer of neutron poison, such as europium oxide (EuO), or another rare earth compound, adjacent to the fissile core. Rare earth elements have a very large thermal neutron absorption (n, γ) cross section (see section 5.11). In a reactor with both TRISO and QUADRISO particles in the core, energy is produced by induced fission in the TRISO particles. The QUADRISO particles do not contribute significantly to energy production because the neutron poison surrounding the fissile core prevents neutrons from reaching the fissile kernel. Over time, (n, γ) reactions in the neutron poison transmute elements with large neutron cross sections into elements with smaller cross sections. This is referred to as the 'burn-up' of the poison layer. Thus, over time, the kernels of the QUADRISO particles contribute progressively more energy to the reactor's output. This behavior compensates for two features of fission reactor operation. First, as fissile material in the TRISO pellets undergo fission, the net energy output will gradually decrease. Second, as fission by-products accumulate in

Figure 7.3. QUADRISO fuel particle. From the center of the particle the layers are: fissile fuel (light green), burnable poison (orange), porous carbon (light blue), inner pyrolytic carbon layer (purple), silicon carbide (yellow), and outer pyrolytic carbon layer (purple). This QUADRISO particle image has been obtained by the author from the Wikimedia website where it was made available by Alberto Talamo (2008) under a CC BY-SA 3.0 licence. It is included within this chapter on that basis. It is attributed to Alberto Talamo.

the fuel, some of these products will be neutron poisons (i.e., typically rare earth elements) and this will also decrease the energy output form the TRISO pellets.

As noted above, two different geometries are used for the cores of high-temperature reactors: prismatic cores and pebble bed cores. Both reactors make use of fuel pellets as described above. The details of the fuel elements and the design of these two reactors are discussed in detail in the next two sections.

7.3 Prismatic core high-temperature reactors

For the prismatic core high-temperature reactor, the TRISO (or TRISO and QUADRISO) fuel pellets are embedded in a graphite matrix cylinder about 1 cm in diameter and 2.5 cm in length (sometimes referred to as the fuel pin), as illustrated in figure 7.4. These fuel pins are assembled to produce the reactor core, as shown in figure 7.5. Stacks of fuel pins are arranged along with cooling channels into hexagonal fuel blocks and these fuel blocks are then assembled into the reactor core.

Prismatic core high-temperature reactors typically use uranium fuel enriched to about 6% in ^{235}U and are cooled with helium gas because water is not suitable at the operating temperatures of these reactors. Figure 7.6 shows the reactor core of a prismatic reactor in its containment vessel along with cooling by helium gas. Typically, heat is transferred from the helium to a secondary water/steam loop through a heat exchanger.

Figure 7.4. Idaho National Laboratory Fellow Dave Petti, technical director for TRISO fuel research, holding a cylinder (fuel compact) of TRISO containing fuel designed for a prismatic core high-temperature reactor. Idaho National Laboratory (2009) CC BY 2.0.

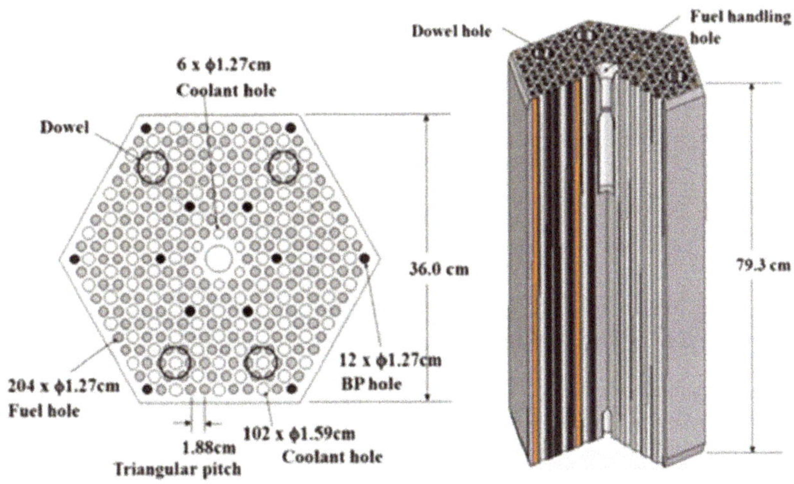

Figure 7.5. Design of a fuel block for the core of a prismatic core high-temperature reactor. Reprinted from Tak *et al* (2014), Copyright (2014), with permission from Elsevier.

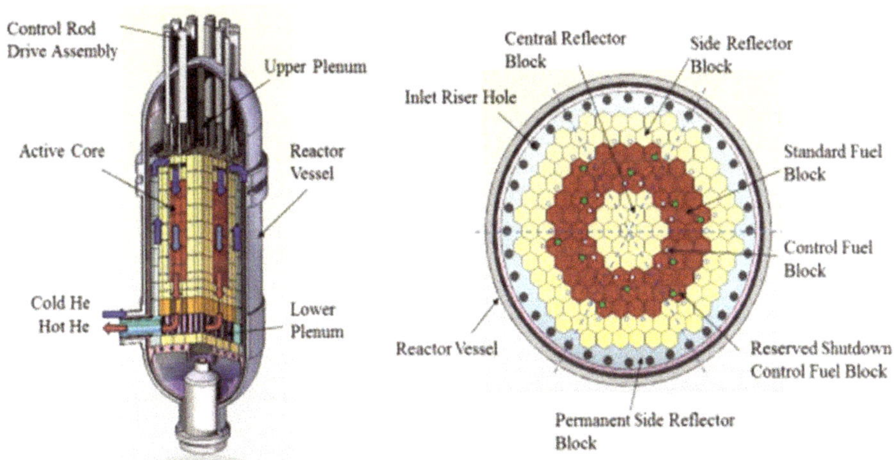

Figure 7.6. Prismatic core high-temperature reactor core and pressure vessel showing the flow of helium gas for cooling. Reprinted from Tak *et al* (2014), Copyright (2014), with permission from Elsevier.

7.4 Pebble bed reactors

For the pebble bed reactor, TRISO or TRISO and QUADRISO fuel pellets are embedded in a graphite sphere approximately 6 cm in diameter, as shown in figure 7.7. The fuel pellets are encapsulated in the spherical fuel element as illustrated in figure 7.8 and a cross section through the graphite sphere is shown in figure 7.9. Pebble bed reactors typically use uranium fuel enriched to about 8.6% in ^{235}U.

The general layout of a pebble bed reactor is shown in figure 7.10. Fuel pebbles are injected at the top of the reactor core and add to the pebble bed inside the core.

Figure 7.7. 60 mm diameter graphite fuel element for a pebble bed reactor. https://commons.wikimedia.org/wiki/File:Pebble_Bed_Graphite_Ball.jpg This Graphite Pebble for Pebble Bed Reactor, Diameter 60mm image has been obtained by the author from the Wikimedia website where it was made available by Weirdmeister (2017) under a CC BY-SA 4.0 licence. It is included within this chapter on that basis. It is attributed to Weirdmeister.

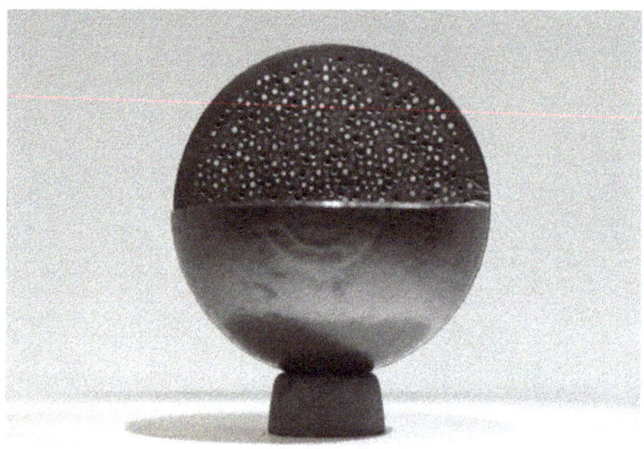

Figure 7.8. Internal structure of a spherical fuel element for a pebble bed high-temperature fission reactor. Reproduced from U.S. Department of Energy (2021). Image stated to be in the public domain.

Pebbles are extracted from the bottom of the core and are recirculated back to the top of the reactor. Pebbles undergo this circuit about a dozen times before being replaced by new pebbles. The fuel cycle is typically open. Heat is extracted, as in the prismatic design, by circulating helium gas through the core. This heat is transferred through a heat exchanger to a secondary water loop which produces steam to operate a turbine/generator.

Further details of a typical pebble bed reactor design are illustrated in figure 7.11. The pebble bed core is surrounded by a graphite reflector. This reflects neutrons that escape from the core back into the core by elastic scattering to increase fission in the fuel pebbles. The reactor is controlled using control rods that are inserted not into the core but into the graphite reflector surrounding the core. Inserting the control rods decreases the flux of neutrons reflected back into the core and reduces the fission

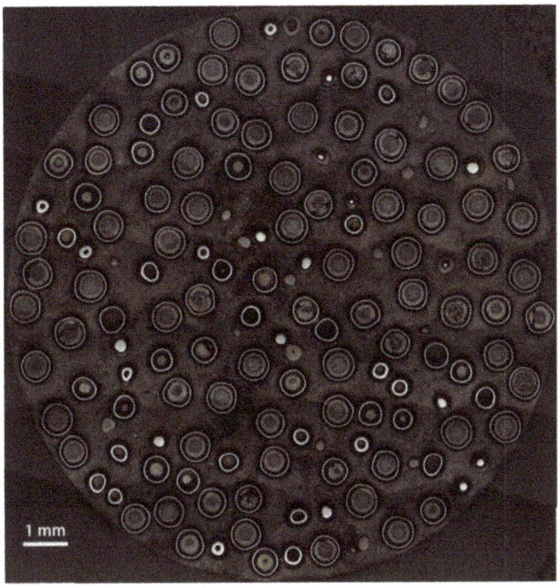

Figure 7.9. Cross section of a spherical fuel element for a pebble bed reactor showing the distribution of TRISO fuel pellets. Reproduced from Idaho National Laboratory (2013c). CC BY 2.0.

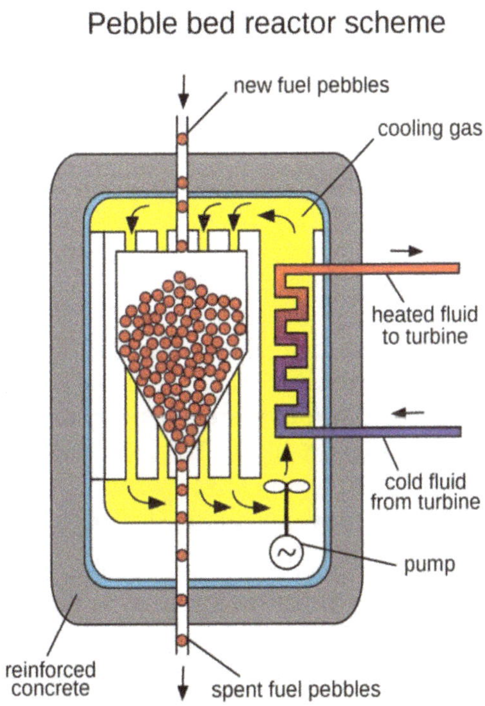

Figure 7.10. Simplified schematic of a pebble bed reactor. This Pebble Bed Reactor scheme image has been obtained by the author from the Wikimedia website where it was made available by Picoterawatt (2019) under a CC0 licence. It is included within this chapter on that basis. It is attributed to Picoterawatt.

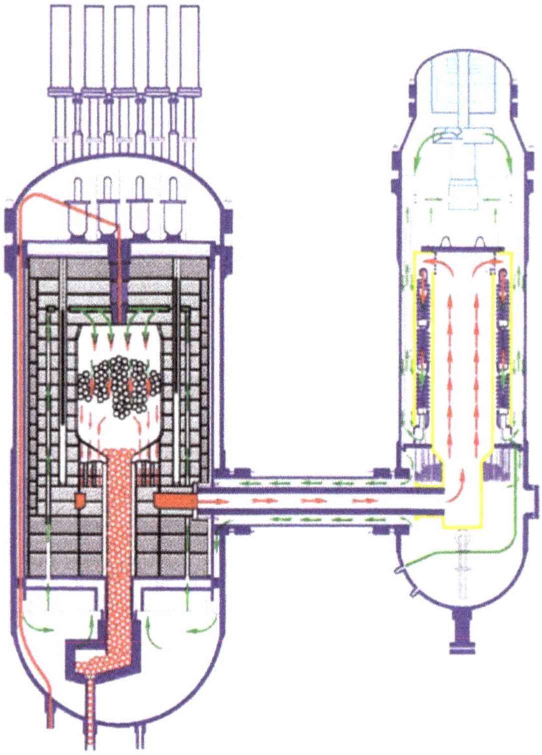

Figure 7.11. Schematic of a typical modular pebble bed reactor, the HTR-10 reactor (see section 7.5.7). Reprinted from Zhang *et al* (2009), Copyright (2009), with permission from Elsevier.

rate in the fuel. As a secondary emergency control system, balls of neutron absorbing material can be dropped into absorber ball channels in the reflector.

Figure 7.12 shows a modular pebble bed reactor with a once-through steam generator. Helium gas is circulated through the reactor by a blower. Cold gas is injected into the reactor vessel and travels upward through channels in the graphite reflector. It then flows downward to the pebble bed and flows through spaces between the loosely packed spherical graphite fuel pebbles. The hot helium gas then flows through the heat exchanger (once-through steam generator; OTSG) where it transfers heat to the water in the secondary loop.

7.5 Progress towards the development of a high-temperature reactor

Interest in high-temperature reactors results from their improved efficiency and the possibility of cogeneration. An additional attractive aspect of high-temperature reactors is the inherent safety of their design. The encapsulation of the fissile fuel prevents the dispersal of radioactive material, even in the event of overheating. During a test of the Arbeitsgemeinschaft Versuchsreaktor (AVR) pebble bed reactor (see section 7.5.3), all control rods were removed and the helium blower was shut off. There was no resulting damage to the fuel pebbles. The use of helium as a coolant

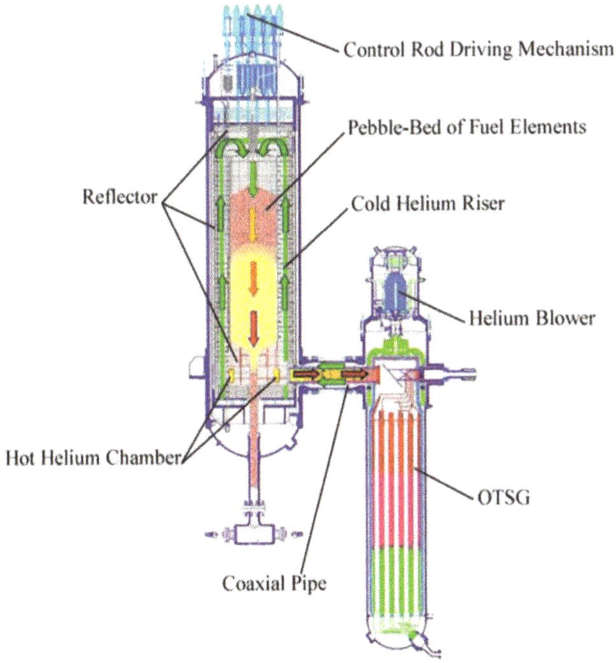

Figure 7.12. Diagram of a modular pebble bed reactor showing helium flow from the reactor core to the OTSG. Reproduced from Dong (2014). CC BY 4.0.

also reduces the risk of serious accident because the use of water as a coolant can lead to the production of combustible hydrogen through thermal decomposition. In the typical modular pebble bed reactor design, as shown in figure 7.12, the steam generator is placed lower than the reactor core so that any potential water leaks will not come into contact with the nuclear fuel.

The development of high-temperature reactors began in the early-1960s. To date, eight test or prototype high-temperature reactors have been built worldwide. Table 7.1 summarizes some of the properties of these reactors. Electrical power is listed for those reactors that included a turbine/generator. The remainder of this section provides some details of these reactors.

7.5.1 Dragon reactor

The Dragon reactor was an experimental high-temperature gas-cooled fission reactor constructed at Winfrith in Dorset, UK, see figure 7.13. Winfrith was the site of nine different experimental nuclear reactors constructed primarily in the 1950s and 1960s. The Dragon reactor was the world's first high-temperature reactor and used a prismatic core geometry containing TRISO fuel pellets cooled by helium gas. The reactor was constructed for the purpose of testing the TRISO fuel geometry and high-temperature reactor materials. The TRISO fuel was highly enriched. The initial fuel pellets contained uranium enriched to 93% in ^{235}U. Later fuel pellets used

Table 7.1. Some properties of test and prototype high-temperature reactors. The start date is commission date or date of first criticality.

Reactor	Country	Start date	Shutdown	Core geometry	Outlet temperature (°C)	Thermal power (MW_{th})	Electrical power (MW_e)
Dragon	UK	1965	1976	Prismatic	750	21.5	—
Peach Bottom	USA	1967	1974	Prismatic	725	115	40
AVR	Germany	1967	1988	Pebble bed	950	46	13
FSV	USA	1979	1989	Prismatic	775	842	330
THTR-300	Germany	1985	1989	Pebble bed	750	750	308
HTTR	Japan	1998	Operational	Prismatic	950	30	—
HTR-10	China	2000	Operational	Pebble bed	700	10	2.5
HTR-PM	China	2021	Operational	Pebble bed	750	2×250	210

Figure 7.13. Photo of a cutaway model of the Dragon reactor at Winfrith. This photo of a cutaway model of the Dragon reactor at Winfrith image has been obtained by the author from the Wikimedia website where it was made available by Geni (2023) under a CC BY-SA 4.0 licence. It is included within this chapter on that basis. It is attributed to Geni.

a lower enrichment of around 20% ^{235}U. Overall, the Dragon reactor operated without any significant problems.

7.5.2 Peach Bottom Nuclear Generating Station, Unit 1

Peach Bottom Nuclear Generating Station in Peach Bottom Township, York County, Pennsylvania was the site of three nuclear reactors. Unit 1 (commissioned 1967) was an experimental high-temperature gas-cooled reactor, as described below. Units 2 and 3 (commissioned 1974) were conventional boiling water reactors and are still in operation as of 2024.

Peach Bottom Unit 1 was a high-temperature graphite-moderated reactor with a prismatic core and helium cooling. Figure 7.14 shows an external view of the reactor building. The fuel consisted of mixed uranium and thorium carbides in a TRISO-like graphite matrix. The initial fuel consisted of 14% uranium (enriched to 93% ^{235}U) and 86% thorium (^{232}Th). In contrast to the fast breeder reactors described in section 5.8, Peach Bottom was a thermal breeder reactor. Neutrons from the fissile ^{235}U produced fissile ^{233}U from fertile ^{232}Th. The breeding reactions are as follows:

$$n + {}^{232}\mathrm{Th} \rightarrow {}^{233}\mathrm{Th} + \gamma \qquad (7.1)$$

followed by the β^- decays,

$${}^{233}\mathrm{Th} \rightarrow {}^{233}\mathrm{Pa} + e^- + \bar{\nu} \qquad (7.2)$$

Figure 7.14. The Peach Bottom high-temperature reactor. Reproduced from U.S. Nuclear Regulatory Commission (2015). CC BY 2.0.

Figure 7.15. Some of the 825 graphite-clad fuel elements fabricated by General Atomic Division of General Dynamics for the Peach Bottom high-temperature reactor. This photo shows some of the 825 graphite-clad fuel elements fabricated by General Atomic Division of General Dynamics in San Diego, Calif., for the Peach Bottom High Temperature Gas-cooled Reactor nuclear power station in Pennsylvania image has been obtained by the author from the Wikimedia website where it was made available by ENERGY.GOV (c. 1967). Image stated to be in the public domain. It is included within this chapter on that basis. It is attributed to ENERGY.GOV and Bomazi. Transferred from Flickr via Flickr2commons.

and

$$^{233}\mathrm{Pa} \rightarrow {}^{233}\mathrm{U} + \mathrm{e}^{-} + \bar{\nu} \tag{7.3}$$

The TRISO fuel particles were contained in fuel elements, as illustrated in figure 7.15, which were inserted into the prismatic graphite core.

Overall, the Peach Bottom high-temperature reactor operated successfully from 1967 to 1974. It provided a research platform to refine high-temperature reactor designs, including the structure of the fuel particles. It also produced a maximum electrical output of 40 MW_e and over its lifetime supplied more than 1.2×10^6 MWh of electricity to the Philadelphia Electric Company grid. The operation of the Peach Bottom high-temperature reactor has been reviewed by Everett and Kohler (1978).

7.5.3 Arbeitsgemeinschaft Versuchsreaktor

The AVR was the world's first operational pebble bed reactor. It was constructed adjacent to the Forschungszentrum Jülich (Jülich Research Centre) and was the first reactor that resulted from Germany's pebble bed reactor program that paralleled the development of prismatic core high-temperature reactors in the United States and the United Kingdom. It operated and was connected to the grid from 1967 to 1988. The reactor building is shown in figure 7.16.

Figure 7.17 shows a cross section of the reactor. The core is about 3.0 m in diameter and about 4.3 m high. It contains about 100 000 fuel pebbles. The fuel

Figure 7.16. The AVR high-temperature reactor at Forschungszentrum, Jülich, Germany. This AVR High Temperature Reactor at Forschungszentrum Jülich in the dismantling phase image has been obtained by the author from the Wikimedia website where it was made available by Maurice van Bruggen (2002) under a CC BY-SA 3.0 licence. It is included within this chapter on that basis. It is attributed to Maurice van Bruggen.

consisted of a mixture of uranium and thorium oxides and carbides, analogous to the fuel described above for the Peach Bottom reactor. Further design details are discussed in Yamashita *et al* (1994). It is interesting to note that the steam generator in this design is located above the reactor core inside the reactor vessel. This is in contrast to the more contemporary modular design, as shown in figure 7.12, where the steam generating module is separate from the reactor and reduces the possibility of hydrogen production in the event of a water leak.

7.5.4 Fort Saint Vrain Nuclear Power Plant

The Fort Saint Vrain (FSV) Nuclear Power Plant was located in the town of Platteville, Colorado. It was the first and only nuclear power plant in that state. The plant operated commercially and provided power to the grid from 1979 to 1989. The reactor building is shown in figure 7.18.

The FSV nuclear reactor was a prismatic core high-temperature reactor. It used a graphite moderator and was cooled by helium gas. An image of the reactor refueling floor is shown in figure 7.19. The typical prismatic geometry of the fuel elements is seen in the photograph. The fuel was a mixture of fissile uranium and fertile thorium.

From a commercial standpoint, the FSV high-temperature reactor was not successful. High operating costs, primarily due to excessive maintenance problems, made the plant economically unviable and eventually led to its shutdown. One of the persistent major problems was the infiltration of water into the helium cooling

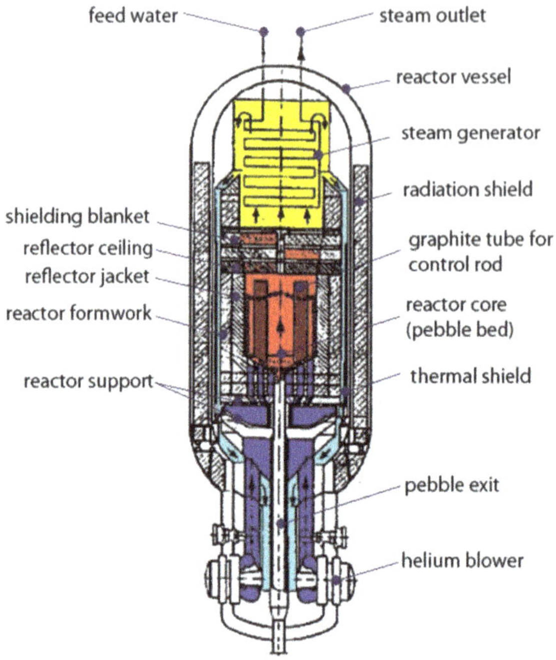

Figure 7.17. Schematic of the AVR high-temperature reactor. This schematic of the AVR-pebble-bed reactor image has been obtained by the author from the Wikimedia website where it was made available by Cschirp (2009) under a CC BY-SA 3.0 licence. It is included within this chapter on that basis. It is attributed to Cschirp.

Figure 7.18. FSV Nuclear Power Plant. This Fort St. Vrain Generating Station image has been obtained by the author from the Wikimedia website where it was made available by ENERGY.GOV (2014). Image stated to be in the public domain. It is included within this chapter on that basis. It is attributed to ENERGY.GOV and Bomazi. Transferred from Flickr via Flickr2commons.

Figure 7.19. Refueling floor of the FSV high-temperature reactor. U.S. National Archives and Records Administration (412-DA-2333). Image stated to be in the public domain.

system. This was caused by the difficulty of making adequate seals to prevent helium from escaping from the cooling system. The presence of water in the high-temperature reactor cooling system led to corrosion issues. Therefore, although it was not a commercial success, the FSV reactor provided valuable data for future high-temperature reactor materials and designs.

After the shutdown of the reactor in 1989, the radioactive fuel was removed and decommissioning was completed in 1992. This represented the first decommissioning of a commercial nuclear power reactor in the United States. Following the removal of the nuclear reactor, three natural gas fired combustion turbines were installed in the FSV facility. This plant operated in a combined cycle mode where excess heat from the combustion turbines was used to generate steam to operate the turbine and generator from the original nuclear generating facility.

7.5.5 Thorium Hocktemperatur Reaktor-300

The Thorium Hocktemperatur Reaktor-300 (THTR-300) reactor was a helium-cooled pebble bed reactor in Hamm-Uentrop, Germany, see figures 7.20 and 7.21. It contained about 670 000 fuel pebbles consisting of ^{235}U and ^{232}Th particles in a graphite matrix. It became operational and was connected to the grid in 1985. The reactor experienced a number of technical problems related to fuel pebble issues.

Figure 7.20. Reactor building of the THTR-300 high-temperature reactor at the Westphalia Hamm Power Plant. This Ausschnitt aus dem Bild File:Luftbild Kraftwerk Westfalen image has been obtained by the author from the Wikimedia website where it was made available by Tim Reckmann and ChNPP (2007) under a CC BY-SA 3.0 licence. It is included within this chapter on that basis. It is attributed to Tim Reckmann and ChNPP.

Figure 7.21. Dry cooling tower of the THTR-300 high-temperature reactor. This Trockenkühlturm des THTR-300 in Hamm-Uentrop kurz vor der Fertigstellung des Kraftwerks. Der Turm wurde 1991 abgerissen image has been obtained by the author from the Wikimedia website where it was made available by Smial (1980) under a CC BY-SA 2.0 licence. It is included within this chapter on that basis. It is attributed to Smial.

These included a fuel pebble jamming in a fuel feed pipe and fuel pebbles breaking. These incidents were detrimental to the operation of the THTR-300 reactor for several reasons. The required maintenance and repairs, along with reactor downtime, reduced economic viability. In addition, because these incidents occurred shortly after the Chernobyl accident in April 1986, public sentiment shifted away from nuclear power and nuclear regulatory agency policies in Germany became stricter. As a result of these factors, the THTR-300 was permanently shutdown in 1989.

7.5.6 High Temperature Engineering Test Reactor

The High Temperature Engineering Test Reactor (HTTR) is a helium-cooled prismatic core high-temperature research reactor in Ōarai, Ibaraki, Japan. The HTTR became operational in 1998 and is designed for testing reactor operations and materials. It does not include a turbine/generator for electricity production. The design of the reactor core is shown in figure 7.22. Details of the HTTR have been described by Shiozawa *et al* (2004), Bess *et al* (2009), and Fujiwara *et al* (2021).

The tests that have been performed using the HTTR include studies of the fabrication of fuel particles from uranium dioxide enriched to 6% in ^{235}U, the study of high strength graphite moderators and the investigation of corrosion resistant reactor metals. In 2022, the Japan Atomic Energy Agency (JAEA) in collaboration with Mitsubishi Heavy Industries, Ltd (MHI) initiated a program to use heat from the HTTR to produce hydrogen.

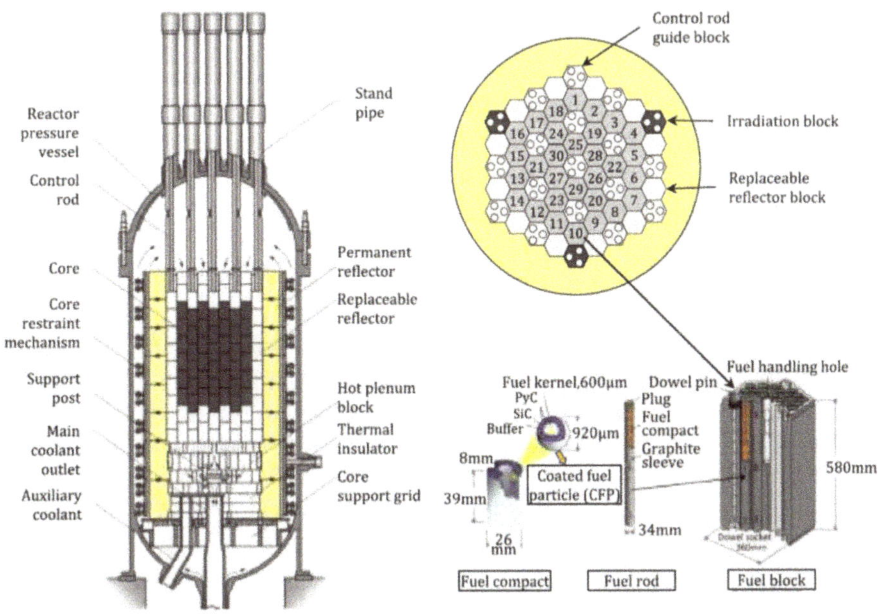

Figure 7.22. Schematic of the HTTR reactor showing the prismatic core design. Reproduced from Fujimoto *et al* (2021). CC BY 4.0.

7.5.7 HTR-10

The HTR-10 is a prototype high-temperature helium-cooled pebble bed that was constructed by the Institute of Nuclear and New Energy Technology in China at Tsinghua University. The details of the HTR-10 modular reactor are shown in figure 7.11. The core, which is 1.8 m in diameter and 1.97 m high, contains 27 000 fuel pebbles. The pebbles contain TRISO fuel particles with uranium dioxide enriched to 17% in ^{235}U.

The HTR-10 reactor is a prototype reactor and was designed to test various reactor principles prior to the construction of the commercial-scale HTR-PM, as described below. The reactor design features that will be investigated include the following:

- The use of TRISO containing fuel pebbles that are continuously circulated through the core to ensure uniform burn-up.
- A reactor core that can be cooled passively in the event of an accident and that under such conditions will not exceed a maximum safe fuel temperature.
- A modular design that separates the reactor from the water/steam containing secondary loop.
- A reactor core constructed only of graphite with no metallic components that are susceptible to corrosion.
- Control systems consisting of control rods, as well as emergency shutdown absorber balls (see figure 7.11).
- The use of a conventional steam turbine/generator for electricity production.
- An additional gas turbine that makes use of the high-temperature helium gas output from the reactor.

7.5.8 HTR-PM

The HTR-PM reactor is a commercial-scale follow-up to the prototype HTR-10 reactor (Zhang *et al* 2016). Two 250 MW_{th} high-temperature pebble bed HTR-PM reactors with a total electrical output of 210 MW_e are located at the Shidao Bay Nuclear Power Plant in Shidao Bay (also known as Shidaowan), Shandong Province, China. The reactors were reported to have begun commercial service on 6 December 2023 and are the world's first Generation IV reactors in commercial operation.

The modular design of the HTR-PM reactor is shown in figure 7.23. The reactor core is 3.0 m diameter by 11.0 m high and contains approximately 420 000 fuel pebbles with TRISO uranium dioxide fuel enriched to 8.6% ^{235}U. Two reactors are utilized in this facility where the thermal power of each reactor is limited to 250 MW_{th}. It has been determined that limiting the reactor thermal power to 250 MW_{th} ensures that in the event of a cooling system failure the reactor will shutdown, residual heat will be dissipated, and the fuel temperature will not exceed 1600 °C. Figure 7.24 shows the results of a model for HTR-PM core temperature as a function of time after a loss of coolant accident (i.e., depressurization of the helium cooling loop). The TRISO fuel particles are designed to withstand this temperature without suffering any damage.

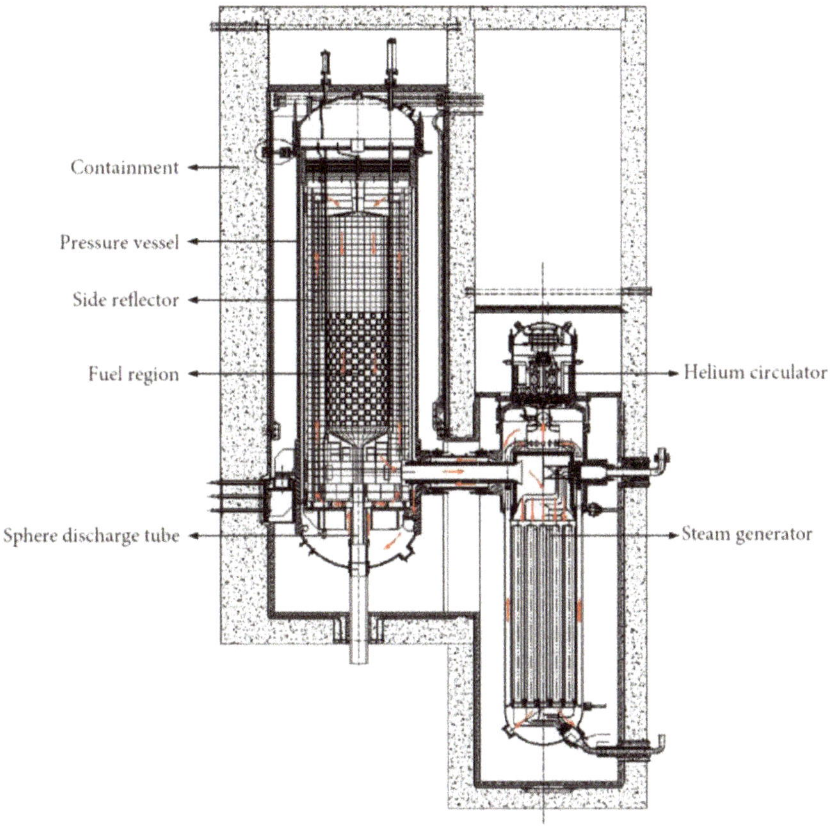

Figure 7.23. HTR-PM reactor schematic. Reproduced from Hao *et al* (2020). CC BY 4.0.

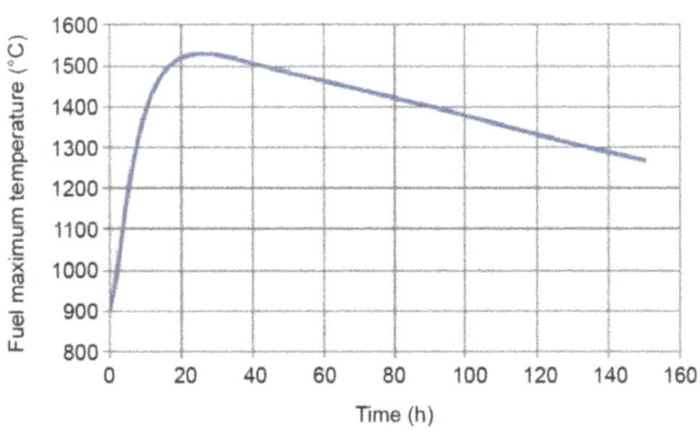

Figure 7.24. HTR-PM core temperature as a function of time after a loss of coolant accident (i.e., depressurization of the helium cooling loop). Reproduced from Zhang *et al* (2016). CC BY 4.0.

7.6 Cogeneration with a high-temperature reactor

Increased thermodynamic efficiency and improved reactor safety are two potential advantages of high-temperature gas-cooled reactors. Another potential benefit is the ability to use heat output for cogeneration. The high temperature of the output, compared to traditional gas or water-cooled reactors, increases the potential cogeneration applications. Figure 7.25 shows the typical range of temperatures accessible from the output of a high-temperature gas-cooled reactor and the temperature ranges of some useful industrial processes.

Two particular processes have attracted interest as applications of cogeneration using nuclear reactors: seawater desalination and hydrogen production. These two possibilities are discussed below.

7.6.1 Nuclear desalination

The desalination of seawater can be accomplished either by the use of a membrane (e.g., reverse osmosis) or by thermal methods. Reverse osmosis is effective for small desalination plants, but thermal methods are typically used for commercial plants with high desalinated water output (Şahin and Şahin 2020). The most common methods used for commercial thermal desalination are multiple effect distillation (MED), multi-stage flash (MSF), and vapor compression distillation (VCD). As illustrated in figure 7.25, desalination does not require the high temperatures that are available from a high-temperature gas-cooled reactor. In fact, the use of heat from a

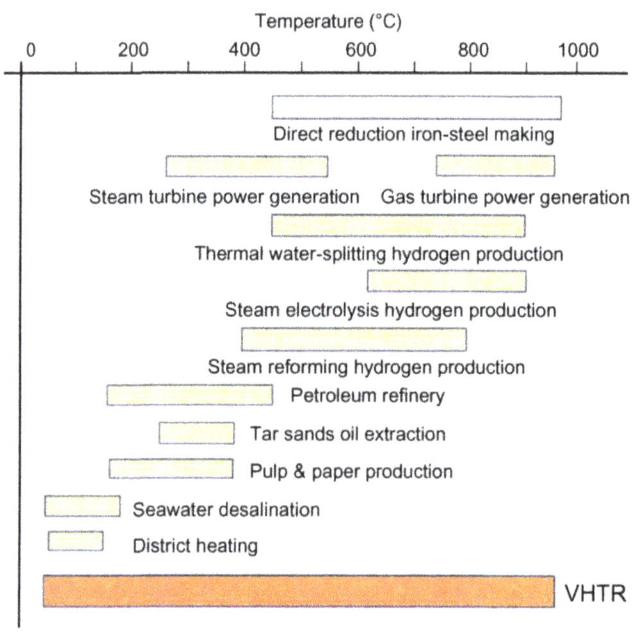

Figure 7.25. Range of temperatures needed for various industrial processes. The range of temperatures available from a high-temperature gas-cooled reactor are also shown. Reprinted from Yan (2023), Copyright (2023), with permission from Elsevier.

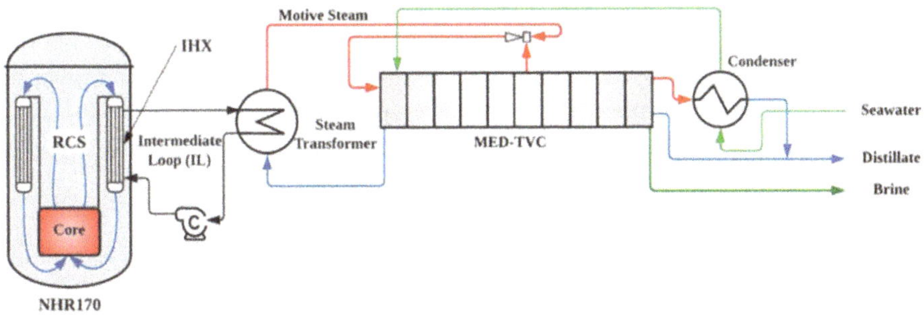

Figure 7.26. Schematic diagram showing the connection of a nuclear reactor (NHR170) and a desalination unit (MED-TVC) as described in the text. RSC = reactor coolant system, IHX = intermediate heat exchanger, MED-TVC = multi-effect distillation thermal vapor compression. Reproduced from Alessi and Al-Rabiah (2022). CC BY 4.0.

lower temperature nuclear reactor has been considered as a means of seawater desalination in the past.

The BN-350 was a sodium-cooled fast neutron reactor located at the Mangyshlak Nuclear Power Plant in Aktau, Kazakhstan that operated from 1964 to 1992. It produced steam at 450 °C, that was used to drive turbine/generators with a capacity of 350 MW_e. Residual steam at 230 °C was used for thermal desalination and could produce up to 120 000 m^3 day^{-1} of desalinated seawater. The reactor was also designed to produce military grade plutonium. Further design details of this type of reactor are presented in chapter 11.

In more recent years, the use of small nuclear reactors dedicated to heat production has been considered as a possible method of seawater desalination. The NHR-200 is a 200 MW_{th} pressurized light-water reactor designed in China for thermal applications such as district heating, seawater desalination, and industrial processes (Zheng and Wang 1995). The proposed application of a modified version of the NHR-200, the NRH-170 with a thermal output of 169 MW_{th}, for thermal desalination using multiple effect distillation is shown in figure 7.26. Steam from the reactor is provided to the MED unit while seawater is used to cool the condenser and is then directed to the MED unit for desalination.

The higher temperature steam available from the secondary loop of a high-temperature gas-cooled reactor can be used for electricity generation and the residual heat can then supply a thermal desalination facility.

7.6.2 Hydrogen production

As shown in figure 7.25, temperatures produced by high-temperature reactors are suitable for the production of hydrogen by various processes. These processes include steam reforming, thermal water splitting and high-temperature electrolysis. Steam reforming is the principal method that is used for the commercial production of hydrogen. It involves the thermal breakdown of methane (natural gas) by the process,

$$CH_4 + 2H_2O \rightarrow 4H_2 + CO_2.$$

This method releases carbon dioxide and contributes to greenhouse gas emissions (to the same extent as the combustion of methane) unless the carbon dioxide that is produced is sequestered. The thermal splitting of water (or thermal decomposition of water) occurs at high temperatures and represents the process,

$$2H_2O \rightarrow 2H_2 + O_2. \tag{7.4}$$

At about 3000 °C, about half of water molecules will dissociate. At a temperature of 2200 °C, the fraction is about 3%. At lower temperatures it is correspondingly smaller, although this can be increased in the presence of a catalyst. At ambient temperatures, it is customary to use electrolysis to produce hydrogen. This uses the process in equation (7.4) except that electrical energy rather than thermal energy drives the reaction. High-temperature electrolysis increases the efficiency of the electrolysis process by using thermal energy to decrease the required electrical energy (Fujiwara *et al* 2008). A typical high-temperature electrolysis arrangement using heat from a high-temperature nuclear reactor is shown in figure 7.27. Hot helium from the reactor can generate electricity using a gas turbine/generator. It also produces steam through a heat exchanger and the steam produces hydrogen through high-temperature electrolysis using electrical energy from the generator. The higher the temperature of the steam that is input into the electrolysis cell, the lower the electricity requirement to produce hydrogen.

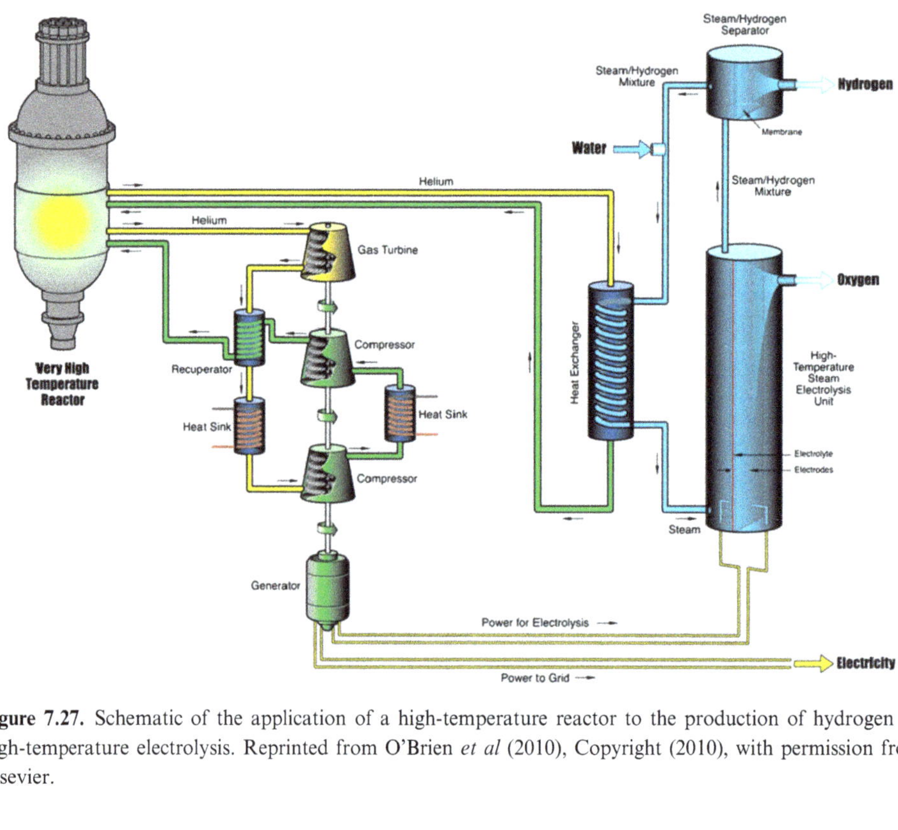

Figure 7.27. Schematic of the application of a high-temperature reactor to the production of hydrogen by high-temperature electrolysis. Reprinted from O'Brien *et al* (2010), Copyright (2010), with permission from Elsevier.

References

Alessi Y M and Al-Rabiah A A A 2022 Feasibility study of utilizing nuclear energy for an existing MED-TVC desalination plant *Appl. Sci.* **12** 9506

Bess J D, Fujimoto N, Dolphin B H, Snoj L and Zukeran A 2009 *Evaluation of the Start-Up Core Physics Tests at Japan's High Temperature Engineering Test Reactor (Fully-Loaded Core)* NEA/NSC/DOC(2006)1 (Idaho Falls: Idaho National Laboratory)

Cschirp 2009 Schematic of the AVR-pebble-bed-reactor https://commons.wikimedia.org/wiki/File:AVR_Reaktor.png

Dong Z 2014 Saturated adaptive output-feedback power-level control for modular high temperature gas-cooled reactors *Energies* **7** 7620–39

ENERGY.GOV c. 1967 Photo shows some of the 825 graphite-clad fuel elements fabricated by General Atomic Division of General Dynamics https://commons.wikimedia.org/wiki/File:HD.6D.290_(10824493836).jpg

ENERGY.GOV 2014 Fort St. Vrain Generating Station https://commons.wikimedia.org/wiki/File:HD.6D.154_(10822121114).jpg

Everett J and Kohler E J 1978 Peach bottom unit no. 1: a high performance helium cooled nuclear power plant *Ann. Nucl. Energy* **5** 321–35

Fujimoto N, Tada K, Ho H Q, Hamamoto S, Nagasumi S and Ishitsuka E 2021 Nuclear data processing code FRENDY: a verification with HTTR criticality benchmark experiments *Ann. Nucl. Energy* **158** 108270

Fujiwara S *et al* 2008 Hydrogen production by high temperature electrolysis with nuclear reactor *Prog. Nucl. Energy* **50** 422–6

Fujiwara Y *et al* 2021 Design of high temperature engineering test reactor (HTTR) ed T Takeda and Y Inagaki High Temperature Gas-Cooled Reactors *High Temperature Gas-Cooled Reactors* (Amsterdam: Elsevier) pp 17–177

Geni 2023 Photo of a cutaway model of the Dragon reactor at Winfrith https://commons.wikimedia.org/wiki/File:Winfrith_Dragon_reactor_model.JPG

Hao C, Li P, She D, Zhou Z and Yang R 2020 Sensitivity and uncertainty analysis of the maximum fuel temperature under accident condition of HTR-PM *Sci. Technol. Nucl. Install.* **2020** 9235783

Idaho National Laboratory 2009 David Petti, TRISO fuel pellet www.flickr.com/photos/inl/3640364326/

Idaho National Laboratory 2013a Individual TRISO particles at 500 um scale www.flickr.com/photos/inl/9017944822/in/album-72157634070807282/

Idaho National Laboratory 2013b TRISO fuel pellet www.flickr.com/photos/inl/9713145192/

Idaho National Laboratory 2013c Cross-section of fuel pellet containing TRISO particles at 10 mm scale www.flickr.com/photos/inl/9017971338/

O'Brien J E, McKellar M G, Harvego E A and Stoots C M 2010 High-temperature electrolysis for large-scale hydrogen and syngas production from nuclear energy—summary of system simulation and economic analyses *Int. J. Hydrogen Energy* **35** 4808–19

Picoterawatt 2019 Pebble bed reactor scheme https://commons.wikimedia.org/wiki/File:Pebble_bed_reactor_scheme_(English).svg

Reckmann T and ChNPP 2007 Ausschnitt aus dem Bild File:Luftbild Kraftwerk Westfalen.jpg, THTR-300 https://commons.wikimedia.org/wiki/File:Luftbild_THTR-300.jpg

Şahin S and Şahin H M 2020 Generation-IV reactors and nuclear hydrogen production *Int. J. Hydrogen Energy* **46** 28936–48

Shiozawa S, Fujikawa S, Iyoku T, Kunitomi K and Tachibana Y 2004 Overview of HTTR design features *Nucl. Eng. Des.* **233** 11–21

Smial 1980 Trockenkühlturm des THTR-300 in Hamm-Uentrop kurz vor der Fertigstellung des Kraftwerks https://commons.wikimedia.org/wiki/File:Thtr300_kuehlturm.jpg

Tak N I, Lee S N, Kim M H, Lim H S and Noh J M 2014 Development of a core thermo-fluid analysis code for prismatic gas cooled reactors *Nucl. Eng. Tech.* **46** 641–54

Talamo A 2008 QUADRISO particle https://commons.wikimedia.org/wiki/File:Quadriso.png

U.S. Department of Energy 2021 X-energy is developing a pebble bed reactor that they say can't melt down www.energy.gov/ne/articles/x-energy-developing-pebble-bed-reactor-they-say-cant-melt-down

U.S. Nuclear Regulatory Commission 2015 Peach bottom atomic power station, Unit 1 www.flickr.com/photos/nrcgov/17671649350/

U.S. National Archives 1972 Refueling floor at St. Vrain Nuclear Power Plant, Image 412-DA-2333 https://catalog.archives.gov/id/544826

van Bruggen M 2002 AVR high temperatur reactor at Forschungszentrum Jülich https://commons.wikimedia.org/wiki/File:Hogetemperatuurreactor.JPG

Weirdmeister 2017 Graphite pebble for pebble bed reactor https://commons.wikimedia.org/wiki/File:Pebble_Bed_Graphite_Ball.jpg

Yamashita K, Murata I, Shindo R, Tokuhara K and Werner H 1994 Analysis of control rod reactivity worths for AVR power plant at cold and hot conditions *J. Nucl. Sci. Technol.* **31** 470–8

Yan X L 2023 Very high temperature reactor *Handbook of Generation IV Nuclear Reactors* 2nd edn I L Pioro (Cambridge, MA: Woodhead Publishing) ch 3 pp 133–65

Zhang Z, Wu Z, Wang D and Xu Y 2009 Current status and technical description of Chinese 2 × 250 MW_{th} HTR-PM demonstration plant *Nucl. Eng. Des.* **239** 1212–9

Zhang Z *et al* 2016 The Shandong Shidao Bay 200 MW_e high-temperature gas-cooled reactor pebble-bed module (HTR-PM) demonstration power plant: an engineering and technological innovation *Engineering* **2** 112–8

Zheng W and Wang D 1995 NHR-200 nuclear energy system and its possible applications *Prog. Nucl. Energy* **29** 193–200

IOP Publishing

Generation IV Nuclear Reactors
Design, operation and prospects for future energy production
Richard A Dunlap

Chapter 8

Molten salt reactors (MSRs)

Chapter 8 provides a review of the basic operation, history, and current development of molten salt reactors (MSRs). These reactors are distinct from other nuclear fission reactors because, in many designs, the fuel is part of the liquid reactor coolant. These reactors may be thermal neutron reactors that use a graphite moderator or they may be fast reactors with no moderator. The only operational MSRs were constructed in the United States in the 1950s and 1960s. More recently, they have been designated as a design for Generation IV reactors and several MSR development projects are underway in the United States, Canada, and the United Kingdom. These reactors fall into three categories: those with fuel that circulates in the molten salt coolant, those with fuel in static molten salt, and those that use solid fuel with molten salt as the coolant. The present chapter reviews some of the development projects that fall into these three categories.

8.1 Introduction

Conceptually, the MSR is quite distinct from other reactor designs because the fissile fuel is actually contained in a high-temperature molten salt. They were first envisioned by the U.S. military as a power source for aircraft, analogous to the nuclear reactors used for submarines and aircraft carriers. These reactors can either be thermal neutron reactors or fast neutron reactors. In the most common design, the nuclear fuel contained in the molten salt circulates in the primary loop of the reactor and serves as both fuel and coolant, transferring its heat through a heat exchanger to a secondary loop. Such reactors, where the fuel is distributed uniformly throughout the coolant, are referred to as homogeneous reactors. An alternative approach, the so-called stable salt reactor, uses molten salt containing the fissile fuel that is contained in sealed tubes, analogous to the solid fuel tubes in a traditional reactor. Non-radioactive molten salt is used as a primary coolant, similar to water or gas coolant. A final design utilizing molten salt is the fluoride salt-cooled high-temperature reactor. This reactor uses a pebble bed design with tristructural-

doi:10.1088/978-0-7503-6069-2ch8

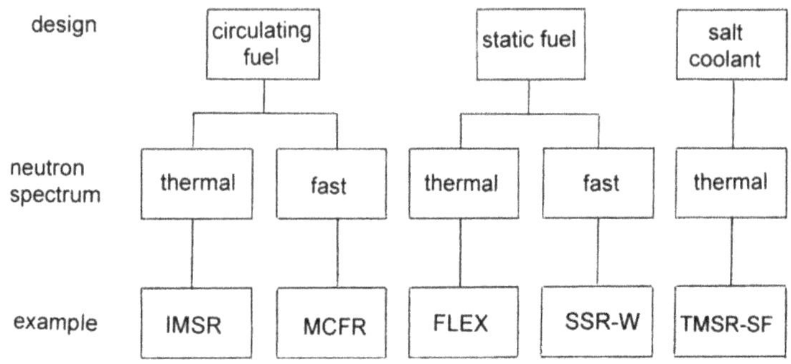

Figure 8.1. Examples of the various designs of MSRs with some examples that have been constructed or are under development.

Table 8.1. Details of the example reactors shown in figure 8.1 along with the organizations that are developing each reactor.

Reactor (abbreviation)	Reactor (name)	Organization
IMSR	Integral molten salt reactor	Terrestrial Energy
MCFR	Molten chloride fast reactor	TerraPower and Southern Company
FLEX	FLEX	MoltexFLEX
SSR-W	Stable salt reactor-wasteburner	Moltex Energy Canada
TMSR-SF	Thorium molten salt reactor-solid fuel	Shanghai Institute of Applied Physics

isotropic (TRISO) particle containing fuel pebbles but uses a molten salt rather than helium as the primary coolant. A breakdown of the various MSR designs is shown in the graph in figure 8.1, along with examples of reactors that have been constructed or are under development. The reactor names and the organizations that are developing them are given in table 8.1.

The next three sections provide an overview of the design and operation of circulating fuel MSRs. This is followed by an historical review of previously constructed MSRs. The final section describes how MSRs fit in with the Generation IV reactor initiative and provides details of the designs of the reactors listed in table 8.1.

8.2 Molten salt reactor fuel requirements

Actinides that serve as fissile or fertile materials in nuclear fission reactors form fluorides. These are uranium (UF_4), plutonium (PuF_3), and thorium (ThF_4). In principle, these could serve as fuels for a MSR and benefit from the low neutron absorption of 100% naturally occurring ^{19}F. In practice, however, their high melting temperatures of more than 1000 °C makes the design of a viable MSR difficult. It is possible, however, to mix the fuel-containing fluoride with other fluorides to lower the melting temperature, e.g., to around 500 °C. Using a low melting temperature salt speeds reactor start-up, reduces the risk of molten salt freezing in the heat exchanger, and reduces the corrosivity of the salt (see below). For the tetravalent fluoride UF_4, there are several possibilities that can effectively lower the melting temperature of the salt. ZrF_4 is one possibility that has a reasonably low neutron absorption cross section. Other possibilities are LiF_4 and BeF_4. Lithium consists of two naturally occurring isotopes: ^{6}Li (7.5%) and ^{7}Li (92.5%). ^{6}Li has a large neutron absorption cross section and produces problematic tritium by the process,

$$n + {}^{6}\text{Li} \rightarrow {}^{4}\text{He} + {}^{3}\text{H}.$$

Therefore, highly enriched ^{7}Li must be used for the LiF_4. Beryllium (100% naturally occurring ^{9}Be) is interesting because it undergoes the neutron reaction,

$$n + {}^{9}\text{Be} \rightarrow 2{}^{4}\text{He} + 2n.$$

It, therefore, acts as a neutron doubler and this can benefit the overall fission reaction.

8.3 Molten salt thermal reactors

Figure 8.2 shows the design of a typical thermal MSR utilizing a graphite moderator. The reactor is controlled by means of conventional control rods. In this design, the salt containing the fuel is circulated through a heat exchanger where heat is transferred to a secondary (nonfuel) molten salt loop. This secondary loop produces steam by means of a steam generator, as shown in the illustration, and this is used to power the turbine/generator.

The three important features of the reactor in figure 8.2 are the design of the primary salt loop, the presence of the emergency dump (or drain) tanks, and the fuel chemical processing facility. These three features are discussed below.

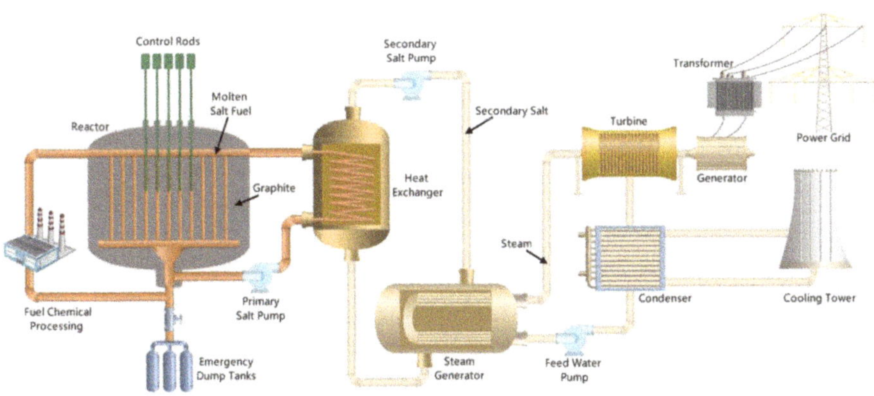

Figure 8.2. Design of a thermal neutron graphite-moderated MSR. Reproduced from Luo *et al* (2021). CC BY 4.0.

Primary salt loop design: The primary loop that carries the fuel-containing molten salt circulates outside the reactor core to an external heat exchanger. This means that the reactor design must include appropriate radiation shielding for the entire primary loop, as well as the heat exchanger.

Emergency dump tanks: The emergency dump tanks provide a means of quickly removing the fuel from the core in the event of an incident. Core meltdown does not have the same meaning as it does for traditional reactors because the fuel is already in a molten state. However, the corrosivity of molten salt increases with increasing temperature. Therefore, it is important in the case of (for example) interruption of coolant in the secondary loop that the reactor temperature does not increase to an unacceptable level. In such a case, the liquid fuel can be quickly transferred to the dump tanks. In the design shown in figure 8.2, there are three dump tanks to ensure that the mass of fuel in each tank remains subcritical.

Fuel chemical processing facility: Unlike most traditional solid fuel nuclear reactors which needs to be shutdown for refueling (which is accomplished by replacing the fuel assemblies), the circulating fuel MSR can be refueled during operation. The fuel chemical processing facility can continuously adjust the chemical content of the fuel as it circulates. For example, fission by-products, some of which can act as neutron poisons and some of which can be gaseous and can increase the pressure in the fuel loop, can be continuously removed from the fuel. The composition of actinides in the fuel can also be adjusted to maintain efficient operation.

The corrosivity of molten salts is a major design concern for MSRs. At present, the nickel-based alloy Hastelloy N developed at Oak Ridge National Laboratory is the customary material for use in systems that are in contact with molten salt. Table 8.2 gives the nominal composition of Hastelloy N. Current MSR designs are limited to an operating temperature of around 700 °C for materials reasons. However, the operating temperature of current designs is sufficiently higher than that of light-water reactors and this leads to improved efficiency and increased

Table 8.2. Nominal composition of Hastelloy N. Data adapted from Haynes International (ND).

Element	Composition (wt%)
Nickel	71 (balance)
Molybdenum	16
Chromium	7
Iron	4 (max)
Silicon	1 (max)
Manganese	0.8 (max)
Aluminum + titanium	0.5 (max)
Tungsten	0.5 (max)
Vanadium	0.5 (max)
Carbon + cobalt + copper	0.43 (max)

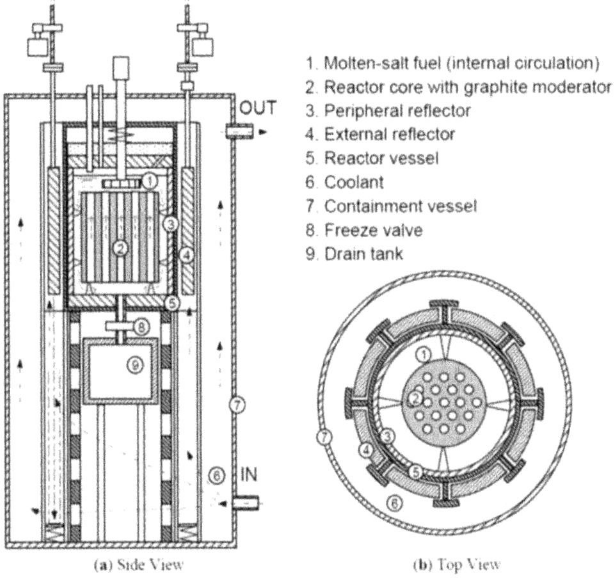

Figure 8.3. Diagram of a thermal neutron MSR core showing heat transfer through a circulating nonfuel coolant. Reproduced from Kamei (2012). CC BY 3.0.

opportunities for thermal applications. The ability to raise the operating temperature to 850 °C would substantially increase the available applications, see chapter 7, and this is a goal of future MSR designs.

Figure 8.3 shows a potential MSR core design where the primary loop is contained within the pressure vessel and coolant from the secondary loop is circulated through the reactor vessel to transport heat to a heat exchanger. This approach eliminates the potential hazards associated with circulating radioactive fuel outside the reactor vessel as in figure 8.2.

The integral molten salt reactor (IMSR), which is being developed by Terrestrial Energy Inc. in Canada, uses a heat exchanger to transfer the heat from the primary loop to the non-radioactive secondary loop that is contained within the reactor vessel. This arrangement is illustrated in figure 8.4 and is discussed in further detail in section 8.6.

8.4 Molten salt fast reactors

A basic schematic of a molten salt fast reactor (MSFR) is shown in figure 8.5. The molten salt fuel is contained in the core of the reactor and is circulated through heat exchangers on the sides. A portion of the molten salt is diverted through the liquid-gas separation and sampling system for salt reprocessing. This system serves the purpose of the fuel chemical processing facility as described above. The safety tanks (i.e., emergency dump tanks) allow the fuel to be rapidly removed from the core in the event of a nuclear incident.

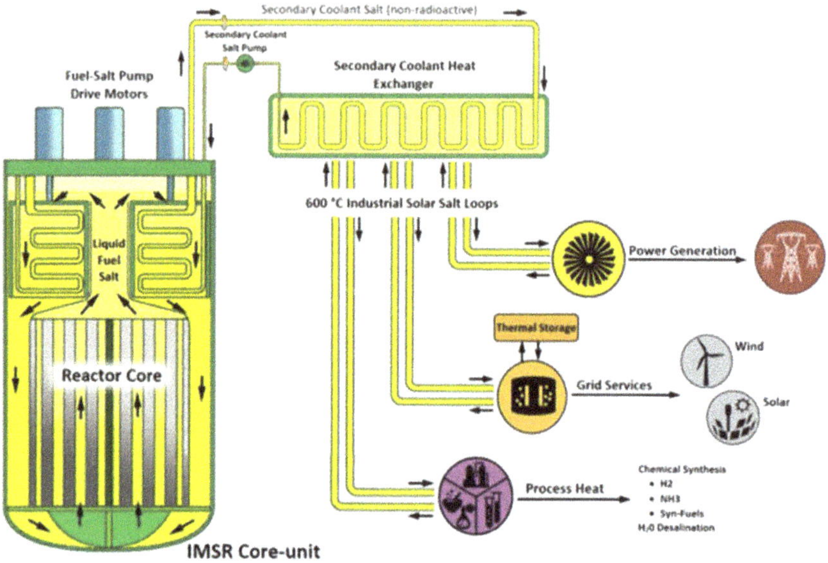

Figure 8.4. Diagram of the IMSR showing possible applications. This schematic describes the different possible heat applications for the IMSR, image has been obtained by the author from the Wikimedia website where it was made available by Terrestrial Energy (2017) under a CC BY-SA 4.0 licence. It is included within this chapter on that basis. It is attributed to Terrestrial Energy.

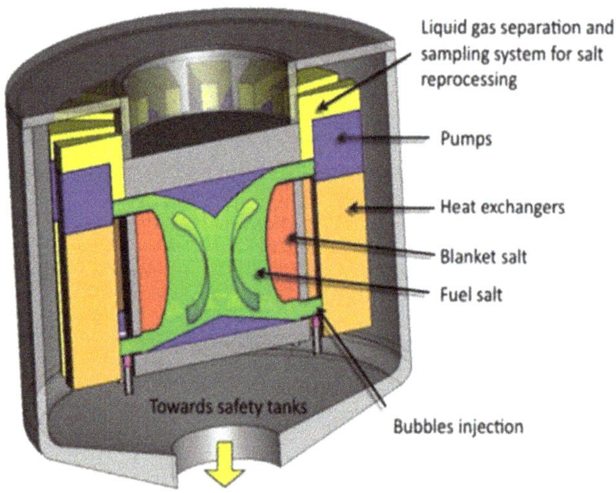

Figure 8.5. Schematic of the core of a fast neutron MSR. Reprinted from Serp *et al* (2014), Copyright (2014), with permission from Elsevier.

Since the reactor operates using fast neutrons, there is no moderator and the blanket salt contains fertile material to breed new fissile material. There are no control rods to regulate the reactor's output. Instead, two approaches are used to regulate the reactor power output. First, the core temperature is related to the rate at which the molten salt fuel is pumped through the reactor core and the rate at which cooling fluid is pumped through the heat exchanger. Second, the reactor's output can be controlled by the injection of helium bubbles into the molten salt in the reactor core. As the bubble density increases, the fuel density decreases and the energy production per unit volume in the core decreases due to the large negative void coefficient that is characteristic of MSRs. Bubbles can be introduced using the bubble injector and removed using the liquid-gas separator, as shown in the figure.

8.5 The history of molten salt reactor development

Two operational MSRs have been constructed, one in the 1950s (the Aircraft Reactor Experiment; ARE) and one in the 1960s (the Molten Salt Reactor Experiment; MSRE). Both were located at Oak Ridge National Laboratory in Tennessee and were circulating fuel thermal reactors. Brief descriptions of these two reactors follow.

8.5.1 Aircraft Reactor Experiment

In 1946, the U.S. Army Air Forces established the aircraft nuclear propulsion program to consider the possibility of using a nuclear reactor to power long-range hypersonic aircraft. Initial plans considered a sodium-cooled beryllium oxide-moderated reactor using traditional fuel rods. Due to concerns that xenon gas, produced as a fission by-product, would accumulate in the solid fuel rods, this approach was abandoned and replaced with a design utilizing a circulating liquid fuel. The resulting ARE was designed and constructed at Oak Ridge National Laboratory beginning in 1950. Figure 8.6 shows the ARE building at Oak Ridge. Details of the reactor design are given in Cottrell *et al* (1952, 1955, 1958).

Figure 8.6. ARE building. This Aircraft Reactor Experiment Building, Oak Ridge National Laboratory image has been obtained by the author from the Wikimedia website where it was made available by Oak Ridge National Laboratory (2007). Image stated to be in the public domain. It is included within this chapter on that basis. It is attributed to Oak Ridge National Laboratory.

This reactor was designed for a thermal output of 3 MW_{th} and was a prototype for a potential 350 MW_{th} reactor that would be appropriate for aircraft propulsion.

The design of the ARE MSR is shown in cross section in figure 8.7. The reactor core was about 1 m in diameter by about 1 m in height and consisted of an array of hexagonal beryllium oxide-moderator blocks (which also served as neutron reflectors) with tubing for the molten salt fuel to pass through. This is illustrated in figure 8.8. The reactor was cooled with liquid sodium, which transferred heat through a heat exchanger to helium gas and then to water. The fuel consisted of 53.09 mol% NaF, 40.73 mol% ZrF_4, and 6.18 mol% UF_4, with the uranium enriched to 93.4% ^{235}U and had a melting temperature of 537 °C (Bettis *et al* 1957). The salt containment system

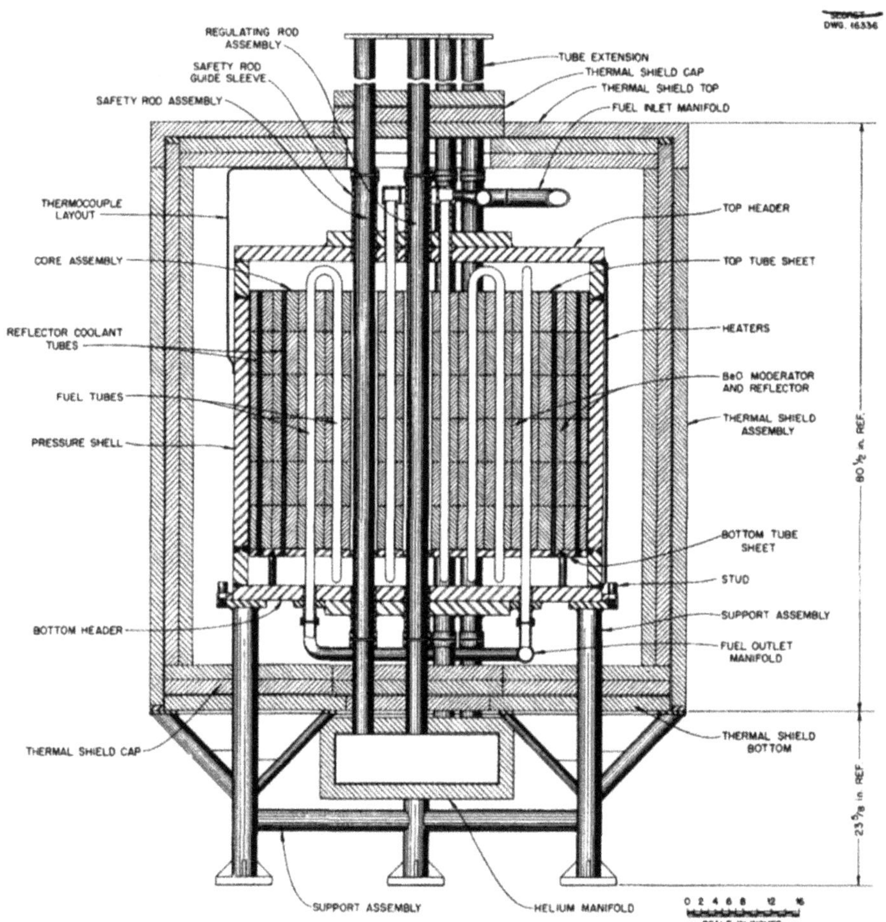

Figure 8.7. Cross section of the reactor core of the ARE. This cross section of the fluid-fueled Aircraft Reactor Experiment, an experimental nuclear reactor that operated in 1954 image has been obtained by the author from the Wikimedia website where it was made available by Oak Ridge National Laboratory (1952). Image stated to be in the public domain. It is included within this chapter on that basis. It is attributed to Oak Ridge National Laboratory.

Figure 8.8. Top view of the core of the ARE. The core is about 1 m in diameter. Fuel tubes which carry the molten salt fuel up and down through the core are seen projecting from the top of the core. The light color hexagonal elements are the beryllium oxide-moderator, and the dark color hexagonal elements are instrument and control tubes. The small holes in the hexagonal moderator blocks are for the liquid sodium coolant. Reproduced from Oak Ridge National Laboratory (2018). CC BY 2.0.

was made of Inconel, a corrosion-resistant nickel-chromium alloy that predated Hastelloy N. Under normal operating conditions the fuel flowed through the reactor core at a rate of 170 liters per minute. The reactor contained a 15 Ci radioactive starter source consisting of ^{210}Po in beryllium. The ^{210}Po decays by the process

$$^{210}\text{Po} \rightarrow {}^{206}\text{Pb} + \alpha$$

and the alpha particle interacts with the beryllium to produce a neutron,

$$\alpha + {}^{9}\text{Be} \rightarrow {}^{12}\text{C} + \text{n}.$$

The neutron produced by the starter source can start the reactor from the cold condition by inducing fission in the fissile fuel. This approach can be used when the spontaneous fission processes in the fuel are not sufficient to initiate a chain reaction. Although the reactor has a negative coefficient of reactivity (see section 3.7), additional safety controls were incorporated into the reactor design. A boron containing regulating rod ran down the center of the reactor core. This can be seen in figures 8.7 and 8.8. Three boron carbide safety rods surrounded the regulating rod, as seen in figure 8.7. These safety rods are the three dark hexagonal elements around the center of the core in figure 8.8. They could be dropped into place in the core in the event that it was necessary to scram the reactor.

The ARE first reached criticality on 3 November 1954. It operated from 8 November 1954 to 12 November 1954, and produced a maximum sustained power of 2.5 MW_{th}. Over the duration of its operation, the reactor produced 96 MWh of thermal energy. The core reached a steady operating temperature of 860 °C (compared to a design temperature of 820 °C).

Although the reactor only operated for a short period of time (about four days), it was used to undertake a variety of tests on reactor operation and stability, and thus provided valuable information for future reactor designs. A follow-up project planned to construct a larger (60 MW_{th}) MSR, the aircraft reactor test (ART), based on the experience gained with the ARE. However, this project was canceled, and the ARE building was subsequently used to house the MSRE, as described below.

8.5.2 Molten Salt Reactor Experiment

The MSRE was constructed at Oak Ridge National Laboratory. The reactor produced 7.4 MW_{th} and first attained criticality in 1965. It operated until 1969. The design and operation have been reviewed by Rosenthal (2010).

Figure 8.9 shows a layout of the MSRE. Fuel salt is pumped through the reactor core and heat is transferred to non-radioactive molten salt in a heat exchanger. Heat is dissipated through an air-cooled radiator connected to an exhaust chimney.

The MSRE used a graphite-moderated core (see figure 8.10) that was cooled by molten salt with a composition of 66 mol% $^{7}LiF_4$ and 34 mol% BeF_4. The core was about 1.4 m in diameter and 1.7 m in length, and was constructed of 5 cm square graphite bars with slots machined along their length to allow the molten fuel salt to flow through the core. Note that the lithium was enriched in ^{7}Li as described in

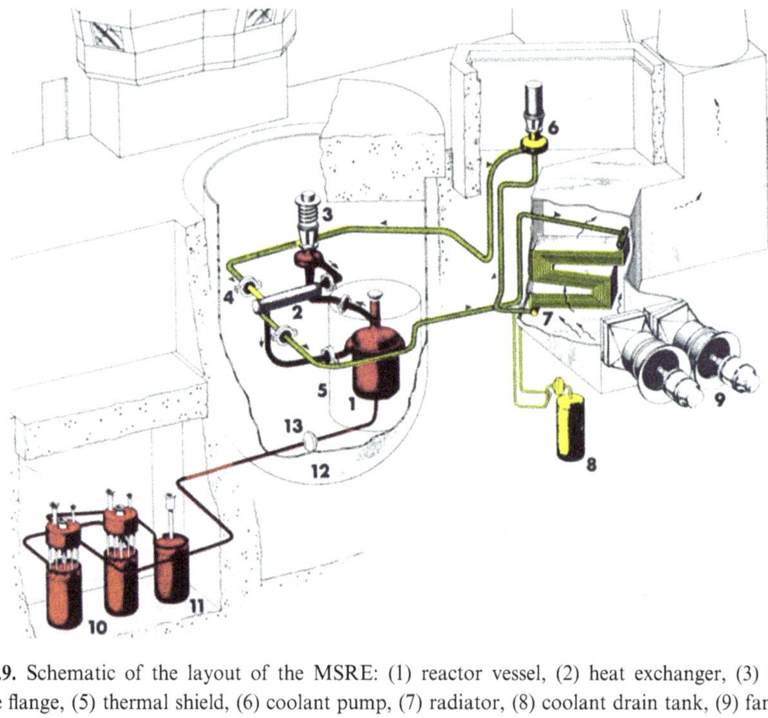

Figure 8.9. Schematic of the layout of the MSRE: (1) reactor vessel, (2) heat exchanger, (3) fuel pump, (4) freeze flange, (5) thermal shield, (6) coolant pump, (7) radiator, (8) coolant drain tank, (9) fans, (10) fuel drain tanks, (11) flush tank, (12) containment vessel, and (13) freeze valve. The control area is in the upper right of the diagram and the chimney is in the upper left. The primary (radioactive) salt loop is shown in red and the secondary (non-radioactive) salt loop is shown in yellow. All components that contacted molten salt were made of Hastelloy N, as discussed above. This diagram showing the internals of the Molten-Salt Reactor Experiment image has been obtained by the author from the Wikimedia website where it was made available by Oak Ridge National Laboratory (2011). Image stated to be in the public domain. It is included within this chapter on that basis. It is attributed to Oak Ridge National Laboratory.

Figure 8.10. Core of the graphite-moderated MSRE. This Molten Salt Reactor Graphite Core image has been obtained by the author from the Wikimedia website where it was made available by Oak Ridge National Laboratory (1964). Image stated to be in the public domain. It is included within this chapter on that basis. It is attributed to Oak Ridge National Laboratory.

section 8.2 to avoid tritium production. The control rods were hollow cylinders of Gd_2Os-Al_2Os and were encased in Inconel.

The fuel salt had a composition of 65 mol% ^7LiF, 29.1 mol% BeF_4, 5 mol% ZrF_4, and 0.9 mol% UF_4 enriched to 33% in ^{235}U (Haubenreich and Engel 1970). The fuel had a melting temperature of 434 °C and at full power the core had an outlet temperature of 663 °C. The fuel flow rate through the reactor core was 1500 liters per minute. Molten salt from the reactor core could be dumped to the storage tanks through the freeze valve, which could be opened or closed by melting or freezing salt in the valve.

Figure 8.11 shows a view looking down on the reactor vessel during the construction of the reactor. A similar view looking down into the reactor vessel of the completed reactor is shown in figure 8.12. The reactor core, fuel pump, and heat exchanger can be seen, as described in the figure caption.

In order to start the reactor, molten salt which was deficient in enriched uranium could be circulated through the reactor. Enriched UF_4–^7LiF with a eutectic composition (61% uranium by weight) could be added, a little at a time, to the fuel until the reactor became critical. The reactor first became critical on 1 June 1965 and operated for periods of up to six months at full power until 26 March 1968, when the reactor was shutdown (Haubenreich and Engel 1970). During that time,

Figure 8.11. The MSRE under construction in 1964. Reproduced from Oak Ridge National Laboratory (2015). CC BY 2.0.

Figure 8.12. Top view of the MSRE showing core (upper center), fuel pump (upper left), and heat exchanger (lower center). ORNL Photo 67051–64. This top down view of the molten salt reactor experiment image has been obtained by the author from the Wikimedia website where it was made available by Oak Ridge National Laboratory (c. 1964). Image stated to be in the public domain. It is included within this chapter on that basis. It is attributed to Oak Ridge National Laboratory.

much was learned about the operation and stability of MSRs and this knowledge formed the basis for much of the current work in this field.

In September 1968 the reactor fuel was replaced with salt containing ^{233}U (91% enrichment) that was bred from ^{232}Th in another reactor. The reactor attained criticality with ^{233}U on 2 October 1968 and operated until it was shutdown in December 1969. It was the first reactor to operate with ^{233}U fuel and during its operation much new information was attained on neutron cross sections and other fission properties of ^{233}U.

8.6 Advanced molten salt reactor designs

In recent years, there has been considerable renewed interest in MSRs and the present section provides details of the different categories of these reactors as given in figure 8.1. The reactors discussed here represent only a fraction of those currently under development worldwide but illustrate the major approaches to a modern MSR. Of these designs, the Generation IV Forum specifically identified the MSFR exemplified by the molten chloride fast reactor (MCFR) and the fluoride salt-cooled high-temperature reactor (FHR) exemplified by the TMSR-SF as of particular interest (Serp *et al* 2014).

8.6.1 Integral molten salt reactor

The IMSR is a derivative of the denatured MSR that was designed by Oak Ridge National Laboratory, but never built (Engel *et al* 1980). The denatured MSR was designed to be operated on denatured (low enrichment, $<5\%$ ^{235}U) uranium with the specific goal of minimizing nuclear proliferation. The low enrichment, as well as the fact that the reactor was not intended to be operated as a breeder, meant that fuel processing costs were minimal.

The IMSR is under development by Terrestrial Energy Inc. in Canada (Terrestrial Energy Inc. 2023). A schematic of the intended design is shown in figure 8.13. The graphite-moderated reactor core along with circulating fuel pumps and the heat exchangers to transfer heat from the primary (radioactive) cooling loop to the secondary (non-radioactive) loop are contained within the sealed integral core

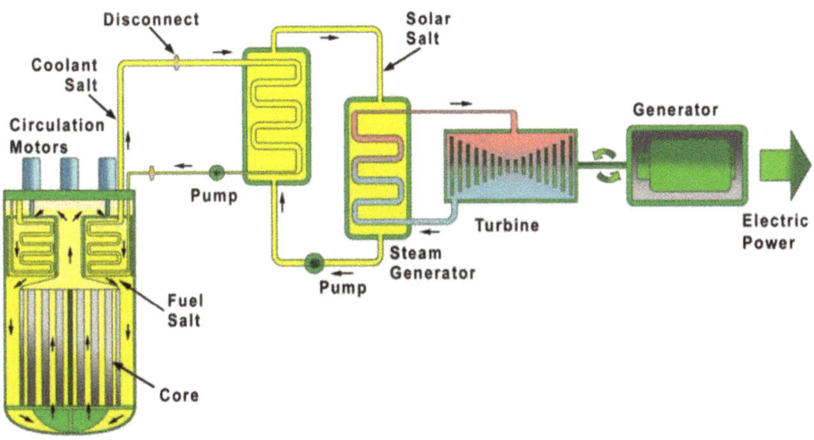

Figure 8.13. Schematic of the graphite-moderated IMSR. This schematic overview of the IMSR heat transfer flow image has been obtained by the author from the Wikimedia website where it was made available by Terrestrial Energy Inc. (2017) under a CC BY-SA 4.0 licence. It is included within this chapter on that basis. It is attributed to Terrestrial Energy.

unit. A major operating problem with graphite-moderated reactors are the defects in the moderator that are caused by neutron irradiation. For commercial operation, periodic graphite moderator replacement is complex and expensive, and results in reactor downtime. The IMSR is designed to operate for seven years, after which the integral core unit is replaced as a whole. The spent reactor core can be stored as an integral unit in a suitable facility. Since no fuel is removed from the core during reactor operation (although small amounts of fresh fuel may be added), the possibility of radioactive contamination is negligible.

Details of the reactor core are shown in the diagram in figure 8.14. There are no control rods since the reactor is self-regulating as a result of a large negative coefficient of reactivity. The reactor contains a neutron-absorbing shutdown rod that can be lowered into the core in the event that the reactor needs to be shutdown or if there is an operational anomaly. The fuel is a mixture of UF_4 with LiF, NaF, and/or BeF_2. The composition will be designed to optimize the fuel's heat capacity

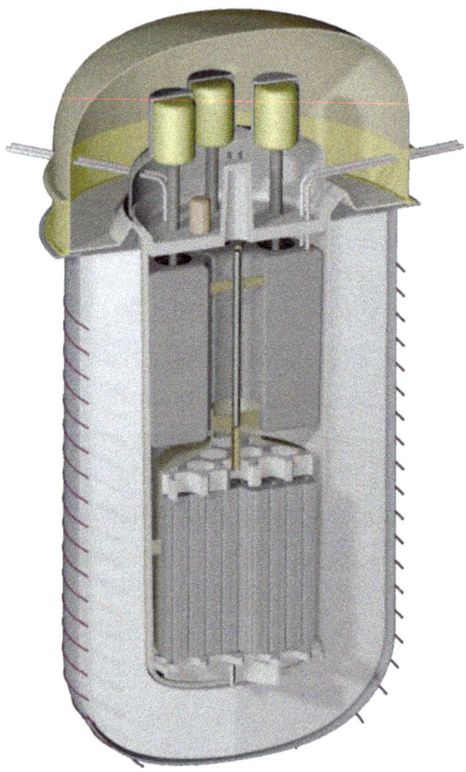

Figure 8.14. Cut-away view of the core of the IMSR reactor. This cut-away view of the IMSR core unit, showing the core unit placed inside a surrounding solid buffer salt tank, itself sitting in a below grade silo image has been obtained by the author from the Wikimedia website where it was made available by Terrestrial Energy (2014) under a CC BY-SA 3.0 licence. It is included within this chapter on that basis. It is attributed to Terrestrial Energy.

and maintain a reasonable melting temperature. As noted above, low enrichment uranium with less than 5% ^{235}U will be used.

As illustrated in figure 8.13, the primary coolant (and nuclear fuel) will transfer heat to salt in a secondary loop through a heat exchanger within the reactor core vessel. This secondary coolant will transfer heat through a heat exchanger to a tertiary salt loop. This arrangement will facilitate utilization of the reactor thermal output for electricity generation, thermal storage, or heat for industrial processes (see figure 8.4). A typical IMSR will provide a thermal output of 440 MW_{th} at a temperature of around 700 °C. At an efficiency of 44%, the reactor can provide 195 MW_e of electricity. Terrestrial Energy aims to have an operational commercial reactor by the end of the 2020s.

Another circulating MSR development project, the Thorium Molten Salt Reactor-Liquid Fuel (TMSR-LF), is underway in China. This reactor uses a fuel salt consisting of FLiBe (a mixture of LiF enriched in ^{7}Li and BeF_2) and ZrF_4 that contains fissile uranium (UF_4 enriched to 19.75% in ^{235}U) and fertile ThF_4 (for breeding new fissile material). A small (2 MW_{th}) experimental version, TMSR-LF1, became operational in 2023.

8.6.2 Molten chloride fast reactor

The first proposal for a MSR that utilized chloride salts appeared in the early-1950s (Goodman *et al* 1952). This work was motivated by the possibility of constructing a fast neutron reactor for the purpose of breeding weapons-grade plutonium (see e.g., Mausolff *et al* 2021). In 1956, researchers at Oak Ridge National Laboratory proposed a commercial fast neutron molten chloride salt reactor to produce heat for industrial processes (Bulmer *et al* 1956). Further analysis of the operation of a MCFR has been reported by Nelson *et al* (1967), Taube and Ligou (1974), and Mourogov and Bokov (2006).

A schematic of a generic MCFR is shown in figure 8.15. Typical dimensions are indicated. The chloride-based molten salt fuel is pumped through the active core from bottom to top. Neutron reflectors surround the core to increase the fission energy production.

As noted in table 8.1, TerraPower, in partnership with Southern Company, is developing a MCFR as part of the U.S. Department of Energy initiative to develop Generation IV reactors (American Nuclear Society 2021, U.S. Department of Energy Office of Nuclear Energy 2021). Figure 8.16 shows a schematic of the proposed design of the MCFR. This reactor includes neutron reflectors around the core and an integral heat exchanger to transfer heat from the primary (radioactive) salt loop to the secondary (non-radioactive) loop. The initial output of this demonstration reactor is designed to be 30 MW_{th} (Southern Company 2021).

The reactor will be inherently stable as the result of a large negative coefficient of reactivity. This negative response is primarily the result of the fact that as the core heats, the thermal expansion of the fuel decreases the power density, and this prevents the core from heating further. The negative coefficient of reactivity also provides a means of effective load following. Reducing the heat extracted from the

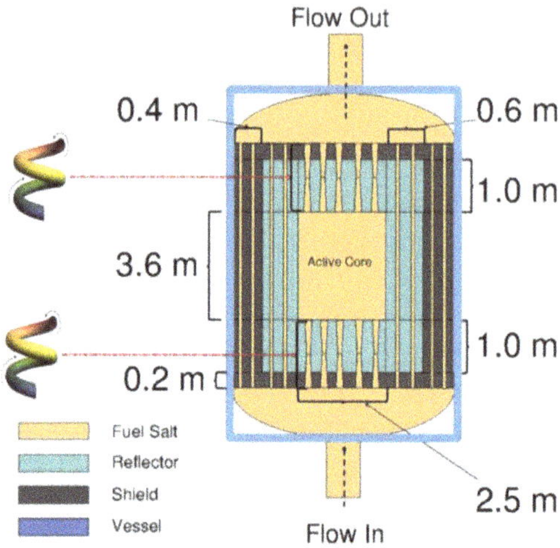

Figure 8.15. Schematic of a generic fast neutron molten chloride salt reactor showing the core surrounded with neutron reflectors and a radiation shield. The fuel flows from bottom to top and the inlet and outlet tubes for the reactor fuel are helical, as shown in the diagram. Reprinted from Mausolff *et al* (2021), Copyright (2021), with permission from Elsevier.

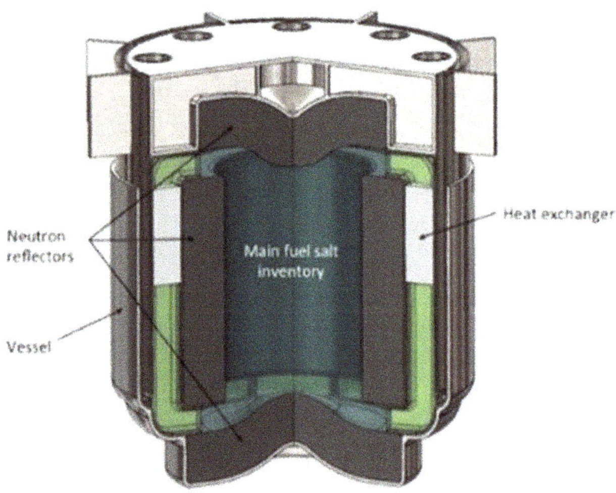

Figure 8.16. Core of the molten chloride salt reactor being developed by TerraPower and Southern Company under contract from the U.S. Department of Energy. Reproduced from U.S. Department of Energy Office of Nuclear Energy (2021). U.S. Department of Energy (2021). Image stated to be in the public domain.

secondary coolant loop causes an increase in coolant temperature and a reduction in reactor output. Conversely, increasing the heat extracted from the secondary coolant loop results in an increase in reactor output (World Nuclear Association 2021).

TerraPower and Southern Company are in the process of constructing a small version of the MSFR at the Idaho National Laboratory (Post Register 2021). This reactor, the Molten Chloride Reactor Experiment, has a thermal output of 200 kW_{th}.

The fuel is a eutectic mixture of NaCl and UCl_3 with a melting temperature of 525 °C. The core is designed to operate at 600 °C with a fuel flow rate through the core of between 25 and 100 kg s^{-1} (American Nuclear Society 2023).

8.6.3 Stable salt reactor-wasteburner

The stable salt reactor-wasteburner (SSR-W) is a stable salt fueled, molten salt-cooled fast neutron reactor being developed by Moltex Energy Canada, Ltd. Moltex has been approved by NB Power to build a 300 MW_{th} demonstration reactor (SSR-W 300) at the Point Lepreau Nuclear Generating Station in New Brunswick, Canada. This reactor is expected to be operational by the early-2030s.

Figure 8.17 shows a schematic of the proposed reactor. The core consists of approximately 300 fuel tubes, each 10 mm in diameter and about 1.8 m long. The fuel tubes are made of steel with a zirconium coating to resist corrosion from the salt. The design of the fuel tube is shown in figure 8.18. A gas vent is included at the top of each tube to avoid build-up of pressure due to gaseous fission products, particularly Kr and Xe. The fuel salt transfers heat to the cooling loop containing molten sodium-based salt. The cooling salt contains a redox potential reducing agent

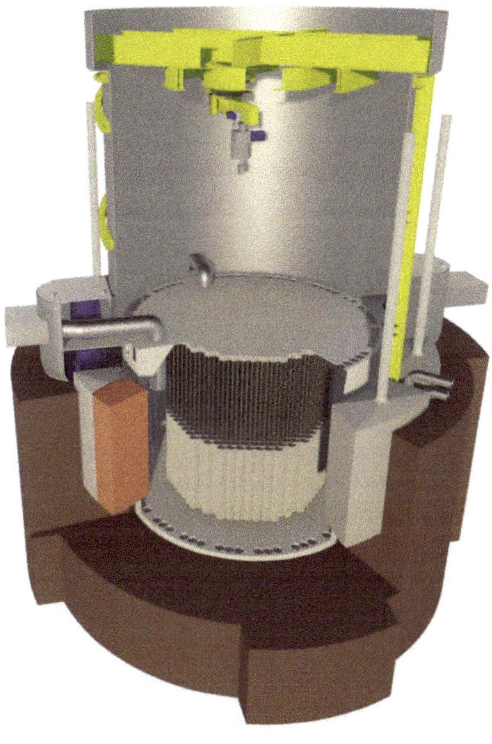

Figure 8.17. A cutout of a SSR-W core. This cutout of a stable salt reactor (SSR) core image has been obtained by the author from the Wikimedia website where it was made available by Rybr00159 (2022) under a CC BY-SA 4.0 licence. It is included within this chapter on that basis. It is attributed to Rybr00159.

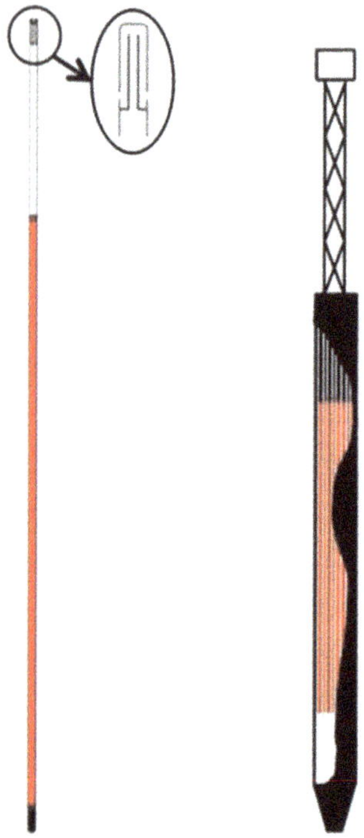

Figure 8.18. (Left-hand side) A fuel rod with gas vent (see inset) and (right-hand side) the complete fuel assembly. This crayon de combustible avec évacuation de gaz « cloche de plongée », et un assemblage complet image has been obtained by the author from the Wikimedia website where it was made available by Moltex Energy (2016) under a CC BY-SA 4.0 licence. It is included within this chapter on that basis. It is attributed to Moltex Energy.

to reduce its corrosivity so that it will be safe to use with standard grade stainless steel (316L). The heat output from the SSR-W reactor will be stored as thermal energy in molten salt tanks and can be used to produce electricity as needed to respond to load demands.

The fuel is a mixture of two-thirds sodium chloride and one-third lanthanide/actinide trichlorides. The reactor is designed so that the actual composition of the lanthanide/actinide trichlorides is not important as long as there is sufficient fissile material to achieve a critical reaction within the core of the reactor. This feature allows the SSR-W reactor to utilize waste fuel from conventional nuclear fission reactors and this is the reason it is referred to as a wasteburner. The ability to utilize radioactive waste from other reactors makes the SSR-W economical to fuel and also reduces the amount of radioactive spent fuel that needs to be directed to disposal.

The SSR-W reactor is inherently safe as a result of a large negative coefficient of reactivity, along with the ability to continuously remove heat from the core through the cooling system. As such, the reactor does not utilize control rods to regulate the reaction. However, an array of boron carbide rods can be released to fall under gravity into the reactor to shut it down in the event of any serious incidents.

Analyzes have shown that the capital cost to construct the SSR-W reactor is estimated to be 1.95 USD/W (for a 1 GW plant). This may be compared to the estimated cost of a coal-fired plant and a conventional nuclear plant of 3.25 USD/W and 5.50 USD/W, respectively (Reuters News & Media Ltd 2016). Because of the low capital construction cost, low maintenance, and low fuel costs resulting from the use of waste fuel, the levelized cost of electricity from the SSR-W reactor is estimated to be about 45 USD MWh^{-1}. This can be compared to the values given in figure 1.20, particularly traditional nuclear at about 180 USD MWh^{-1}.

8.6.4 FLEX

The FLEX reactor is a graphite-moderated reactor that uses static molten salt fuel and is being developed by MoltexFLEX Ltd (MoltexFLEX 2024) in the United Kingdom. The fuel is a mixed salt containing UCl_3 enriched to 5% in ^{235}U (Nuclear Engineering International 2023). The salt fuel is contained in a fuel rod, as described previously for the SSR-W reactor. The top is vented to release fission gases (e.g., Xe) to avoid pressure build-up in the fuel rod. The primary coolant is molten salt that transfers heat to a secondary salt loop through a heat exchanger external to the reactor vessel. Output heat is stored in molten salt tanks at 700 °C and can be used as needed for district heating, industrial processes, or electricity generation. In this way, the reactor can be used as a top-up to base load electricity when required by demand.

The FLEX reactor is designed to produce 60 MW_{th} and uses modular components to minimize construction and maintenance costs. It is designed to operate for five years with minimum maintenance before refueling. As is the case for other MSRs, the FLEX is inherently safe due to a large negative coefficient of reactivity. Development of the FLEX reactor is in its early stages.

8.6.5 Thorium molten salt reactor-solid fuel

The thorium molten salt reactor-solid fuel (TMSR-SF) along with the TMSR-LF, as noted above, are under development at the Shanghai Institute of Applied Physics (SINAP) and form the basis of China's approach to utilizing thorium as a nuclear reactor fuel (Zhang *et al* 2018). The TMSR-SF reactor is designed to be a 100 MW_{th} molten salt cooled pebble bed reactor containing uranium and thorium fuel. The TMSR-SF1, a smaller 10 MW_{th} version of the reactor as described below, was designed to test concepts for the larger demonstration reactor.

A diagram of the TMSR-SF1 is shown in figure 8.19. The fuel consists of 6 cm diameter graphite pebbles containing TRISO particles (see chapter 7) with either ^{232}Th or uranium enriched to 17% ^{235}U. The 3 m diameter by 2.85 m high core contains 14 650 pebbles (Ho 2021, World Nuclear Association 2021) and is surrounded by a graphite reflector. Control rods in the center of the core and in the reflector, as shown in figure 8.20, allow for reactor control. Molten salt consisting of FLiBe (a mixture of 66.7 mol% LiF enriched in ^7Li and 33.3 mol% BeF_2) is used as the coolant and transfers heat through a heat exchanger for electricity generation or industrial heat applications (Sun *et al* 2018).

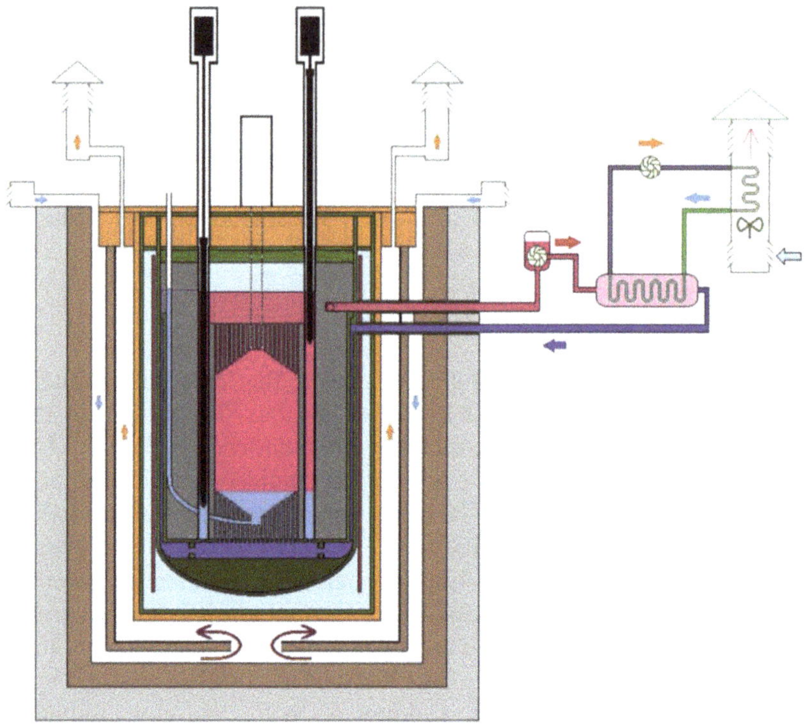

Figure 8.19. Schematic diagram of the TMSR-SF1 reactor. Reprinted from Sun *et al* (2018), Copyright (2018), with permission from Elsevier.

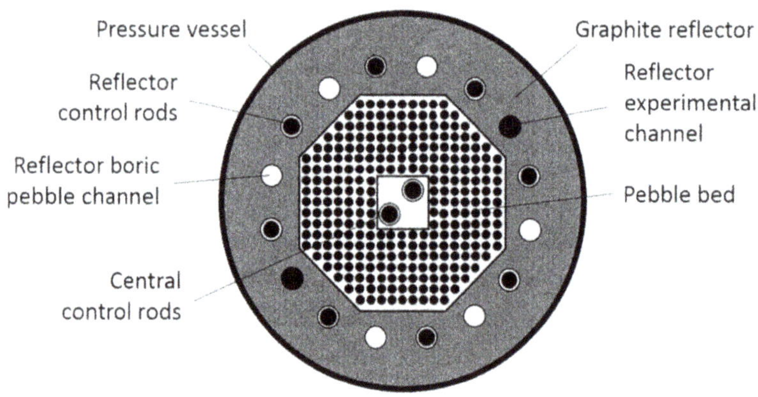

Figure 8.20. Cross section through the core of the TMSR-SF1 molten salt-cooled reactor. Reproduced from Schulenberg (2022), with permission from Springer Nature.

The fuel cycle includes partial utilization of the energy content of the fertile ^{232}Th component of the fuel. Thorium energy is utilized along the lines of the Peach Bottom Reactor (section 7.5) and the breeding of fissile ^{233}U is through the reactions given in equations (7.1)–(7.3). This approach, where there is a once-though fuel cycle that extracts energy from ^{232}Th, is referred to as self-breeding or as a breed and burn fuel cycle. Fuel can also be reprocessed to extract the fissile component for new fuel fabrication.

The use of molten salt cooling for a pebble bed reactor containing uranium and thorium fuel provides a means of using at least a portion of the energy from ^{232}Th. It also provides a high temperature thermal output that optimizes reactor efficiency for electricity production and increases opportunities for thermal applications. However, at present, the Shanghai Institute of Applied Physics has discontinued development of the TMSR-SF1 in favor of the TMSR-LF project which, although providing a lower temperature output, makes more efficient use of the thorium energy content.

References

American Nuclear Society 2021 ARDP recipient Southern announces molten salt fast reactor demonstration plans *Nuclear Newswire* (19 November 2021) www.ans.org/news/article-3452/ardp-recipient-southern-announces-molten-salt-fast-reactor-build-plans/

Bettis E S, Cottrell W B, Mann E R, Meem J L and Whitman G D 2021 1957 The aircraft reactor experiment-operation *Nucl. Sci. Eng.* **2** 841–53

Bulmer J, Gift E, Holl R, Jacobs A, Jaye S, Kassman E, McVean R, Oehl R and Rossi R 1956 *Fused Salt Fast Breeder* (Oak Ridge, TN: Oak Ridge National Laboratory) ORNL CF-56-8-204 www.osti.gov/servlets/purl/4319127

Cottrell W B (ed) 1952 *Aircraft Reactor Experiment Hazards Summary Report* (Oak Ridge National Laboratory ORNL-1407) www.osti.gov/servlets/purl/4704625

Cottrell W B, Hungerford H E, Leslie J K and Meem J L 1955 *Operation of the Aircraft Reactor Experiment* (Oak Ridge, TN: Oak Ridge National Laboratory) ORNL-1845 www.osti.gov/servlets/purl/4237975

Cottrell W B, Crabtree T E, Davis A L and Piper W G 1958 *Disassembly and Postoperative Examination of the Aircraft Reactor Experiment* (Oak Ridge National Laboratory) ORNL-1868 www.osti.gov/servlets/purl/4223435

Engel J R, Bauman H F, Dearing J F, Grimes W R, McCoy H E and Rhoades W A 1980 *Conceptual Design Characteristics of a Denatured Molten-Salt Reactor with Once-Through Fueling* (Oak Ridge, TN: Oak Ridge National Laboratory) ORNL/TM-7207 http://www.osti.gov/servlets/purl/5352526

Goodman C, Greenstadt J L, Kiehn R M, Klein A, Mills M M and Tralli N 1952 *Nuclear problems of non-aqueous fluid-fuel reactors* Technical Report MIT-5000 (Cambridge, MA: Massachusetts Institute of Technology) https://moltensalt.org/references/static/downloads/pdf/MIT-5000.pdf

Haubenreich P N and Engel J R 1970 Experience with the molten-salt reactor experiment *Nucl. Appl. Technol.* **8** 118–36

Haynes International ND Hastelloy® N alloy https://haynesintl.com/en/datasheet/hastelloy-n-alloy/#nominal-composition

Ho M 2021 Molten salt reactor review www.nuclearaustralia.org.au/wp-content/uploads/2021/04/Mark_Ho_20210512.pdf

Kamei T 2012 Recent research of thorium molten-salt reactor from a sustainability viewpoint *Sustainability* **4** 2399–418

Luo R, Liu C and Macián-Juan R 2021 Investigation of control characteristics for a molten salt reactor plant under normal and accident conditions *Energies* **14** 5279

Mausolff Z, DeHart M and Goluoglu S 2021 Design and assessment of a molten chloride fast reactor *Nucl. Eng. Des.* **379** 111181

Moltex Energy 2016 Un crayon de combustible avec évacuation de gaz «cloche de plongée», et un assemblage complet https://commons.wikimedia.org/wiki/File:RSS_-_crayon_%26_assemblage.jpg

MoltexFLEX 2024 FLEX Reactor www.moltexflex.com/flex-reactor/

Mourogov A and Bokov P M 2006 Potentialities of the fast spectrum molten salt reactor concept: REBUS-3700 *Energy Convers. Manag.* **47** 2761–71

Nelson P A, Butler D K, Chasanov M G and Meneghetti D 1967 Fuel properties and nuclear performance of fast reactors fueled with molten chlorides *Nucl. Appl.* **3** 540–7

Nuclear Engineering International 2023 MoltexFLEX refines design of its FLEX reactor (13 September 2023) www.neimagazine.com/news/newsmoltexflex-refines-design-of-its-flex--reactor-11143421

Oak Ridge National Laboratory 1952 A cross-section of the fluid-fueled Aircraft Reactor Experiment https://commons.wikimedia.org/wiki/File:Aircraft_Reactor_Experiment_cross_section.png

Oak Ridge National Laboratory c. 1964 A top down view of the molten salt reactor experiment https://commons.wikimedia.org/wiki/File:MoltenSaltReactor.jpg

Oak Ridge National Laboratory 1964 Molten salt reactor graphite core https://commons.wikimedia.org/wiki/File:MSRE_Core.JPG

Oak Ridge National Laboratory 2007 Aircraft reactor experiment building https://commons.wikimedia.org/wiki/File:ARE_Building.JPG

Oak Ridge National Laboratory 2011 Diagram showing the internals of the Molten-Salt Reactor Experiment https://commons.wikimedia.org/wiki/File:MSRE_Diagram.png

Oak Ridge National Laboratory 2015 View of MSRE reactor pit under construction in 1964 www.flickr.com/photos/oakridgelab/21575353623/in/album-72157618874649144/

Oak Ridge National Laboratory 2018 Aircraft reactor experiment www.flickr.com/photos/oakridgelab/45763884494/in/album-72157618874649144/

Post Register 2021 INL is targeted site for world's first fast-spectrum salt reactor—Southern Company will lead effort in collaboration with TerraPower, Idaho National Laboratory *Post Register* (18 November 2021) www.postregister.com/news/inl/inl-is-targeted-site-for-world-s-first-fast-spectrum-salt-reactor/article_a58ef52a-a409–509e-97a9-36eeda907f2b.html

Reuters News & Media Ltd 2016 Moltex energy sees UK, Canada SMR licensing as springboard to Asia *Reuters Events Nuclear* (28 June 2016) www.reutersevents.com/nuclear/moltex-energy-sees-uk-canada-smr-licensing-springboard-asia

Rosenthal M W 2010 *An Account of Oak Ridge National Laboratory's Thirteen Nuclear Reactors* (Oak Ridge, TN: Oak Ridge National Laboratory) ORNL/TM-2009/181

Rybr00159 2022 A cutout of a Stable Salt Reactor core (SSR) core https://commons.wikimedia.org/wiki/File:SSR-W-cutaway-no-labels-1464x2048.png

Schulenberg T 2022 *The Fourth Generation of Nuclear Reactors—Fundamentals, Types, and Benefits Explained* (Berlin: Springer)

Serp J *et al* 2014 The molten salt reactor (MSR) in generation IV: Overview and perspectives *Prog. Nucl. Energy* **77** 308–19

Southern Company 2021 *Technology Inclusive Content of Application Project For Non-Light Water Reactors TerraPower Molten Chloride Reactor Experiment TICAP Tabletop Exercise Report* Document Number SC-16166-202 Rev 0 www.nrc.gov/docs/ML2122/ML21228A222.pdf

Sun K, Wilson J, Hauptman S, Ji R, Dave A J, Zou Y and Hu L W 2018 Neutronics modeling and analysis of the TMSR-SF1 fuel lattice and full core with explicit fuel particle distribution and random pebble loadings *Prog. Nucl. Energy* **109** 171–9

Taube M and Ligou J 1974 Molten plutonium chlorides fast breeder reactor cooled by molten uranium chloride *Ann. Nucl. Sci. Eng.* **1** 277–81

Terrestrial Energy Inc. 2014 Cut-away view of the IMSR core unit https://commons.wikimedia.org/wiki/File:Terrestrial_Energy_IMSR_reactor_module_and_buffer_salt_tank.jpg

Terrestrial Energy Inc. 2017a This schematic describes the different possible heat applications for the IMSR https://commons.wikimedia.org/wiki/File:IMSR_heat_applications.png

Terrestrial Energy Inc. 2017b Schematic overview of the IMSR heat transfer flow https://commons.wikimedia.org/wiki/File:IMSR_INFO-GRAPHIC_Colored.jpg

Terrestrial Energy Inc. 2023 Integral Molten Salt Reactor: Carbon-free, Low-cost, High-impact. Flexible and Resilient www.terrestrialenergy.com/technology/

U.S. Department of Energy Office of Nuclear Energy 2021 *Southern Company and TerraPower Prep for Testing on Molten Salt Reactor* www.energy.gov/ne/articles/southern-company-and-terrapower-prep-testing-molten-salt-reactor

World Nuclear Association 2021 Molten salt reactors https://world-nuclear.org/information-library/current-and-future-generation/molten-salt-reactors.aspx

Zhang D *et al* 2018 Review of conceptual design and fundamental research of molten salt reactors in China *Int. J. Energy Res.* **42** 1834–48

IOP Publishing

Generation IV Nuclear Reactors
Design, operation and prospects for future energy production
Richard A Dunlap

Chapter 9

Supercritical water-cooled reactors (SCWRs)

Supercritical water-cooled reactors (SCWRs) are discussed in this chapter, which begins with an overview of the properties of supercritical water and the benefits of utilizing water in this phase for cooling a fission reactor. In the 1950s and 1960s, several projects developed designs for SCWRs, but none was ever constructed. SCWRs have attracted attention as potential Generation IV reactors, largely because of the high enthalpy gain by the coolant in the core along with the lack of a phase transition in the cooling system, which means that reactor designs can be simpler and more economical than those for traditional pressurize water reactors. These reactors may utilize either thermal or fast spectrum neutrons. This chapter reviews some of the Generation IV reactor designs being developed in Canada, Asia, and the European Union.

9.1 Introduction

Light-water pressurized water reactors are the most common type of nuclear fission reactor (see table 5.1), with over 300 units in operation as of 2023. A major drawback of pressurized water reactors is their relatively low thermodynamic efficiency, less than 35%, for generating electricity. This low efficiency is a direct result of the temperature of the steam that is input into the turbine. Typically, pressurized water reactors produce steam at a pressure of about 15.5 MPa (153 atm) and a maximum temperature of around 315 °C (see section 5.4). This same situation applied to coal-fired generating stations in the past. Although the method of producing heat is different, both nuclear and coal-fired power plants use the same technology for converting heat to electricity. Since the 1990s, virtually all new coal-fired generating station have utilized supercritical water to drive the turbines, thereby increasing the temperature of the turbine input and, correspondingly, increasing the efficiency from about 35% to around 45%. The concept of the

doi:10.1088/978-0-7503-6069-2ch9

SCWR follows along the lines of modern coal-fired generating stations and intends to similarly increase the efficiency of electricity generation. Thus, this new Generation IV fission reactor makes use of technologies from traditional fission reactors and fossil fuel infrastructure. SCWRs can be thermal neutron reactors, which are either light water or heavy water moderated, or they can be fast neutron reactors. We begin with a brief discussion of the properties of supercritical water.

9.2 Properties of supercritical water

The phase diagram of water is shown in figure 9.1. The transition line separating the gas and liquid phases as a function of temperature and pressure terminates at the critical point, 373.946 °C and 22.064 MPa. For temperatures and pressures above the critical point, there is no distinction between the gas and liquid phases of water. Figure 9.2 shows some properties of water at a pressure above the critical pressure as a function of temperature as it undergoes the transition to the supercritical regime. It can be seen in the figure that minor changes in temperature around the critical temperature can yield substantial changes in the physical properties of water. A particularly important feature in figure 9.2 is a substantial increase in enthalpy above the transition temperature. This feature means that for a given target turbine power output, a smaller steam mass flow rate at the turbine's input is required. This, in turn, reduces the required size of the turbine, as well as the condenser and other associated steam system components, and reduces the capital construction costs as a function of reactor power output. Since amortized capital construction costs are a major component of the levelized cost of electricity for a nuclear power plant, reductions in capital expenses for SCWRs can be as important as the increased efficiency.

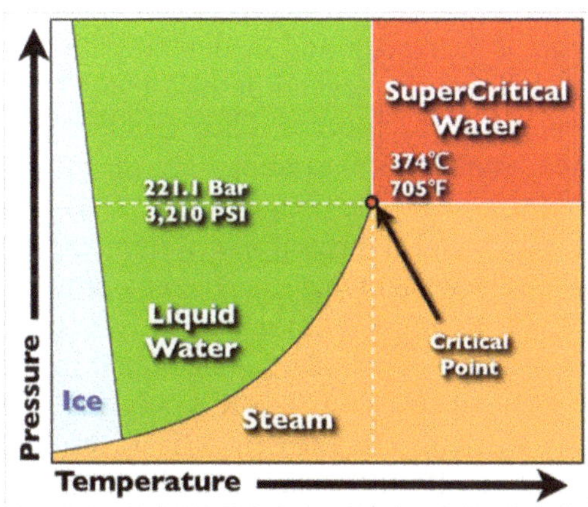

Figure 9.1. Phase diagram for water showing the location of the critical point. This Supercritical Water (Red Area) image has been obtained by the author from the Wikimedia website where it was made available by Jonathan Kamler (2010) under a CC BY-SA 4.0 licence. It is included within this chapter on that basis. It is attributed to Jonathan Kamler.

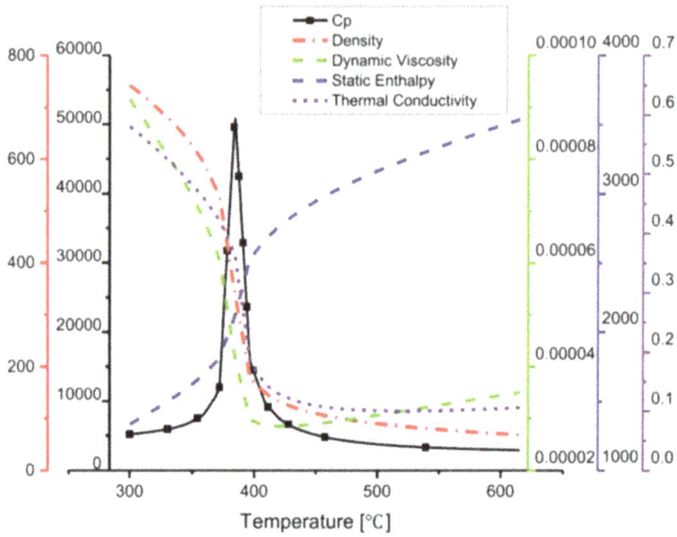

Figure 9.2. Properties of water as a function of temperature at a pressure of 25 MPa. Vertical axes are specific heat (black scale) in J (kg °C)$^{-1}$, density (red scale) in kg m^{-3}, viscosity (green scale) in Pa·s, enthalpy (blue scale) in kJ kg^{-1}, and thermal conductivity (purple scale) in W (m °C)$^{-1}$. Reprinted from Ma *et al* (2018), Copyright (2018), with permission from Elsevier.

9.3 General design of supercritical water-cooled reactors

The cores of SCWRs fall into two general categories: those with pressure vessels and those with pressure tubes. Pressure vessel reactors are similar to conventional light pressurized water reactors. Figure 9.3 shows a generic diagram of a pressure vessel-type SCWR. Typical operating pressure is around 25 MPa and feedwater enters the reactor vessel at around 280 °C. After passing through the core, the supercritical water exits the reactor vessel at a temperature of around 500 °C, or higher, depending on the design of the reactor and the limitation imposed by the properties of the reactor materials. There are several designs related to the geometry of the cooling water flow through the reactor core. These include, for example, single-pass, two-pass, three-pass, etc, some typical examples are shown in figure 9.4.

An important consideration for the design of a SCWR is the neutron spectrum. Normally in a light-water cooled pressurized water reactor, the cooling water functions effectively as the neutron moderator, thereby providing a thermal neutron spectrum. However, it can be noted that the density of superfluid water above the transition temperature drops significantly, as illustrated in figure 9.2. In fact, the density of supercritical water in a typical SCWR is only about 10% of that of cooling water in a typical pressurized water reactor. As a result of this low density, supercritical water in not an effective neutron moderator. Therefore, reactors that are intended to utilize thermal neutrons typically employ other approaches to neutron moderation, as described in some of the examples later in this chapter. In some reactors, the neutron spectrum may not be completely thermalized, and these reactors are sometimes referred to as epithermal.

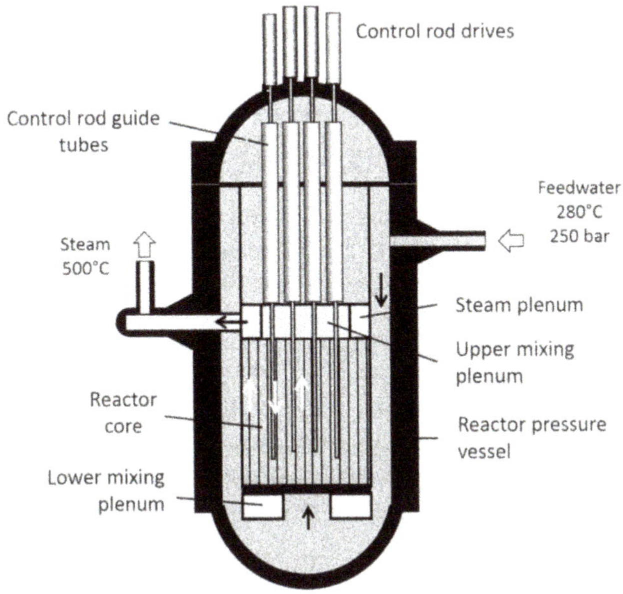

Figure 9.3. Cross section of the reactor vessel for a light SCWR. Reproduced from Schulenberg (2022), with permission from Springer Nature.

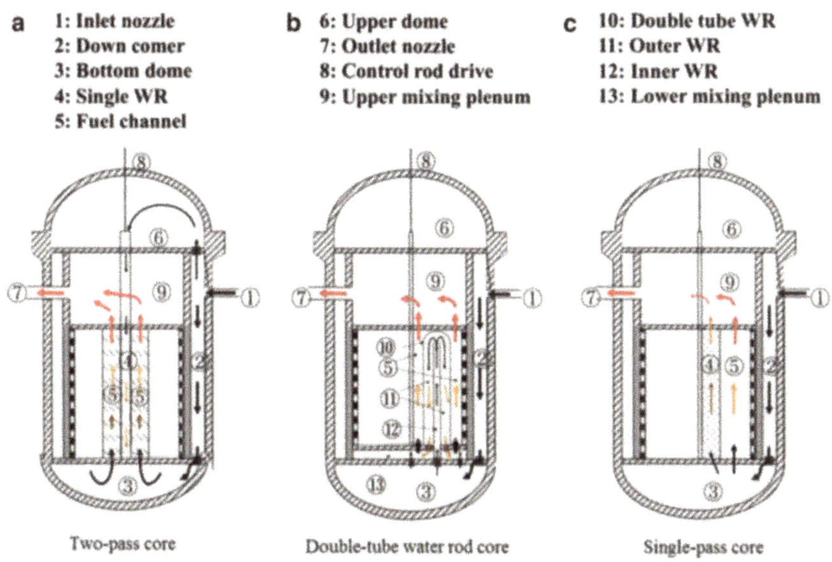

Figure 9.4. Cooling water flow schemes for SCWRs incorporating a pressure vessel. Reproduced from Oka and Morooka (2014), with permission from Springer Nature.

Reactors that do not include additional means of neutron moderation utilize fast spectrum neutrons and can be configured as breeder reactors. Several possible designs of supercritical water-cooled breeder reactors are shown in figure 9.5.

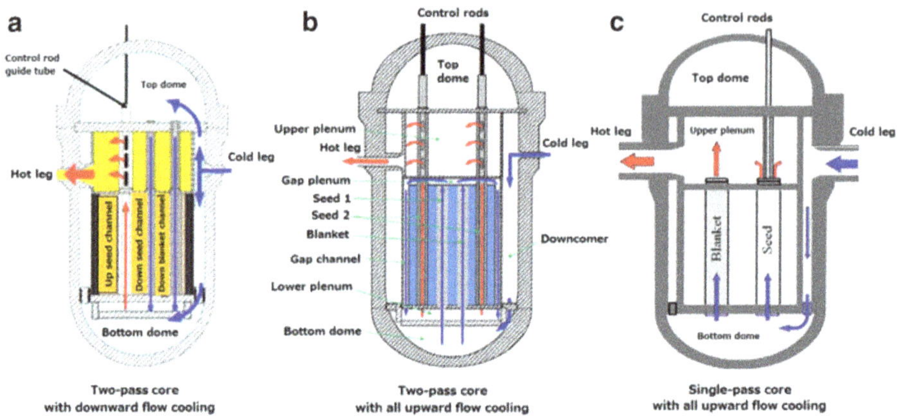

Figure 9.5. Reactor core designs and cooling water flow schemes for supercritical water-cooled breeder reactors. Reproduced from Oka and Morooka (2014), with permission from Springer Nature.

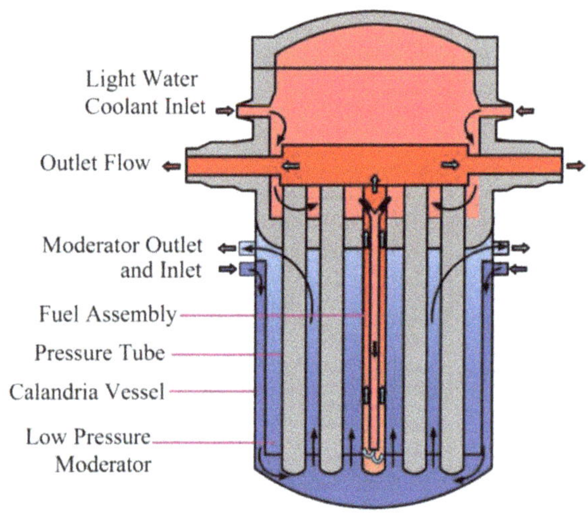

Figure 9.6. Schematic of the Canadian SCWR design incorporating pressure tubes. Reproduced from Wu *et al* (2022). CC BY 4.0.

The seed-blanket design, as shown in the figure, is commonly used. The seed contains fuel with the fissile component, e.g., ^{233}U or ^{235}U, enriched to about 6%, while the blanket has a lower enrichment, around 2%. The Shippingport Atomic Power Station (see section 4.6.4) was one of the earliest reactors to experiment with the seed-blanket configuration.

The second design of a SCWR utilizes a pressure tube, as illustrated in figure 9.6. Cooling is accomplished by supercritical light water inside the pressure tubes and neutron moderation is accomplished by heavy water between the pressure tubes. Pressure tubes containing the fuel assemblies are cooled by circulating supercritical light water, as illustrated. Light water enters the reactor vessel at the top and is distributed to the pressure tubes containing the fuel assemblies through the inlet plenum. The cooling water flows downward through the center channel of the fuel assembly and, after exiting at the bottom of the fuel assembly, flows upward in the pressure tube around the outside of the assembly, as shown in the figure. At the top, the cooling water from the various pressure tubes is collected in the outlet plenum from which it exits at a temperature of around 625 °C and flows to the turbine system. Further details of this design are presented below in the section on the Canadian SCWR reactor.

A simplified diagram of the layout of a supercritical water-cooled reactor is shown in figure 9.7. Supercritical water flows from the reactor to the turbine to drive the generator. The output from the turbine is cooled by the condenser, which is necessary to improve the thermodynamic efficiency of the turbine. The condenser can dissipate heat through conventional means, such as a body of water acting as a heat reservoir or a cooling tower.

A more detailed diagram of a SCWR layout is shown in figure 9.8. A series of three turbines, a high-pressure (HP) turbine, an intermediate-pressure (IP) turbine, and a low-pressure (LP) turbine, are used. A portion of the heat at the output of the HP turbine is directed to the input of the IP turbine. The remaining heat from the output of the HP turbine is directed to the reheater to be combined with the output from the IP turbine, which is then input into the LP turbine. Output from the condenser is directed to the feedtrain, which preheats the water using excess heat from the turbine outputs for input back into the reactor. Much of this technology is borrowed from three decades of experience using supercritical water in coal-fired generating stations.

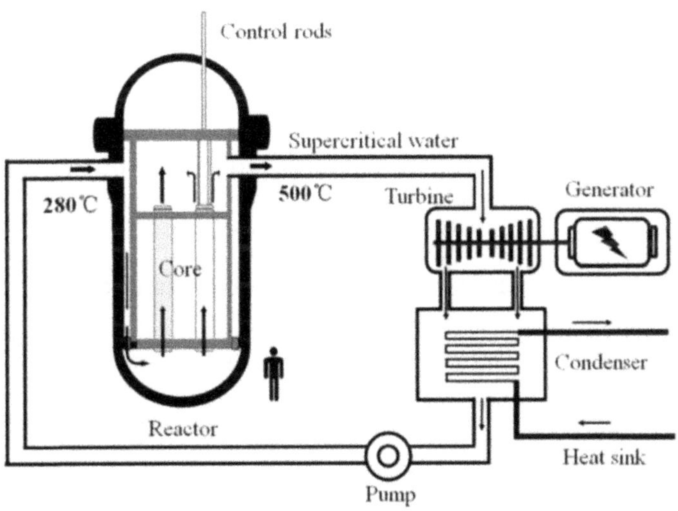

Figure 9.7. Simple layout of a SCWR with turbine, generator, and condenser. Reproduced from Oka (2014), with permission from Springer Nature.

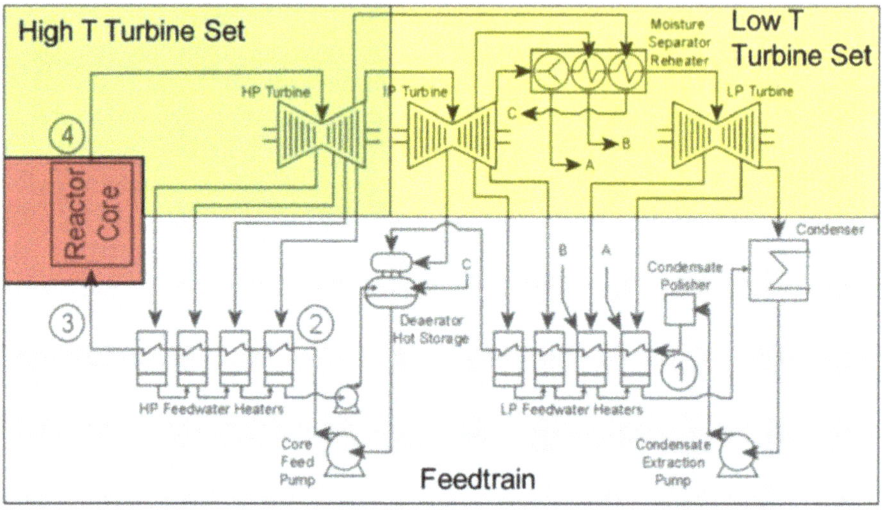

Figure 9.8. Typical layout of a SCWR power plant showing the HP turbine, IP turbine, and LP turbine. Reprinted from Guzonas and Novotny (2014), Copyright (2014), with permission from Elsevier.

9.4 Previous reactor development related to supercritical water-cooled reactors

In the 1950s and 1960s, there was an interest in developing a SCWR. The high output temperature from the reactor was an attractive feature because it allowed for increased thermodynamic efficiency of the turbines. Although there were several plans for supercritical water reactors at that time, no reactor was ever constructed. Over time, traditional boiling water reactors and pressurized water reactors were developed, and successfully constructed and operated. However, interest in supercritical water reactors waned until it was revived in the 1990s. The supercritical water reactor concepts that were considered in the 1950s and 1960s have been overviewed by Oka (2000) and Oka *et al* (2010). Here, we provide a brief description of some of the notable designs that were being considered during that period but were never constructed.

9.4.1 Westinghouse light-water moderated supercritical water-cooled reactor (1959)

One of the earliest designs of a SCWR was the Westinghouse (WH) light-water moderated from 1957. Some of the basic properties of this design are given in table 9.1. The reactor uses fuel assemblies in HP tubes cooled by supercritical water. Each tube contains seven fuel rods. LP (i.e. subcritical) light water fills the space between the HP tubes and acts as the neutron moderator. The reactor uses an indirect cycle where heat from the supercritical water in the primary loop is transferred to a secondary loop through a heat exchanger to drive the turbine. This approach was taken to avoid the possibility of distributing radioactive material

Table 9.1. Properties of some SCWR from the late 1950s and early 1960s. Data adapted from Oka *et al* (2010). MOX = mixed oxides.

Property	WH light water	GE heavy water	SCOTT-R	B&W FR
Reactor type	Pressure tube	Pressure tube	Pressure tube	Pressure vessel
Neutron spectrum	Thermal	Thermal	Thermal	Fast
Moderator	H_2O	D_2O	Graphite	None
Fuel	UO_2	UO_2	UO_2	MOX
Coolant pressure (MPa)	27.6	37.9	24.1	25.3
Outlet temperature (°C)	538	621	566	538
Thermal power (MW_{th})	70	300	2297	2326
Thermal efficiency (%)	30.3	40	43.5	42.2

outside the reactor. The choice of an indirect steam cycle is the principal reason for the low thermodynamic efficiency given in table 9.1.

9.4.2 General electric heavy-water moderated supercritical water-cooled reactor (1959)

The General Electric (GE) heavy-water moderated supercritical reactor also uses a pressure tube design. Some of its properties are given in table 9.1. The spherical reactor vessel has 300 fuel assemblies contained in HP supercritical water-cooled tubes. The spaces between the tubes contain the LP heavy water moderator. It can be noted from the data in table 9.1 that this reactor operates at a significantly higher pressure than the other supercritical water reactors that have been discussed. This design followed from the design of the supercritical coal-fired generating station in Philo, Muskingum County, Ohio. Unit 6 at the Philo Power Plant became operational in 1957 and was the first supercritical water-cooled coal-fired generating station in the United States. Although some early supercritical coal-fired generating stations operated at such high pressures, current systems typically operate more in range of 25 MPa. The GE heavy-water moderated reactor operates on direct cycle where the primary loop supercritical water drives the turbine directly. It was felt that this approach was necessary in order to achieve a sufficiently high thermodynamic efficiency. However, potential radioactive contamination in the turbine was a major concern.

9.4.3 Westinghouse graphite-moderated supercritical water-cooled reactor (1962)

The 1962 WH design, commonly referred to as the supercritical once-through tube reactor (SCOTT-R) was a pressure tube reactor with a graphite moderator. This reactor was substantially larger than the two designs discussed above and with an efficiency of more than 43% it provided a commercial-scale electricity output of about 1000 MW_e (see table 9.1). The fuel assemblies are contained in HP tubes that are cooled by supercritical water which drives the turbine directly. The HP tubes are located in holes penetrating the graphite moderator block. The moderator is cooled by flowing helium gas.

9.4.4 Babcock & Wilcox supercritical water-cooled fast breeder reactor (1967)

The supercritical water-cooled fast reactor that was designed by Babcock & Wilcox (B&W) uses a pressure vessel. The reactor operates on a direct cycle utilizing the Loeffler boiler design in which the heat outlet from the reactor is divided, where one portion drives the turbine directly and the second portion is mixed with the feedwater to preheat the reactor input. The use of mixed oxide (MOX) fuel, containing a mixture of fissile and fertile nuclides provides the ability to breed new fissile material, which can then be extracted from the fuel by reprocessing.

Sections 9.5–9.8 consider the details of some recent SCWR development projects. It should be noted, however, that these are 'in-progress' designs and will continue to be refined before prototype reactors are constructed.

9.5 Canadian supercritical water-cooled reactor

The Canadian SCWR is a pressure tube type heavy-water moderated reactor designed along the lines of the conventional CANDU heavy water reactors (see section 5.7). The basic design of the Canadian SCWR has been described by Yetisir *et al* (2016).

The general design of the core of the Canadian SCWR is shown in figure 9.6 and a cross section of a fuel assembly is shown in figure 9.9. The relevant design and operational parameters are given in table 9.2. Supercritical cooling light water flows down a central channel and at the bottom of the closed pressure tube is directed upward, where it makes a single pass through the annular ring containing the fuel

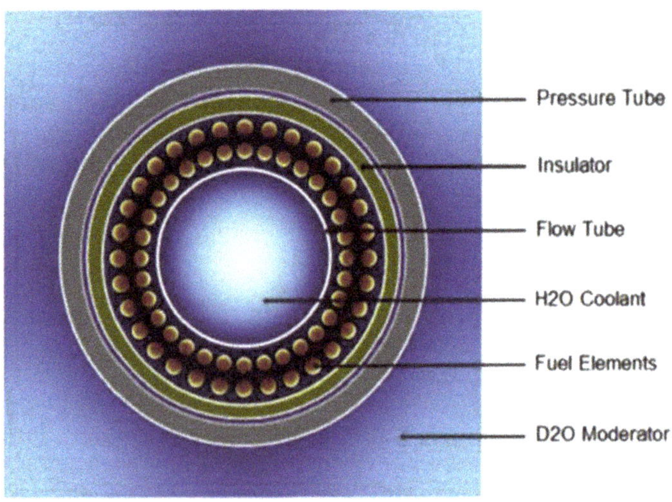

Figure 9.9. Cross-section of the fuel assembly of the Canadian SCWR. Reprinted from Guzonas and Novotny (2014), Copyright (2014), with permission from Elsevier.

Table 9.2. Properties of the Canadian SCWR. Data adapted from Edwards and Leung (2022), Rahman *et al* (2020), and Wu *et al* (2022).

Property	Value
Reactor type	Pressure tube
Neutron spectrum	Thermal
Moderator	D_2O
Fuel	$(Pu,Th)O_2$
Fuel enrichment (%)	9
Pressure (MPa)	25
Inlet temperature (°C)	350
Outlet temperature (°C)	625
Active core height (m)	5.0
Number of flow passes	1
Thermal power (MW_{th})	2540
Thermal efficiency (%)	48

elements. The current design uses 64 fuel elements in each fuel assembly (Nava-Domínguez *et al* 2016) and 336 assemblies in the reactor core (Yetisir *et al* 2016). As shown in figure 9.6, the cooling water from the pressure tubes is collected in an output plenum and directed to a direct drive turbine system.

The pressure tubes that contain the fuel elements and carry the supercritical cooling light water are surrounded by low-pressure heavy water that acts as the neutron moderator. The heat that is transferred to the heavy water moderator is removed by both active and passive systems, as illustrated in figure 9.10. The reactor is designed with a 'no-core-melt' concept, meaning that if all cooling water flow to the core is stopped, then the heat will be dissipated from the fuel assemblies to the moderator at a rate that is sufficient to prevent the core from melting. Computer studies of the Canadian SCWR have shown that under these conditions the reactor core temperature will remain within acceptable limits (Wu and Novog 2015). A key component of this design is the insulator that separates the supercritical water channel and the heavy water moderator. On the one hand, the insulation must be sufficient to minimize heat loss from the supercritical light water to the LP heavy water. On the other hand, sufficient heat transfer should occur in the event of the loss of circulation in the cooling water in order to prevent core meltdown. The insulating material must have an appropriately low thermal conductivity, a low neutron absorption cross section, a resistance to radiation damage from neutron irradiation, and a high corrosion resistance at elevated temperatures. Yttria stabilized Zirconia has been shown to be suitable and is used as a benchmark for the evaluation of other potential materials.

A final point to consider for the Canadian SCWR is the choice of fuel. As noted in table 9.2, the fuel is a mixture of PuO_2 and ThO_2. The use of mixed actinide fuel containing a fissile nuclide and fertile ^{232}Th is an approach to closing the fuel cycle (Morreale *et al* 2017) and is effective at reducing activity levels in spent fuel. It

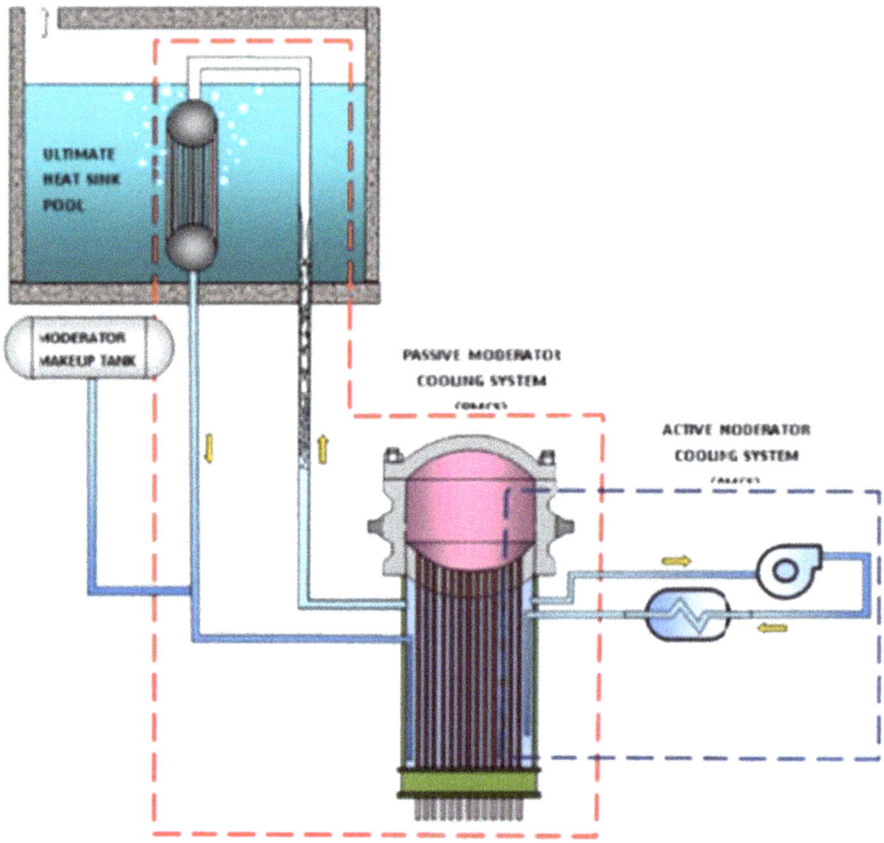

Figure 9.10. Diagram showing the active and passive components of the moderator cooling system for the Canadian SCWR. Reproduced from Wu *et al* (2022). CC BY 4.0.

should be noted, however, that the inclusion of ^{232}Th and/or ^{239}Pu in the fuel alters the neutron energy spectrum and this must be considered in the reactor design. This feature results from the larger thermal neutron absorption cross section for ^{232}Th and ^{239}Pu compared to ^{238}U and ^{235}U. Figure 9.11 shows an example of this behavior, and it is readily seen that the low energy of the neutron spectrum is depressed for Th and MOX containing fuels, meaning that the average neutron energy is increased compared to fuels containing only uranium.

9.6 CSR1000 (China)

The CSR1000 reactor is being considered as both a thermal reactor and a mixed spectrum reactor. The thermal version of the CSR100, as described in the present section, is being developed by the Nuclear Power Institute of China. This reactor is a two-pass light-water cooled and light-water moderated supercritical reactor. The specifications are summarized in table 9.3.

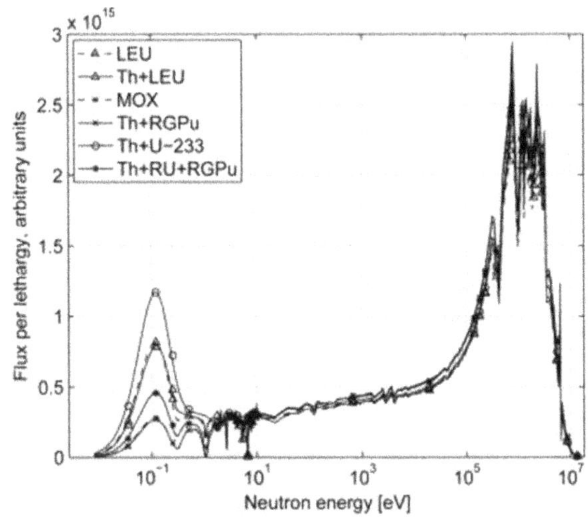

Figure 9.11. Energy spectra for neutrons from different types of nuclear fuel. Reprinted from Björk *et al* (2011), Copyright (2011), with permission from Elsevier.

Table 9.3. Properties of the CSR1000. Data adapted from Edwards and Leung (2022), Rahman *et al* (2020), Wu *et al* (2014), and Wu *et al* (2022).

Property	Value
Reactor type	Pressure vessel
Neutron spectrum	Thermal
Moderator	H_2O
Fuel	UO_2
Fuel enrichment (%)	5.6
Pressure (MPa)	25
Inlet temperature (°C)	280
Outlet temperature (°C)	500
Active core height (m)	3.0
Number of flow passes	2
Thermal power (MW_{th})	2300
Thermal efficiency (%)	43.5

Figure 9.12 shows the design of the CSR1000 core. The core is divided into two regions, as illustrated in figure 9.12(a). An inner region is cooled during the first pass of the cooling water and the outer region is cooled by the second pass. Figure 9.12(b) shows a vertical cross section of the reactor core illustrating the cooling water flow pattern. Cooling water that is input near the top of the vessel is divided into two streams. One stream travels downward on the first cooling pass through the central part of the core (indicated by (1) in figure 9.12(b)). The other stream travels down the outside of the core through the downcomer (labeled (4)), where it joins the stream

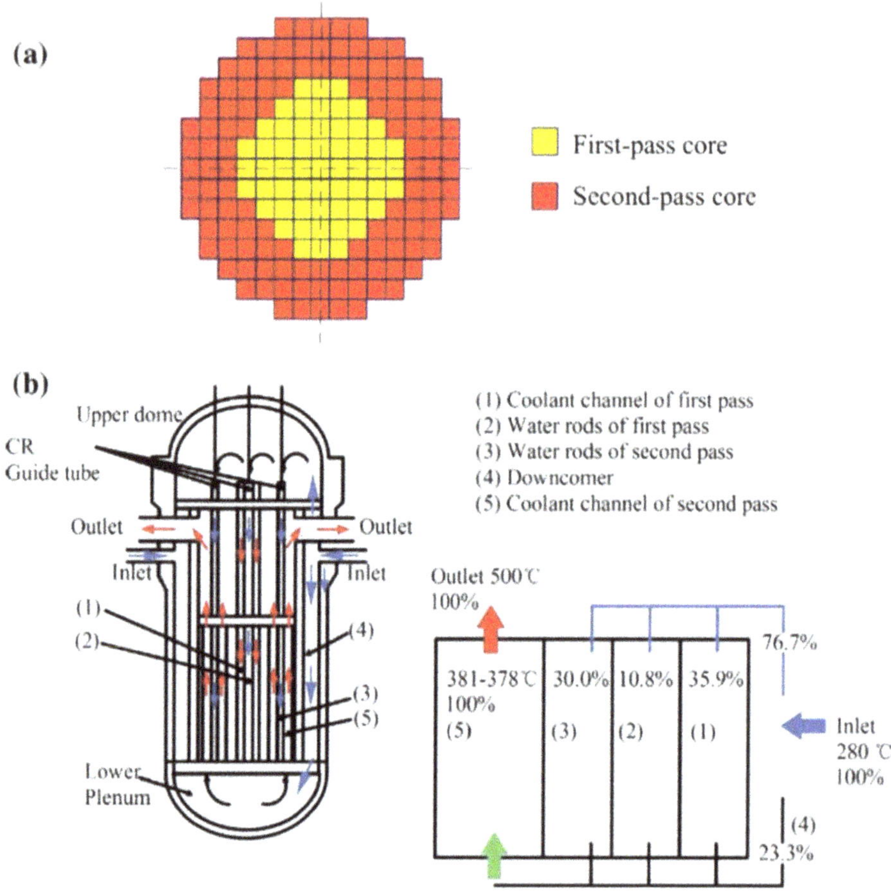

Figure 9.12. (a) Diagram of CSR1000 core showing regions cooled by first and second pass of supercritical light water. (b) Vertical section of reactor core showing the cooling water flow pattern. Reprinted from Wu *et al* (2014), Copyright (2014), with permission from Elsevier.

from the central region after its first pass. The two streams combine and travel upward on the second pass through the outer portion of the core (labeled (5)). In addition to flowing through the coolant channels in the core, cooling water also passes through water rods, as shown in the figure (labeled (2) and (3)). Water rods are commonly used in boiling water reactors to improve cooling (see Vogt 2004). These are typically empty tubes about the size of a fuel rod that are interspersed with the fuel rods in the reactor core. These carry cooling water to increase the heat transfer from the core to the cooling water. Their use in the CSR1000 is described in detail below.

The details of the fuel assembly of the CSR1000 are shown in figure 9.13. Each fuel assembly consists of 56 fuel rods surrounding a central water rod, also referred to as a moderator box in a SCWR. The water rod serves two purposes. First, to increase the rate of heat transfer from the core to the supercritical cooling water.

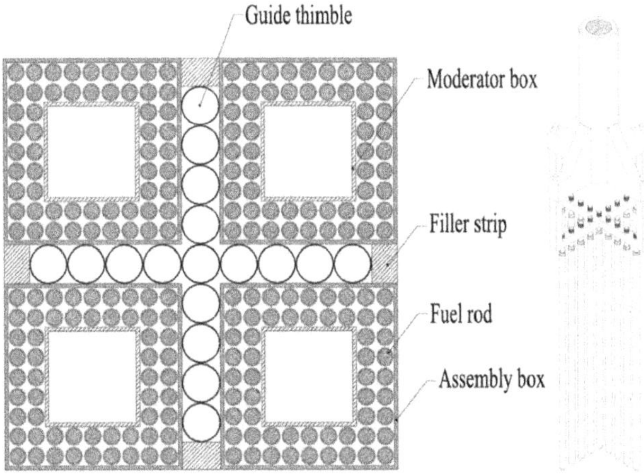

Figure 9.13. (Left-hand panel) Details of a fuel assembly from the CSR1000 reactor and (right-hand panel) proposal for the design of the cruciform control rod. Reproduced from Zhu *et al* (2021). CC BY 3.0.

Second, and most importantly, the supercritical water in the water rod acts as a neutron moderator. As noted previously, the effectiveness of supercritical water as a moderator is significantly reduced compared to subcritical water because of its low density. The large volume of the water rod compensates for the low density of the supercritical water and provides sufficient neutron moderation in the CSR1000 reactor.

The reactor core contains a total of 157 fuel assemblies. These are arranged in groups of four, as shown in figure 9.13. The four fuel assemblies are separated by channels that contain the cruciform control rods. This control rod geometry is commonly used in traditional boiling water reactors (see Westinghouse Electric Company 2021). In the proposed design shown in figure 9.13, the cruciform control rod is composed of several individual circular control rods that fit into guide thimbles, as shown in figure 9.14.

The CSR1000 reactor utilizes uranium dioxide fuel enriched to 5.6% in ^{235}U. This enrichment is typical of light-water moderated supercritical thermal reactors (see Rahman *et al* 2020) but is somewhat greater than that used in traditional light-water moderated boiling water reactors (~2.4% ^{235}U) or pressurized water reactors (~3.2% ^{235}U).

9.7 High-performance light-water reactor (European Union)

The SCWR design from the European Union is a thermal light-water moderated pressure vessel reactor similar in design to the CSR1000. It is designated as the high-performance light-water reactor (HPLWR). The basic design properties are given in table 9.4 and a diagram of the reactor pressure vessel is shown in figure 9.15.

The HPLWR utilizes a three-pass core, and the coolant flow paths are shown in figure 9.16. Analogous to the design of the two-pass CSR1000 core shown in figure 9.12, the three-pass HPLWR core is divided into three sections as illustrated

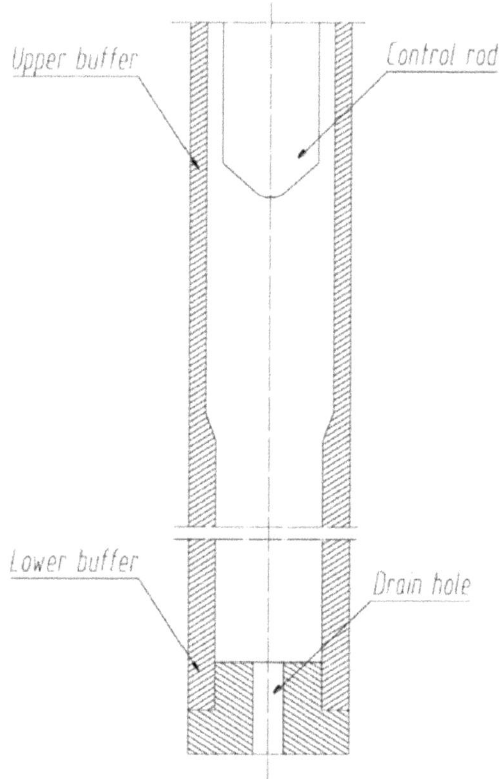

Figure 9.14. Proposed design of the guide thimble for the control rods in the CSR1000 SCWR. Reproduced from Zhu *et al* (2021). CC BY 3.0.

Table 9.4. Properties of the HPLWR. Data adapted from Edwards and Leung (2022), Rahman *et al* (2020), and Wu *et al* (2022).

Property	Value
Reactor type	Pressure vessel
Neutron spectrum	Thermal
Moderator	H_2O
Fuel	UO_2
Fuel enrichment (%)	9.5
Pressure (MPa)	25
Inlet temperature (°C)	280
Outlet temperature (°C)	500
Active core height (m)	4.2
Number of flow passes	3
Thermal power (MW_{th})	2300
Thermal efficiency (%)	43.5

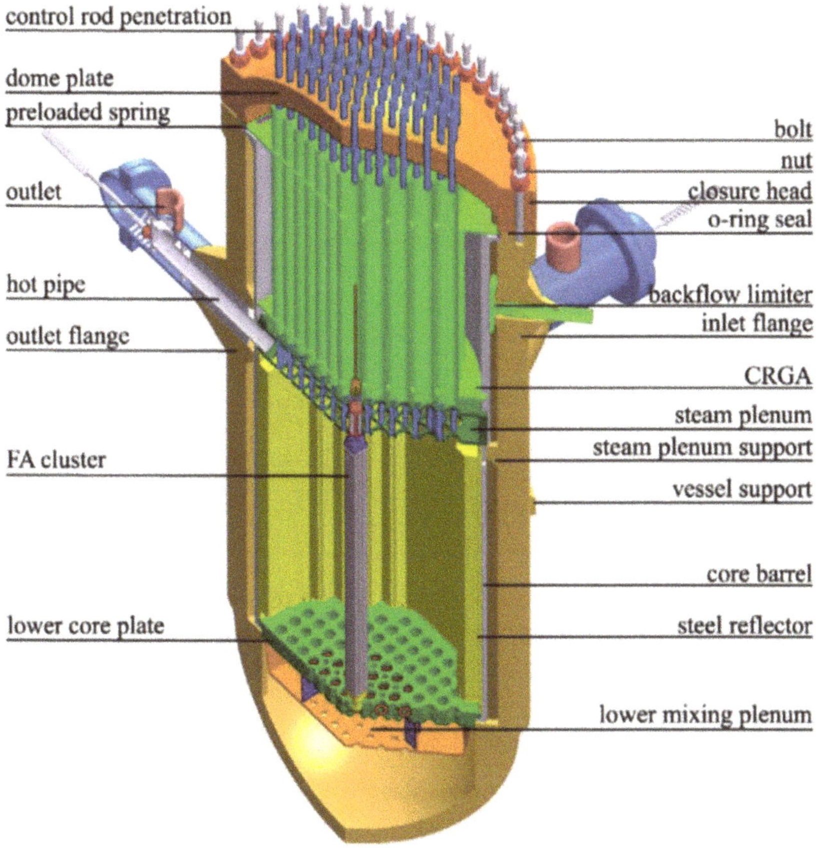

Figure 9.15. Reactor core of the European Union HPLWR. Reprinted from Fischer *et al* (2009), Copyright (2009), with permission from Elsevier.

in figure 9.17. As shown in figure 9.16, the inlet cooling water is split between the downcomer, which directs coolant to the core inlet at the bottom of the reactor, and the upper plenum, which directs coolant to the moderator boxes (labeled water boxes in the figure) from the top. After the first pass of flowing upward through the initial heat transfer stage (the evaporator), the cooling water is directed downward through the core for the second pass, and then upward for the third pass. From the core outlet, the supercritical water is directed to the turbine.

Each fuel assembly cluster, as shown in figure 9.18, has nine sub-assemblies. Each sub-assembly has 40 fuel rods surrounding a central water rod, which acts as the light-water moderator. Fuel rods are wire-wrapped to ensure sufficient space between the rods to allow for adequate cooling water flow through the core. Unlike the cruciform rods in the CSR1000 reactor which are inserted between the fuel assemblies, the control rods in the HPLWR reactor are inserted inside the moderator boxes from the reactor top (Schulenberg 2013, Schulenberg *et al* 2014).

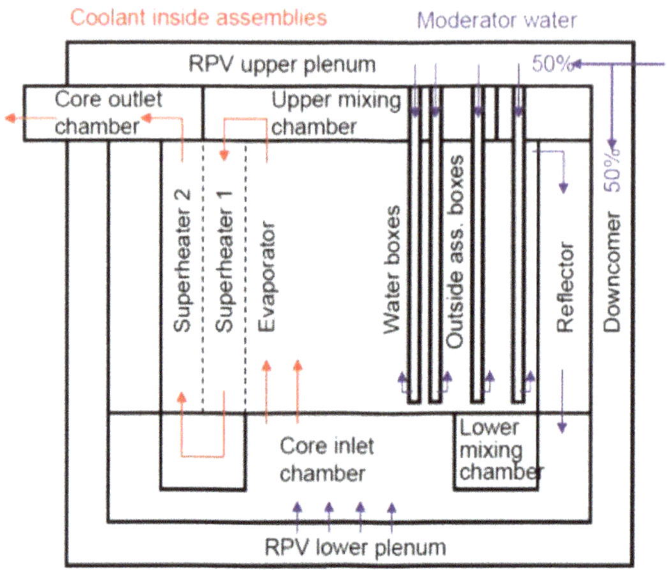

Figure 9.16. Coolant flow path inside the core of the HPLWR. Reproduced from Wu *et al* (2022). CC BY 4.0.

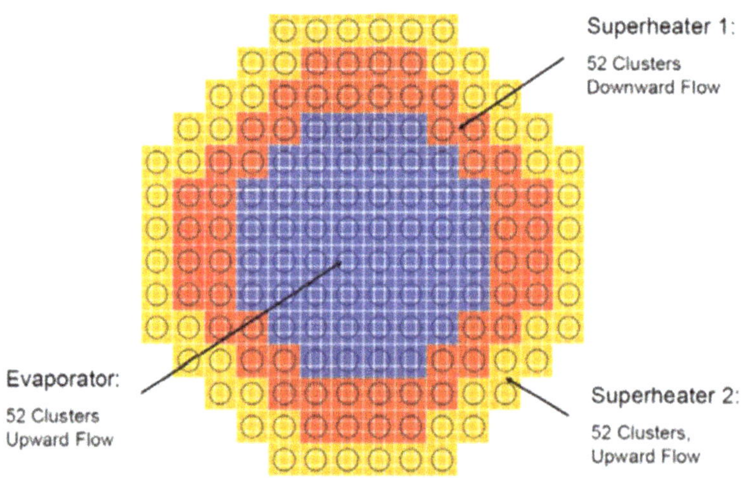

Figure 9.17. Arrangement of clusters of fuel assemblies in the HPLWR core. Reproduced from Wu *et al* (2022). CC BY 4.0.

9.8 Super FR (Japan)

The University of Tokyo has initiated studies of two SCWR concepts: the Super LWR (light water reactor) and the Super FR (fast reactor). The Super LWR is a thermal reactor along the lines of the Chinese CSR1000 and European Union HPLWR, as described in the previous two sections, and the Super FR is a fast neutron spectrum

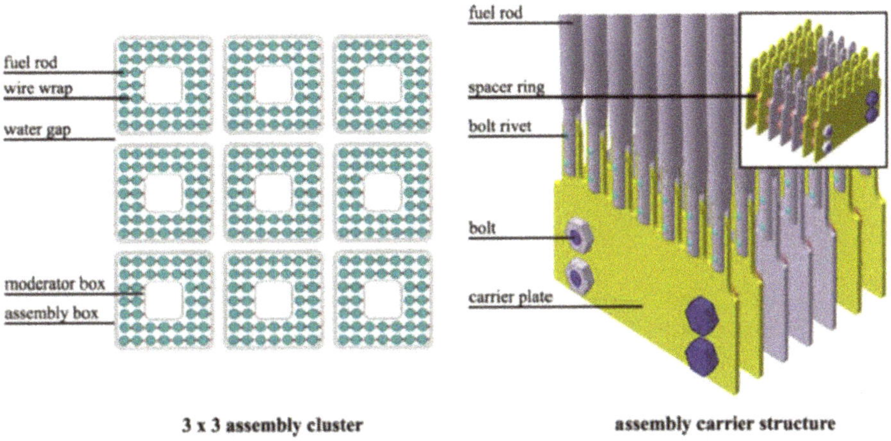

Figure 9.18. Fuel assembly cluster of nine sub-assemblies (left-hand panel) and assembly carrier structure (right-hand panel). Reprinted from Fischer *et al* (2009), Copyright (2009), with permission from Elsevier.

Table 9.5. Properties of the Super FR. MOX = mixed oxide fuel, * = see discussion in text. Data adapted from Edwards and Leung (2022), Rahman *et al* (2020), and Wu *et al* (2022).

Property	Value
Reactor type	Pressure vessel
Neutron spectrum	Fast
Moderator*	$ZH_{1.7}$
Fuel	MOX
Fuel enrichment (%)	—
Pressure (MPa)	25
Inlet temperature (°C)	280
Outlet temperature (°C)	501
Active core height (m)	2.4
Number of flow passes	1
Thermal power (MW_{th})	2337
Thermal efficiency (%)	43

reactor, which is described in the present section. Both Japanese SCWR designs have been described in detail by Oka *et al* (2010), and Oka and Mori (2014).

The basic Super FR reactor properties are summarized in table 9.5. Both one-pass coolant flow and two-pass coolant flow have been considered as possible Super FR designs. The current design utilizes a one-pass core with a single upward flow of coolant, as shown in figure 9.19. The Super FR is designed as a fast breeder reactor with a seed and blanket geometry, where the control rods are inserted from the top into the seed assembly. In order to utilize fast spectrum neutrons, there are no moderator boxes in the core and supercritical cooling water only flows through narrow channels in the tightly packed core, as shown in figure 9.20.

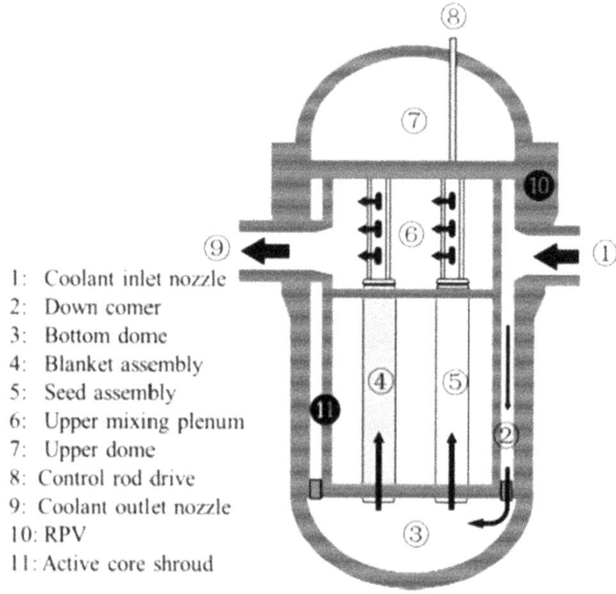

Figure 9.19. Design of the seed and blanket one-pass Super FR. Reprinted from Liu and Oka (2015), Copyright (2015), with permission from Elsevier.

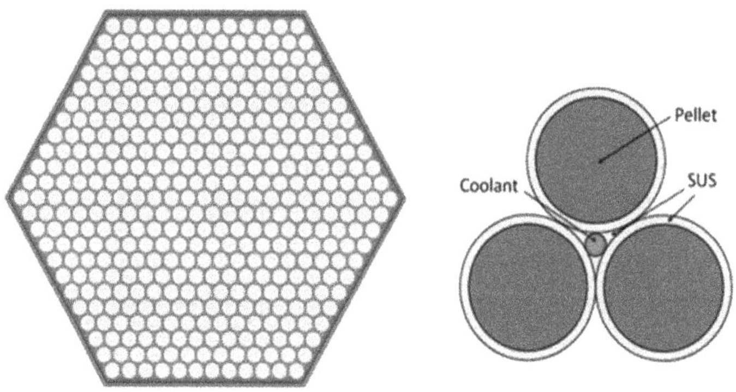

Figure 9.20. Design of the Super FR supercritical water-cooled fast neutron reactor core showing cooling channels. Reproduced from Oka (2014), with permission from Springer Nature.

The seed fuel assemblies are made of a total of 252 fuel rods containing MOX fuel, along with spaces for the control rods (see figure 9.21(a)). The blanket fuel assemblies contain 547 fuel rods. The top 200 cm of the blanket fuel assemblies contain depleted uranium dioxide with 0.2% ^{235}U and the remaining bottom portion of the blanket fuel assemblies contain 10% ^{239}Pu in MOX fuel, as shown in figures 9.21(b) and (c) (Edwards and Leung 2022). Figure 9.21(b) also shows that the upper portion of the blanket fuel assemblies includes a $ZrH_{1.7}$ moderator layer outside of the breeder region. This zirconium hydride layer slows neutrons coming from the seed fuel assemblies, so they are at energies below the ^{238}U fission threshold

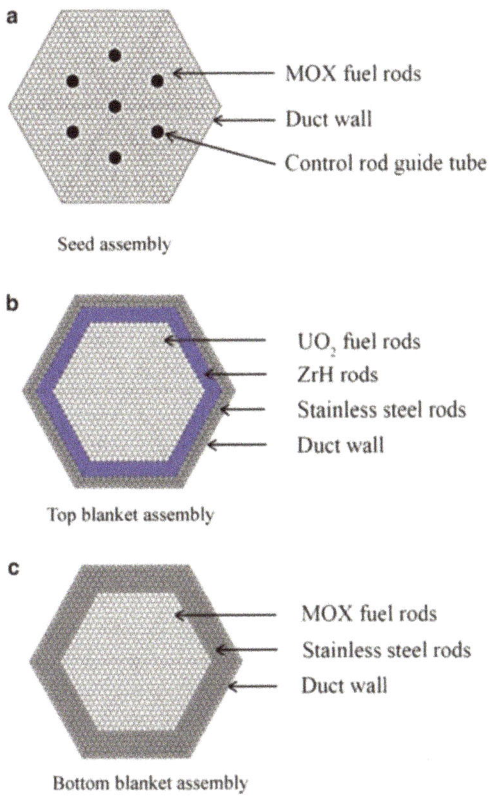

Figure 9.21. Details of the construction of (a) the seed fuel assembly, (b) the upper part of the blanket fuel assembly, and (c) the lower part of the blanket fuel assembly of the Super FR, as described in the text. Reproduced from Oka (2014), with permission from Springer Nature.

of about 1.4 MeV before they interact with the fertile breeder fuel to ensure that they breed new fissile ^{239}Pu fuel rather than inducing fission in ^{238}U (see figure 3.4 and discussion in Dunlap 2023). This also ensures that the reactor has a negative cooling void reactivity, thereby ensuring that fluctuations in the flow of the supercritical cooling water do not lead to operational instabilities. Reactor behavior for different core configurations has been modeled by Oka *et al* (2010).

References

Björk K I, Fhager V and Demazière C 2011 Comparison of thorium-based fuels with different fissile components in existing boiling water reactors *Prog. Nucl. Energy* **53** 618–25

Dunlap R A 2023 *An Introduction to the Physics of Nuclei and Particles section 12.2* (Bristol: IOP Publishing)

Edwards G and Leung L 2022 *GIF Supercritical Water-cooled Reactor—Proliferation Resistance and Physical Protection White Paper* GIF/PRPPWG/2002/002 www.gen-4.org/gif/upload/docs/application/pdf/2022-04/scwr_prpp_white_paper_april_2022_2022-04-25_14-26-21_623.pdf

Fischer K, Schulenberg T and Laurien E 2009 Design of a supercritical water-cooled reactor with a three-pass core arrangement *Nucl. Eng. Des.* **239** 800–12

Guzonas D and Novotny R 2014 Supercritical water-cooled reactor materials—summary of research and open issues *Prog. Nucl. Energy* **77** 361–72

Kamler J 2010 Supercritical water https://commons.wikimedia.org/wiki/File:Supercritical_H2Olr.jpg

Liu Q and Oka Y 2015 Single pass core design for a super fast reactor *Ann. Nucl. Energy* **80** 451–9

Ma D, Zhou T, Feng X and Huang Y 2018 Research on flow characteristics in supercritical water natural circulation: influence of heating power distribution *Nucl. Eng. Tech.* **50** 1079–87

Morreale A C, Ball M R, Novog D R and Luxat J C 2017 Behavior of transuranic mixed-oxide fuel in a CANDU-900 reactor *Nucl. Technol.* **183** 30–44

Nava-Domínguez A, Onder N, Rao Y and Leung L 2016 Evolution of the Canadian SCWR fuel-assembly concept and assessment of the 64 element assembly for thermalhydraulic performance *CNL Nucl. Rev.* **5** 221–38

Oka Y 2000 Review of high temperature water and steam cooled reactor concepts *Proc. 1st Int. Symp. on SCWR Tokyo (6–8 November 2000)*

Oka Y 2014 Introduction and overview ed Y Oka and H Mori *Supercritical-Pressure Light Water Cooled Reactors* (Tokyo: Springer Japan) ch 1 pp 1–20

Oka Y, Koshizuka S, Ishiwatari Y and Yamaji A 2010 *Super Light Water Reactors and Super Fast Reactors—Supercritical-Pressure Light Water Cooled Reactors* (New York: Springer)

Oka Y and Mori H (ed) 2014 *Supercritical-Pressure Light Water Cooled Reactors* (Tokyo: Springer Japan)

Oka Y and Morooka S 2014 Reactor design and safety ed Y Oka and H Mori *Supercritical-Pressure Light Water Cooled Reactors* (Tokyo: Springer Japan) ch 2 pp 21–248

Rahman M M, Dongxu J, Jahan N, Salvatores M and Zhao J 2020 Design concepts of supercritical water-cooled reactor (SCWR) and nuclear marine vessel: a review *Prog. Nucl. Energy* **124** 103320

Schulenberg T 2013 Material requirements of the high performance light water reactor *J. Supercrit. Fluids* **77** 127–33

Schulenberg T, Leung L K H and Oka Y 2014 Review of R&D for supercritical water cooled reactors *Prog. Nucl. Energy* **77** 282–99

Schulenberg T 2022 *The Fourth Generation of Nuclear Reactors—Fundamentals, Types, and Benefits Explained* (Berlin: Springer)

Vogt D K 2004 Nuclear fission reactors: boiling water and pressurized water reactors ed C J Cleveland *Encyclopedia of Energy* (Amsterdam: Elsevier) pp 333–40

Westinghouse Electric Company 2021 BWR control rod CR 82M-1 www.westinghousenuclear.com/Portals/0/flysheets/BWR%20Control%20Rod%20Blade%20CR%2082M-1.pdf?ver=2021-02-26–100539-197#:~:text=The%20Westinghouse%20BWR%20control%20rod,carbide%20(B4C)%20and%20hafnium

Wu P, Gou J, Shan J, Zhang B and Li X 2014 Preliminary safety evaluation for CSR1000 with passive safety system *Ann. Nucl. Energy* **65** 390–401

Wu Y and Novog D R 2015 Prediction of response of the Canadian super critical water reactor to potential loss of forced flow scenarios *Proc. 7th Int. Symp. on Supercritical Water-Cooled Reactors ISSCWR-7 (Helsinki, 15–18 March 2015)*

Wu P, Ren Y, Feng M, Shan J, Huang Y and Yang W 2022 A review of existing SuperCritical Water reactor concepts, safety analysis codes and safety characteristics *Prog. Nucl. Energy* **153** 104409

Yetisir M, Gaudet M, Pencer J, McDonald M, Rhodes D, Hamilton H and Leung L 2016 Canadian supercritical water-cooled reactor core concept and safety features *CNL Nucl. Rev.* **5** 189–202

Zhu F, Zheng L, Ren Q-Y, Yue T, Pang H, Feng L, Li X, Ran R and Huang S 2021 Optimization of a fuel assembly for supercritical water-cooled reactor CSR1000 *Front. Energy Res.* **9** 678741

Chapter 10

Gas-cooled fast reactors (GFRs)

Chapter 10 considers the design of Generation IV gas-cooled fast reactors. In the 1960s and 1970s, several designs were developed in the United States, Europe, and the former Soviet Union for gas-cooled fast reactors but none was ever constructed. These designs are analyzed in the present chapter because they provide insight into the potential difficulties of realizing such a reactor. The major concern in the design of the gas-cooled fast reactor is the low density, and hence low heat capacity, of the coolant. This means that the design of a passive decay heat removal system is a significant challenge. On the positive side, gas-cooled fast reactors have high output temperature, leading to various cogeneration possibilities. In addition, the fuel cycle provides good fuel burn-up, thereby providing good sustainability and minimal radioactive waste. Several Generation IV gas-cooled reactor designs are in progress and representative projects from the European Union and the United States are described in this chapter.

10.1 Introduction

This chapter, along with chapters 11 and 12, present the Generation IV concepts for fast reactors. They differ in the coolant that is used. Here, we consider gas-cooled fast reactors (GFRs). These reactors are the least developed of all of the six Generation IV reactor concepts. While there was considerable interest in this technology between around 1960 and 1980, and several design projects were undertaken, no functioning gas-cooled fast reactor was ever constructed. In 2002, the Generation IV reactor initiative revived interest in these reactors. At present, there are several projects in progress and some of these are discussed in the present chapter.

10.2 History and general design of gas-cooled fast reactors

In the past, there was interest in the development of fast breeder reactors as a means of closing the nuclear fuel cycle. The majority of this work dealt with

doi:10.1088/978-0-7503-6069-2ch10

sodium cooled reactors, as described in detail in the next chapter. Research into a gas-cooled version of a fast breeder reactor was largely motivated by the fact that the lower neutron absorption cross section of the gas coolant compared to a liquid coolant provided a harder (more energetic) neutron spectrum. This, in turn, yielded a higher breeding ratio and a shorter doubling time (see section 3.8). During this early period of interest in gas-cooled fast reactors, several gas coolants were considered, including steam, helium, carbon dioxide, and N_2O_4. Overviews of gas-cooled fast reactors from a historic point of view have been presented by van Rooijen (2009) and Weaver (2012) and several of the more significant designs are discussed below.

10.2.1 General Atomics (United States)

The first gas-cooled fast reactor design project was announced in the United States in 1962 by General Atomics. Designs for a 300 MW_e demonstration reactor, as well as a 1000 MW_e commercial-scale reactor, were considered. The reactors were helium cooled and used $(U,Pu)O_2$ fuel with a highly enriched core and a fertile breeder blanket. Rod-type (sometimes referred to as pin-type) fuel was contained in stainless steel tubing that had its surface roughened to improve heat transfer to the helium gas (see section 10.2.2). The core design was based largely on contemporary liquid metal-cooled fast breeder reactor designs. In 1973, General Atomics announced plans to construct the 300 MW_e demonstration plant, to be operational by 1983, according to the design specification given in table 10.1. However, there were concerns of sufficient decay heat removal in the event of a failure of the coolant circulating blowers. In order to utilize natural convection in the event of such an incident, the coolant flow direction through the core was changed from upward to downward. By the early-1980s, however, the project was discontinued.

10.2.2 The gas breeder reactor association (Europe)

In Europe, an international association, the gas breeder reactor association, was formed to design a gas-cooled fast reactor. In all, the association produced four reactor designs between 1970 and 1974, designated GBR-1 through GBR-4.

Table 10.1. Design specifications for the General Atomics 300 MW_e demonstration gas-cooled fast reactor. Data adapted from van Rooijen (2009).

Parameter	Value
Power, electric (MW_e)	300
Power, thermal (MW_{th})	835
Core height (m)	1.0
Core diameter (m)	2.0
Inlet coolant temperature (°C)	323
Outlet coolant temperature (°C)	550
Coolant pressure (MPa)	8.5

Table 10.2. Summary of the basic design characteristic of the last three reactors designed by the Gas Breeder Reactor Association. CP = coated particle. Data adapted from van Rooijen (2009) and Čížek *et al* (2021).

Property	GBR-2	GBR-3	GBR-4
Year	1972	1972	1974
Coolant	He	CO_2	He
Fuel type (driver)	CP	CP	pin
Fuel	$(U,Pu)O_2$	$(U,Pu)O_2$	$(U,Pu)O_2$
Power, thermal (MW_{th})	3000	3000	3450
Power, electric (MW_e)	1000	1000	1200
Inlet temperature (°C)	260	260	260
Outlet temperature (°C)	700	650	560
Coolant pressure (MPa)	12.0	6.0	12.0

A summary of the basic characteristics for the final three reactor designs is given in table 10.2. As noted, these designs used either helium or carbon dioxide as a coolant and had a driver (fissile) core using either pin-type fuel elements or coated particle fuel elements. The latter are similar to the TRISO fuel particles used in very high-temperature reactors (as described in section 7.2). In all gas-cooled designs, a fertile uranium blanket had a traditional rod fuel geometry.

A schematic of the GBR-4 design is shown in figure 10.1. The reactor vessel contains the reactor core, along with the steam generator, coolant circulating pump, and helium purifier. The details of the fuel assembly are shown in figure 10.2. Each fuel rod contains a central section for the fissile driver fuel and end sections on both the top and bottom that contain the fertile uranium breeder fuel. The image in the lower right-hand corner of the figure shows the design of the roughened surface of the fuel rod tube that is designed to assist with heat transfer between the fuel and the helium coolant. This increases the surface to volume ratio of the fuel rod, analogous to putting fins on a radiator. No gas breeder reactor was ever constructed, and the program was discontinued in the early-1980s.

10.2.3 The former Soviet Union

Around 1970, researchers in the former Soviet Union proposed a design for a gas-cooled fast reactor using dinitrogen tetroxide (N_2O_4) as the coolant. Dinitrogen tetroxide has a melting temperature of −11.2 °C, a boiling temperature of 21.69 °C (at 1 atmosphere) and a density of 1.442 46 g cm^{-3} at 21 °C. The coolant pressure was between 16 MPa and 25 MPa, slightly higher than other gas-cooled fast reactors of that era and the outlet temperature was 677 °C. At these operating conditions, the dinitrogen tetroxide dissociates in the reactor core and then recombines when passing through the heat exchanger. This process is represented by the two reactions,

$$N_2O_4 \leftrightarrow 2NO_2 \leftrightarrow N_2 + 2O_2, \quad (10.1)$$

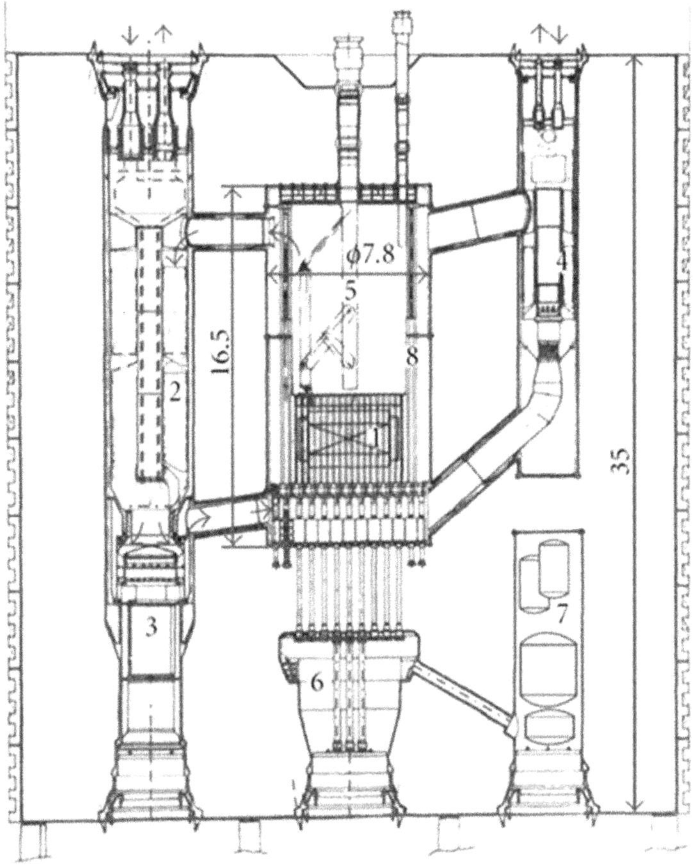

Figure 10.1. Design of the GBR-4 gas-cooled fast reactor. The diagram shows a cross section of the reactor vessel: (1) core, (2) steam generator, (3) gas circulator, (4) emergency cooling loop, (5) fuel manipulator, (6) control cavity, (7) helium purifier, and (8) radiation shield. Reproduced from van Rooijen (2009). CC BY 4.0.

which are endothermic from left to right. Utilizing this dissociation/recombination process has some advantages,

- The combined effects of evaporation and the reaction in equation (10.1) in the reactor core, absorbs additional heat,
- Condensing the coolant in the heat exchanger releases this additional heat, and
- Liquefying the coolant in the heat exchanger reduces the necessary pumping power to return the liquid to the reactor core.

One unfortunate property of dinitrogen tetroxide is that it is very corrosive. In order to deal with this feature of the coolant, chromium dispersion fuel elements were developed. In these, uranium or uranium dioxide fuel is dispersed in a chromium matrix (which is resistant to the corrosive action of dinitrogen tetroxide). Like the gas-cooled fast reactor development programs described above, the Soviet reactor was never constructed and the program ended in the early-1980s.

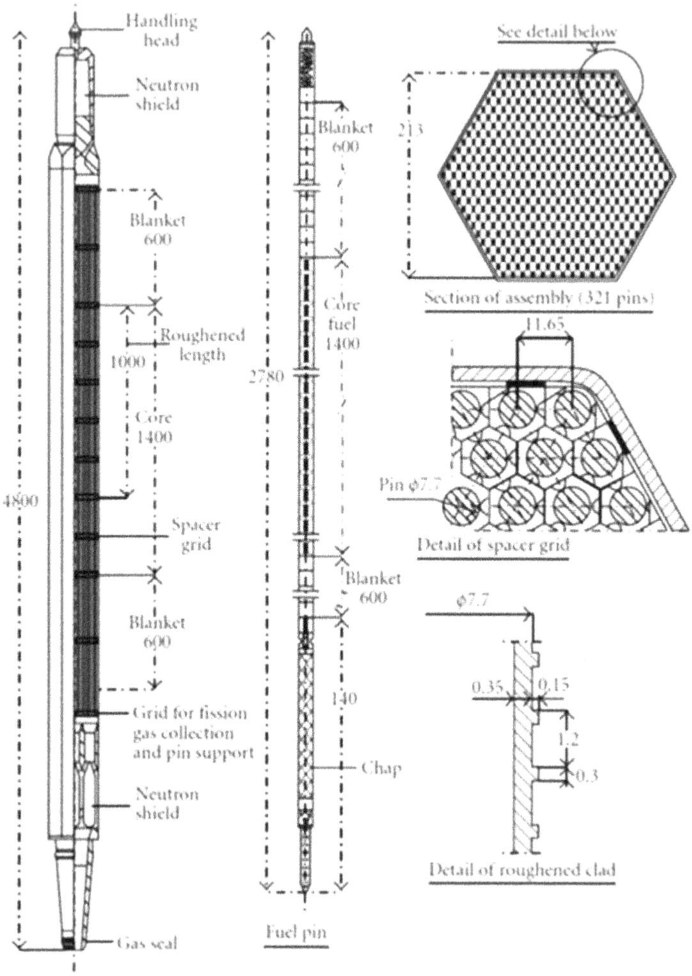

Figure 10.2. Design of fuel element for the GBR-4 gas-cooled fast reactor. (Left-hand side) Diagram of the fuel element, (center) diagram of the fuel pin (rod), (top right-hand corner) design of the reactor core fuel assembly arrangement and (bottom right-hand corner) details of roughening on surface fuel rod tube. Dimensions are in mm. Reproduced from van Rooijen (2009). CC BY 4.0.

10.3 Generation IV gas-cooled fast reactor development

Initial interest in gas-cooled fast reactors stemmed from their ability to produce electricity as well as heat for industrial applications. However, their ability to close the fuel cycle by breeding fissile fuel from fertile material more rapidly than liquid metal-cooled breeder reactors was a major advantage of their gas coolant. This characteristic provided a clear path to greater fuel utilization and radioactive waste reduction. However, as noted above, none of the early programs to develop a gas-cooled fast reactor produced an operational device. The lack of suitable high-temperature structural materials, an appropriate design for fuel elements, and passive safety characteristics all factored against a viable design.

In the early-2000s, the Generation IV forum considered the possibility of an advanced gas-cooled fast reactor. Sustainability and economic operation were potential strong points in its design, and materials advances and research on passive reactor safety subsequent to earlier design attempts made the gas-cooled fast reactor an attractive possibility (see Hervieu *et al* 2022, Idaho National Laboratory 2017, Schulenberg 2022, Tsvetkov 2023, Weaver 2012, Weaver *et al* 2005).

Two different coolants have been considered for possible Generation IV gas-cooled fast reactors: helium and supercritical carbon dioxide. Both coolants have their own advantages and disadvantages, as discussed below.

Helium: The use of helium as a coolant in a gas-cooled fast reactor has many similarities to its use as a coolant in very high-temperature reactors, as discussed in chapter 7. Thus, there has already been considerable research into many aspects of helium-cooled reactors. Helium has a very low neutron absorption cross section, leading to a harder neutron energy spectrum. This, as noted above, provides a higher breeding ratio and a shorter doubling time. In addition, helium is inert and has negligible corrosive effects on reactor components. On the other hand, the faster neutron spectrum leads to greater neutron loss from the core and reduced importance of interactions in the (n, γ) region of the spectrum (see section 3.3). This latter factor reduces the effects of thermal Doppler broadening and, subsequently, yields a less negative temperature coefficient of reactivity (see section 3.7). This characteristic must be considered in the design of suitable safety features. The low heat capacity of helium means that the coolant must be under significant pressure (i.e., 8–10 MPa) in order to improve heat transfer from the core. In the event of a loss of coolant pressure in the core, the helium offers relatively poor removal of decay heat. Thus, appropriate safety systems must be in place in order to avoid overheating of the core.

Carbon dioxide: Contemporary Generation IV designs of gas-cooled fast reactors that utilize carbon dioxide as a coolant use this material in the supercritical phase. Figure 10.3 shows the phase diagram of carbon dioxide. The critical point is at a temperature of 31.0 °C and a pressure of 7.38 MPa. Since pressures in the range of 20–25 MPa, are needed to obtain suitably high thermal efficiencies, the carbon dioxide is naturally in its supercritical state (Idaho National Laboratory 2017, Weaver 2012). The advantages of carbon dioxide over helium as a coolant are that the negative thermal coefficient of reactivity is improved and high thermal efficiency can be maintained at a lower outlet temperature. However, core heat removal in the event of loss of coolant pressure is a concern that still needs to be addressed.

The previous section described the varieties of fuel geometries that have been considered in the past for a viable gas-cooled fast reactor. These are fuel pins (rods), coated fuel particles and fuel dispersions, and these are all potential candidates for advanced gas-cooled fast reactors as well.

Fuel rods follow along the designs described for earlier reactor projects. These typically include fertile (i.e., breeder) sections above and below a central fissile fuel section. These breeder blanket sections at the top and bottom of the core are referred to as axial breeder blankets.

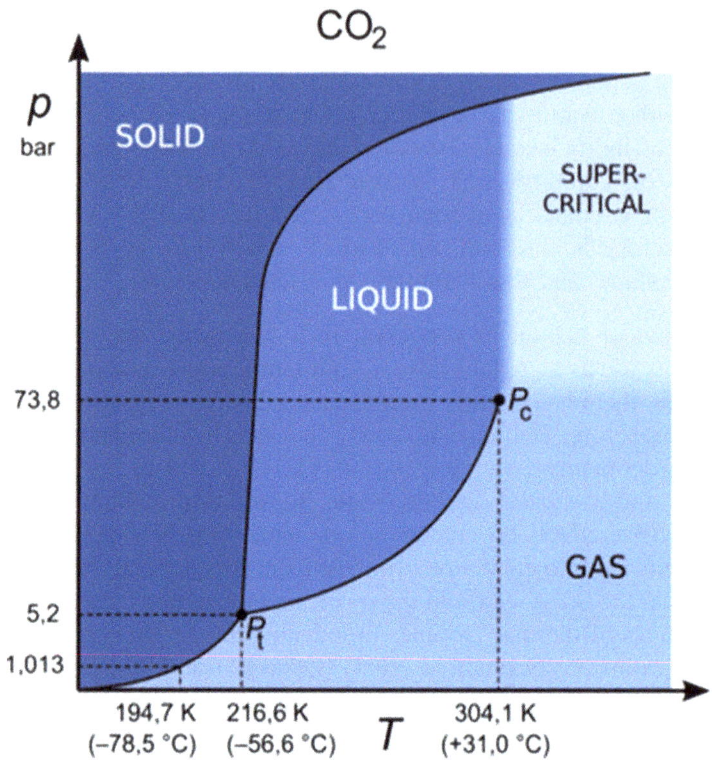

Figure 10.3. Phase diagram of carbon dioxide (CO_2). This colored phase diagram of carbon dioxide (not to scale) image has been obtained by the author from the Wikimedia website where it was made available by Rothwild (2016). Image stated to be in the public domain. It is included within this chapter on that basis. It is attributed to Rothwild.

Coated fuel particles are similar in geometry to the TRISO fuel particles described in detail in chapter 7 and fuel in a dispersion matrix has been discussed above with regard to the Soviet reactor design. In general, suitable properties for coating materials and matrix materials include,

- High temperature resistance
- Low neutron absorption cross sections
- Reasonable economic cost
- Minimal neutron moderation properties

Materials that have attracted interest as possible coating and matrix materials include Nb-based alloys and silicon carbide (SiC). Some typical designs of coated particle and dispersion fuel assemblies are illustrated in figures 10.4 and 10.5, respectively.

Typical core geometries for gas-cooled fast reactors include regions of fissile fuel and also breeder blankets. An example is shown in figure 10.6. The fissile core includes control rods and is surrounded by a reflector. The breeder blankets shown in the radial cross section are referred to as radial blankets. The fissile core may

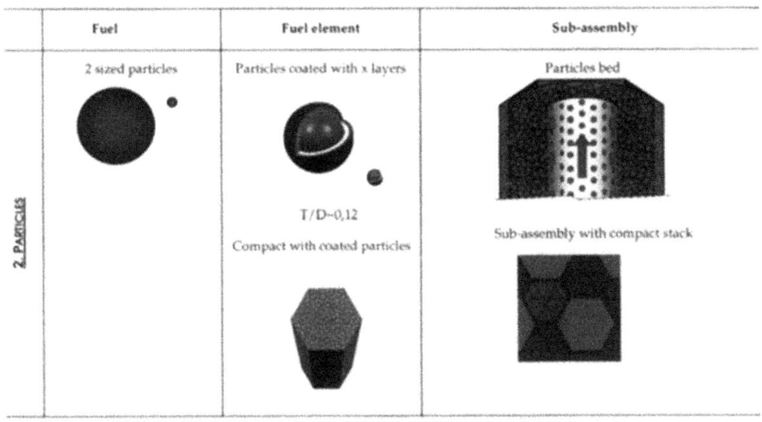

Figure 10.4. Possible designs for particle fuel in a gas-cooled fast reactor. Reproduced from Weaver (2012). CC BY 4.0.

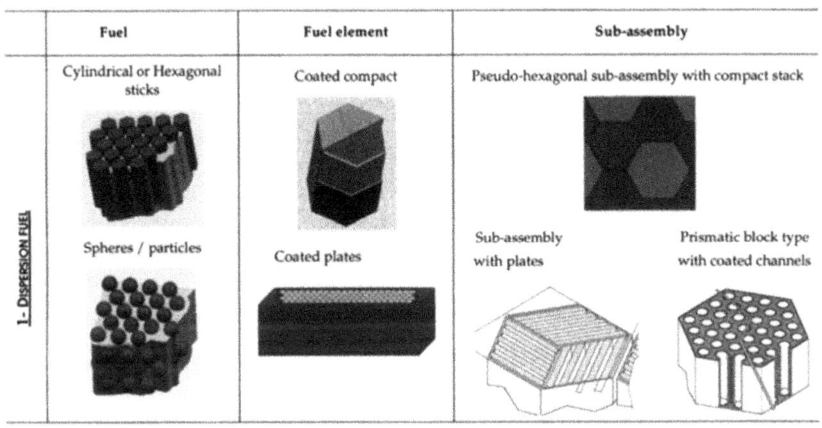

Figure 10.5. Possible designs for fuel dispersion in plates and blocks for a gas-cooled fast reactor. Reproduced from Weaver (2012). CC BY 4.0.

also include fuel assemblies with axial breeder blankets, as shown in the axial cross section.

The gas-cooled fast reactor has several significant design challenges. First, the low neutron absorption cross section of the coolant means greater neutron loss from the core, and this in turn requires a greater amount of fissile fuel and corresponding concerns about nuclear proliferation. Second, the harder neutron spectrum leads to a less negative temperature coefficient of reactivity and potential core overheating. Finally, there need to be provisions for decay heat removal in the event of loss of coolant pressure. While no Generation IV gas-cooled fast reactor has been constructed to date, there are a number of design projects that are currently underway. In sections 10.4 and 10.5, we consider two of these reactor designs.

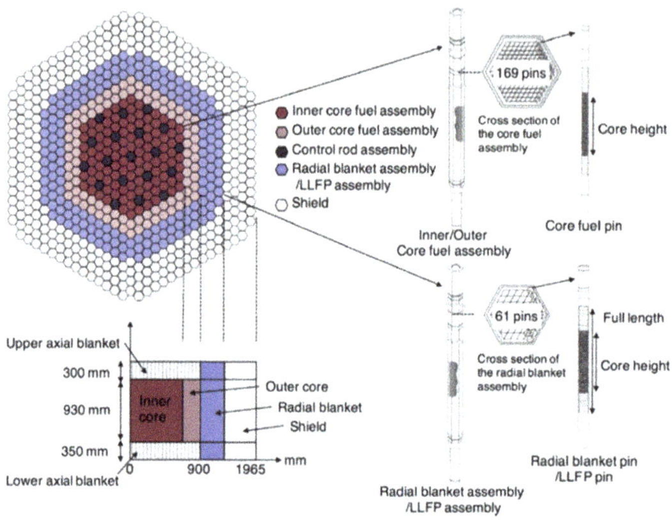

Figure 10.6. (Top) Radial cross section of typical fast reactor core, (bottom) axial cross section of the core, and (right-hand side) hexagonal fuel assembly. Reproduced from Chiba *et al* (2017). CC BY 4.0.

10.4 ALLEGRO

ALLEGRO is an experimental helium-cooled fast reactor being developed by the V4G4 (Visegard 4 for Generation 4) consortium formed by research organizations from the Czech Republic, Hungary, Poland, and Slovakia in association with Commissariat à l'Énergie Atomique et aux Énergies Alternatives, France (see Poette *et al* 2009, Stainsby *et al* 2009, Hervieu *et al* 2022).

ALLEGRO will serve as a 75 MW_{th} demonstration gas-cooled fast reactor and also as a means of testing materials and components in anticipation of the design of a 2400 MW_{th} commercial-scale reactor. There are no plans for electricity Generation by ALLEGRO. The specific goals of the ALLEGRO project are to demonstrate the viability of a commercial gas-cooled fast reactor and to address particular design features. These features include the fuel and design of the fuel elements, as well as safety systems, such as an appropriate decay heat removal system.

The official launch of the project occurred in 2010 and the V4G4 Centre of excellence was established in 2013. The project consists of three phases, a preparatory phase, a realization phase, and an operational phase. The first two phases are subdivided as follows:

Preparatory Phase (2015–25)

- Definition of goals and specifications (2015–16)
- Preconceptual design (2017–20)
- Conceptual design (2021–25)

Realization phase (2025–)

- Basic reactor design
- Detailed reactor design

- Licensing and site selection
- Construction

Figure 10.7 shows a diagram of the present ALLEGRO reactor design. The core is in the center of the image. Three decay heat removal loops (gas-water heat exchangers) are located above the core. Two main heat exchangers (gas-water) are located to the sides of the core with helium circulation blowers attached below. The red module is a prototype (gas-gas) heat exchanger for testing (Orosz *et al* 2022). A major concern for helium-cooled fast reactors is the potential loss of pressure in the

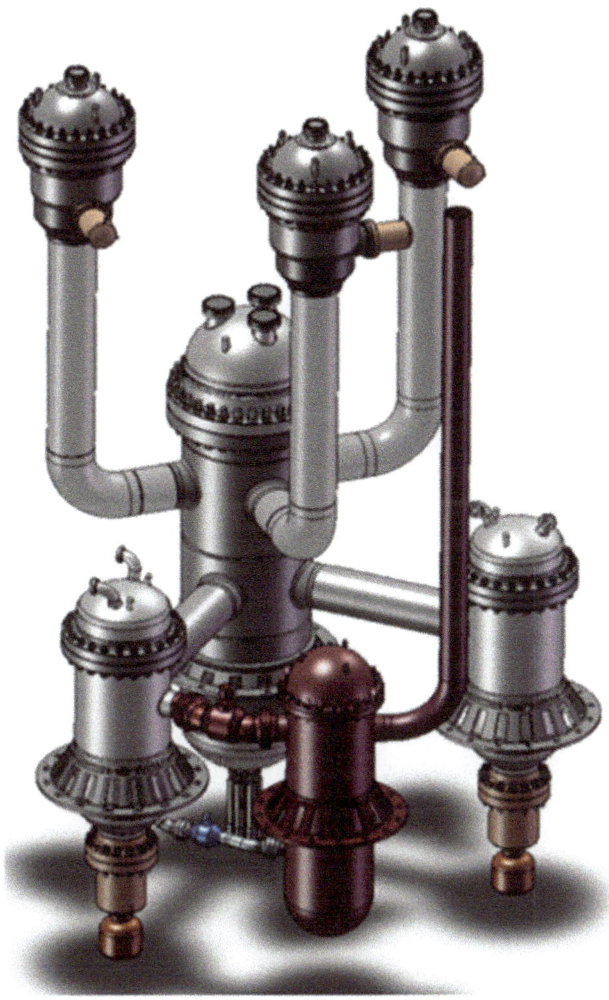

Figure 10.7. Diagram of the ALLEGRO reactor. Reproduced from Gadó (2014). CC BY 4.0.

cooling system and the ability of the reactor to sufficiently eliminate decay heat from the core. As seen in the design in figure 10.7, there are three heat exchangers located above the reactor core to deal with the removal of the decay heat in the event of a primary system cooling loop depressurization. It is essential to ensure that the reactor temperature under such a situation does not exceed the safe operating conditions of the reactor materials. The three decay heat removal heat exchangers are located above the reactor core to provide cooling by means of natural convection. However, natural convection is not sufficient to adequately cool the core and forced gas flow is achieved through the use of electrically driven blowers. If the primary cooling loop depressurizes completely (i.e., to atmospheric pressure), then the decay heat removal system would not be sufficient to adequately cool the core. In order to deal with this problem, the entire reactor is contained within a so-called guard vessel, as shown in figure 10.8. In the event of a depressurization of the cooling system, the helium gas that would be released would increase the pressure inside the guard vessel and prevent the coolant pressure from dropping completely to atmospheric pressure. The guard vessel is contained within a conventional concrete containment building, which acts as a barrier to prevent the spread of radioactive materials to the environment.

The core of the ALLEGRO reactor consists of pin-type fuel assemblies. The design of the fuel rod is shown in figure 10.9. As the figure shows, there are reflector regions above and below the central fuel region of the rod. A major feature of the ALLEGRO reactor is the ability to utilize different fuel configurations. In fact, ALLEGRO is planning to investigate specific fuel designs.

The first core to be investigated will consist of MOX fuel enriched to 18.5%–19.5% that is contained in stainless steel tubes. In this configuration, the reactor will have a moderate outlet temperature, 560 °C, which is compatible with standard reactor materials. The fuel assembly design for the MOX core is shown on the left-hand side of figure 10.10. The fuel rods are wire-wrapped to ensure that there is sufficient space between them for the flow of helium coolant.

The second approach to core testing of the ALLEGRO reactor uses the same MOX core with some of the MOX fuel assemblies replaced with experimental assemblies. This arrangement of the core is illustrated in figure 10.11. The goal is to investigate new materials for fuel assemblies that can withstand high operating temperatures, as well as the high energy neutron flux expected in a fast reactor. A typical test fuel assembly is shown on the right-hand side of figure 10.10. The experimental fuel assemblies contain refractory ceramic composites where the fuel is in the form of (U, Pu)C clad with SiC_f–SiC. This cladding consists of silicon carbide fibers (SiC_f) reinforced by amorphous silicon carbide (SiC).

The third ALLEGRO core design will use only refractory ceramic fuel assemblies. This will be designed on the basis of results from phase two of the reactor tests. It is anticipated that fuels will consist of (U, Pu)C with a SiC_f-SiC composite cladding. This reactor configuration will operate with an outlet temperature of 850 °C and will be a demonstration for the design of the 2400 MW_{th} commercial-scale gas-cooled fast reactor.

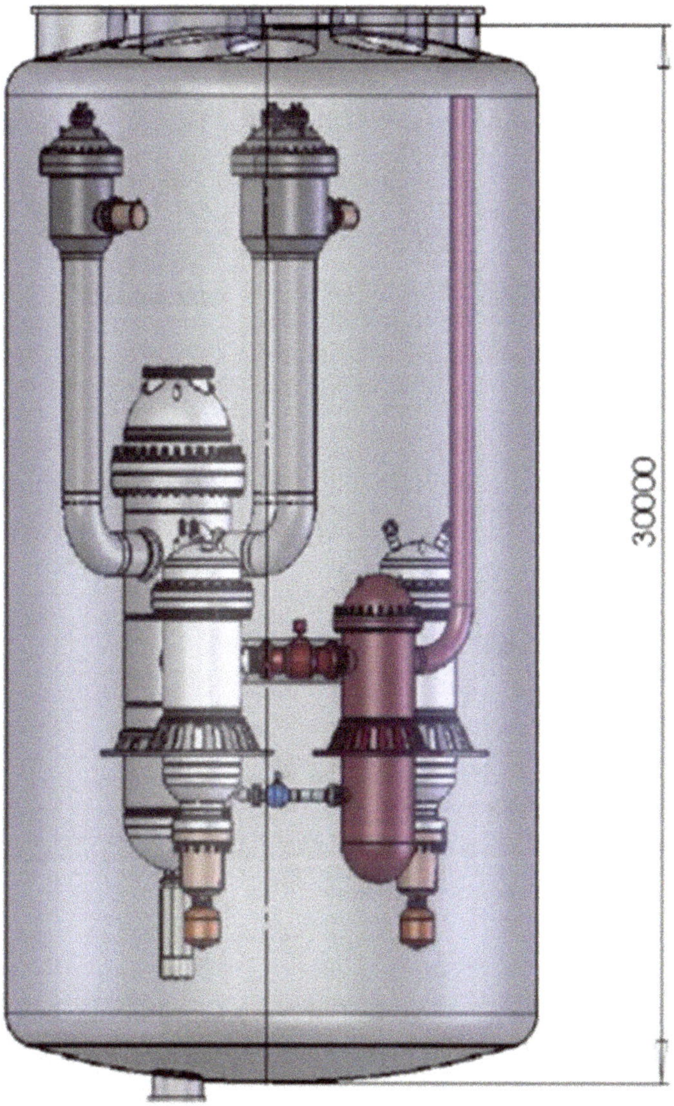

Figure 10.8. Diagram of the ALLEGRO reactor inside its guard vessel. Reprinted from Stainsby *et al* (2011), Copyright (2011), with permission from Elsevier.

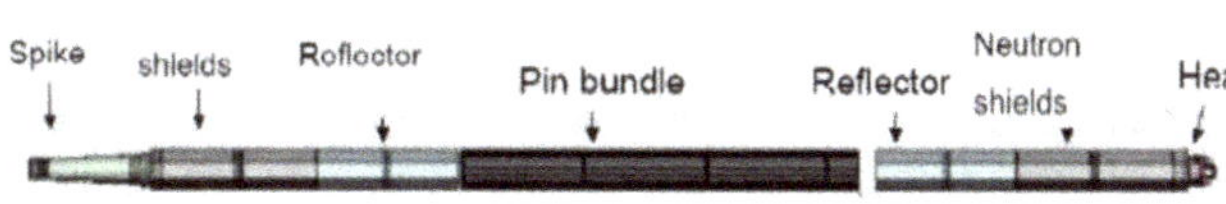

Figure 10.9. Diagram of the ALLEGRO fuel rod. Reproduced from Gadó (2014). CC BY 4.0.

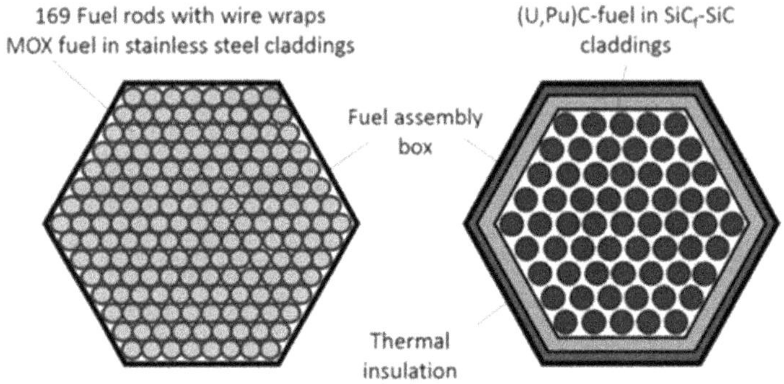

Figure 10.10. Design of the fuel assembly for the ALLEGRO reactor. (Left-hand side) MOX fuel rods in stainless steel tubes and (right-hand side) (U,Pu)C fuel with SiC cladding as described in the text. Reproduced from Schulenberg (2022). CC BY 4.0.

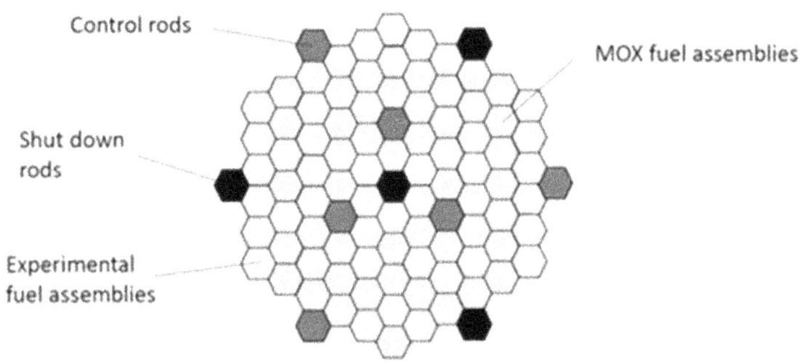

Figure 10.11. Layout of the fuel assemblies in the ALLEGRO reactor core showing fuel assemblies (light gray), control rods (dark gray), shutdown rods (black), and experimental fuel assemblies (white). Reproduced from Schulenberg (2022). CC BY 4.0.

10.5 EM^2

The Energy Multiplier Module (EM^2 or EM2) is a helium-cooled fast reactor that has been designed to produce 500 MW$_{th}$ and 265 MW$_e$. It is being developed by General Atomics (U.S. Department of Energy 2012, 2014, Choi and Schleicher 2017, Idaho National Laboratory 2017, General Atomics 2024). The core contains starter (seed) fissile fuel and fertile blanket fuel. The fissile fuel bred from the fertile blanket by the neutrons produced by the starter fuel is burned in the reactor. This configuration is sometimes referred to as a ‘convert-and-burn’ core. The designation ‘traveling wave reactor’ is also sometimes used as the region of energy production moves from the fissile starter through the blanket as the reactor ages. In this configuration, the core has a long life (i.e., 30 years) without refueling.

Figure 10.12 shows a diagram of the reactor vessel for the EM^2 reactor. This figure illustrates once through cooling, where the helium coolant enters through an outside annulus and flows upward through the core. The outlet temperature is designed to be around 850 °C. Control rods are inserted from the top of the reactor.

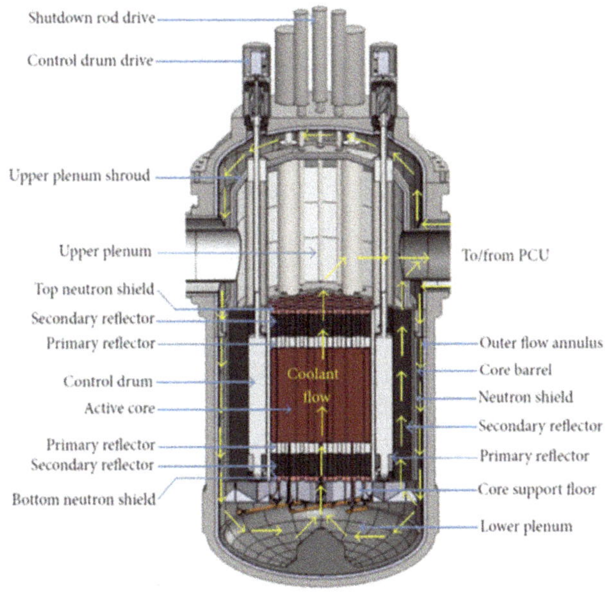

Figure 10.12. Reactor vessel of the EM^2 reactor. The vessel is 4.7 m in diameter and 10.6 m high. Reproduced from Choi *et al* (2013). CC BY 3.0.

The core contains a starter fuel region consisting of pin-type uranium fuel enriched to 14.5% in ^{235}U and a fertile blanket region containing depleted uranium. The average uranium enrichment for the entire core is 7.7% (Choi and Schleicher 2017).

Figure 10.13 shows a cross section of the EM^2 core with a graphite outer reflector. The core contains pin-type uranium carbide fuel with a high-temperature SiC composite cladding. The core contains 85 hexagonal assemblies, 55 of which are standard fuel assemblies, each with 91 fuel rods. Eighteen assemblies contain control rods, and the remaining 12 assemblies contain shutdown rods. The assemblies with control or shutdown rods each contain 84 fuel rods.

As is the case for other gas-cooled fast reactors, the question of decay heat removal in the event of a loss of coolant pressure event must be considered. The design of the EM^2 reactor deals with this concern, as we saw previously for the ALLEGRO reactor, with the addition of a decay heat removal system above the core of the reactor, as shown in figure 10.14. This figure also shows the modular nature of the EM^2 reactor, where the reactor vessel, the decay heat removal system, and the turbine/generator are modular units in separate compartments of the reactor containment structure. It is also important to note that the fuel rods are vented in order to release pressure build-up from fission gases. This ensures that the fuel rods are maintained at a pressure that is lower than the pressure of the of the surrounding helium coolant, so that in the event of a breach of the fuel cladding there will be an ingress of helium into the fuel rod rather than a release of radioactive material into the primary coolant.

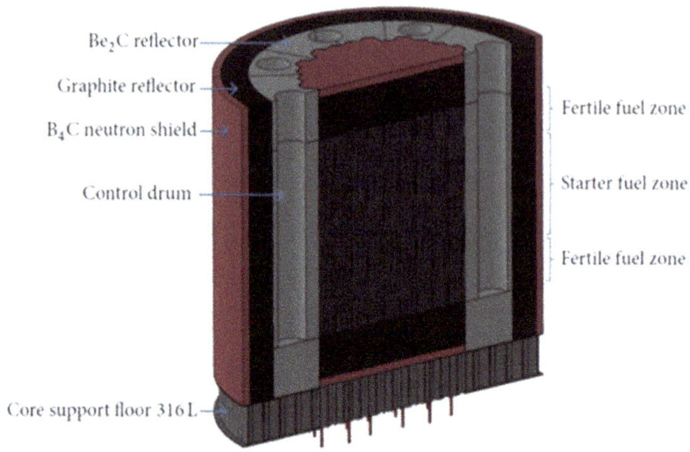

Figure 10.13. Model of the EM^2 core showing the different components as described in the text. The axial reflectors are shown above and below the fuel zones. Reproduced from Choi *et al* (2013). CC BY 3.0.

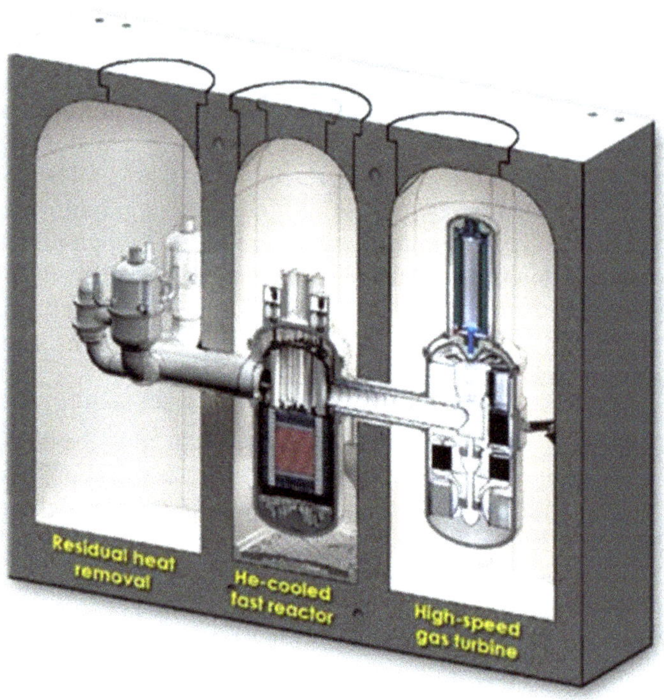

Figure 10.14. Containment structure for the EM^2 reactor showing the reactor vessel (center), residual heat removal system (left-hand side) and direct drive high-speed turbine (right-hand side). Reproduced from U.S. Department of Energy (2012). Image stated to be in the public domain.

The convert-and-burn approach to fuel utilization uses fast neutrons from ^{235}U to breed ^{239}Pu from ^{238}U in the fertile depleted uranium blanket. Thus, over the lifetime of the reactor, the ^{235}U inventory decreases and the ^{239}Pu inventory increases. Figure 10.15 shows the fraction of fission events in the core that result from the different principal fissile nuclides that are present as a function of time. At start-up, the energy production is primarily from fissile ^{235}U and a smaller amount from fast neutron induced fission in ^{238}U (recall uranium fission cross sections in figure 3.4). As time progresses, the amount of starter ^{235}U decreases with a corresponding decrease in the fraction of ^{235}U fission energy. The fission energy from ^{239}Pu, which is bred from fertile ^{238}U according to equations (4.19) to (4.21), increases correspondingly. Fast neutron induced fission in ^{238}U decreases slightly from the burn-up of a small portion of the depleted uranium. There is also a small increasing fission component from ^{241}Pu, which is formed from ^{239}Pu by neutron capture (n, γ) reactions,

$$\mathrm{n} + {}^{239}\mathrm{Pu} \rightarrow {}^{240}\mathrm{Pu}$$

and

$$\mathrm{n} + {}^{240}\mathrm{Pu} \rightarrow {}^{241}\mathrm{Pu}.$$

The convert-and-burn process utilized by the EM^2 makes much better use of natural uranium than typical light-water reactors and results in about a factor of four improvement in fuel utilization (Choi and Schleicher 2017). This has a corresponding decrease in the heavy-metal component of reactor waste. Figure 10.16 shows a comparison of waste per unit energy generated for several different types of reactors.

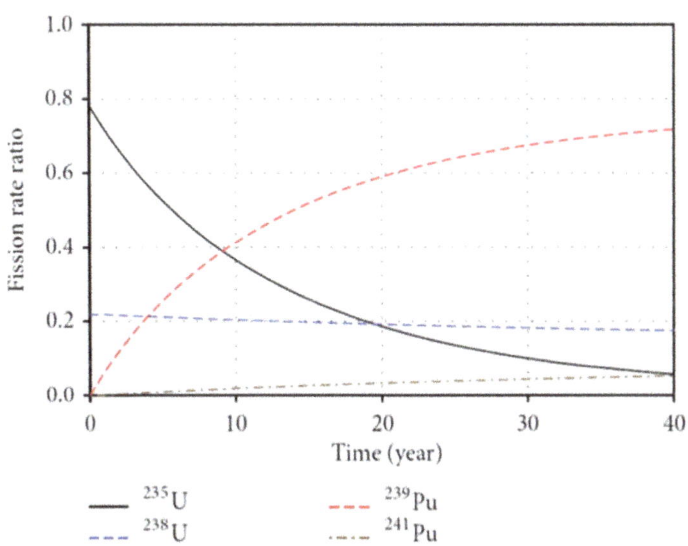

Figure 10.15. Calculated fission rates of major fissile nuclides in the EM^2 core as a function of time after start-up. Reproduced from Choi *et al* (2013). CC BY 3.0.

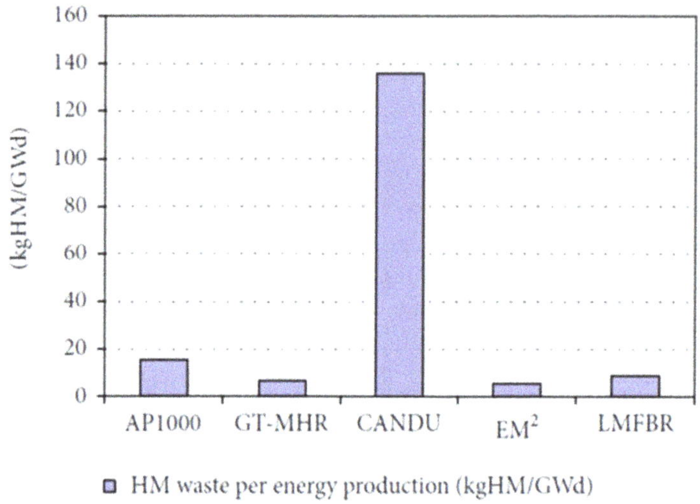

Figure 10.16. Radioactive heavy-metal waste per unit energy generated in (kg GWd^{-1}) for different fission reactor technologies. AP1000 = Advance pressurized water reactor, GT-MHR = gas turbine modular helium reactor, LMFBR = liquid metal-cooled fast breeder reactor. Reproduced from Choi *et al* (2013). CC BY 3.0.

The advantages of the EM^2 reactor are clear. Only fission products need to be removed from spent reactor fuel, leaving a small amount of actinide waste. The fission products will decay to near background in about 500 years, compared to 10 000 years or more (see figure 6.17) from the large amount of actinide waste from a traditional light-water reactor.

As of 2019, the published timescale for the development of the EM^2 reactor (International Atomic Energy Agency 2019) indicated 2029 for the completion of detailed designs, 2030 for the start of construction of a commercial reactor, and 2032 for operation.

References

Chiba S, Wakabayashi T, Tachi Y, Takaki N, Terashima A, Okumura S and Yoshida T 2017 Method to reduce long-lived fission products by nuclear transmutations with fast spectrum reactors *Sci. Rep.* **7** 13961

Choi H and Schleicher R W 2017 The energy multiplier module (EM: status of conceptual design *Nucl. Technol.* **200** 106–24

Choi H, Schleicher R W and Gupta P 2013 A compact gas-cooled fast reactor with an ultra-long fuel cycle *Sci. Technol. Nucl. Install.* **2013** 618707

Čížek J, Kalivodová J, Janeček M, Stráský J, Srba O and Macková A 2021 advanced structural materials for gas-cooled fast reactors—a review *Metals* **11** 76

Gadó J 2014 The reactor ALLEGRO and the sustainable nuclear energy in central Europe *EPJ Web of Conferences* **78** 08001

General Atomics 2024 Advanced Reactors www.ga.com/nuclear-fission/advanced-reactors

Hervieu E, Vasile A and Nguyen F 2022 *GIF Supercritical Water-Cooled Reactor—Proliferation Resistance and Physical Protection White Paper* GIF/PRPPWG/2002/001 www.osti.gov/servlets/purl/1862101

Idaho National Laboratory 2017 *Gas-Cooled Fast Reactor Research and Development Roadmap* INL/EXT-17-41800 https://inldigitallibrary.inl.gov/sites/sti/sti/Sort_1841.pdf

International Atomic Energy Agency 2019 *Status Report—EM2 (General Atomics)* https://aris.iaea.org/PDF/EM2(GeneralAtomics)_2020.pdf

Orosz G I, Magyar B, Szerbák D, Kacz D and Aszódi A 2022 ALLEGRO Gas cooled fast reactor rod bundle investigations with CFD and PIV method *Nucl. Eng. Des.* **400** 112062

Poette C, Brun-Magaud V, Morin F, Pignatel J-F, Stainsby R and Mikityuk K 2009 Allegro: the European gas fast reactor demonstrator project *Volume 2: Structural Integrity; Safety and Security; Advanced Applications of Nuclear Technology; Balance of Plant for Nuclear Applications* (ASMEDC) pp 815–22

Rothwild 2016 Colored phase diagram of carbon dioxide https://commons.wikimedia.org/wiki/File:Colored_phase_diagram_of_carbon_dioxide_(multi_language).svg

Schulenberg T 2022 *The Fourth Generation of Nuclear Reactors—Fundamentals, Types and Benefits Explained* (Berlin: Springer) ch 8, pp 135–46

Stainsby R, Peers K, Mitchell C, Poette C, Mikityuk K and Somers J 2009 Gas cooled fast reactor research and development in the European Union *Sci. Technol. Nucl. Install.* **2009** 238624

Stainsby R, Peers K, Mitchell C, Poette C, Mikityuk K and Somers J 2011 Gas cooled fast reactor research in Europe *Nucl. Eng. Des.* **241** 3481–9

Tsvetkov P 2023 Gas-cooled fast reactors (GFRs) *Handbook of Generation IV Nuclear Reactors* 2nd edn ed I L Pioro (Cambridge, MA: Woodhead Publishing) ch 4 pp 167–72

U.S. Department of Energy 2012 *Advanced Reactor Concepts Technical Review Panel Report—Evaluation and Identification of future R&D on eight Advanced Reactor Concepts, conducted April—September 2012* www.energy.gov/ne/articles/advanced-reactor-concepts-technical-review-panel-report

U.S. Department of Energy 2014 *Advanced Reactor Concepts Technical Review Panel Public Report—Evaluation and Recommendations for Future R&D on Seven Advanced Reactor Concepts, Conducted March through June 2014* www.energy.gov/ne/articles/advance-reactor-concepts-technical-review-panel-public-report

van Rooijen W F G 2009 Gas-cooled fast reactor: a historical overview and future outlook *Sci. Technol. Nucl. Install.* **2009** 965757

Weaver K D 2012 *Gas-Cooled Fast Reactors* Fast *Spectrum Reactors* ed A E Waltar, D R Todd and P V Tsvetkov (Boston: Springer) ch 17 pp 489–511 rights https://link.springer.com/chapter/10.1007/978-1-4419-9572-8_17

Weaver K D *et al* 2005 *Gas-Cooled Fast Reactor (GFR) FY 05 Annual Report* (Idaho National Laboratory Report) INL/EXT-05-00799 https://inldigitallibrary.inl.gov/sites/sti/sti/3480236.pdf

Chapter 11

Sodium-cooled fast reactors (SFRs)

Generation IV sodium-cooled fast reactors (SFRs) will be discussed in this chapter. SFRs have a long history of development, dating back to the late-1950s. In fact, nearly 20 operational reactors have been constructed, which makes SFRs one of the most mature Generation IV technologies. This chapter reviews the general design of an SFR. This is followed by an overview of some of the more notable reactors of this type that have been operational in the past. The operation of these reactors is analyzed to understand the challenges that face the construction of a viable Generation IV reactor. Specifically, the reactivity of liquid sodium and the positive void coefficient that is characteristic of these reactors are difficulties that must be considered. This chapter reviews some current Generation IV SFR projects and discusses the ways in which these reactors form an integral part of India's nuclear energy program.

11.1 Introduction

One of the earliest nuclear reactors, and the world's first fast reactor, was the Los Alamos Fast Plutonium Reactor, commonly known as Clementine (see section 4.5). This reactor utilized ^{239}Pu as a fuel and liquid mercury as a coolant. Clementine first reached criticality in 1946 and operated until 1952. Clementine was not designed to generate electricity but provided valuable experimental information about the operation of a fast neutron reactor and about the neutron cross sections of a wide variety of materials (Jurney *et al* 1954). The operation of Clementine pointed to the need to make use of other liquid metal coolants. Some of the most promising possibilities are summarized in table 11.1. An appropriate coolant must have minimal cross section for neutrons over a wide range of energies but must also have suitable thermal properties, such as heat capacity and thermal conductivity. An important consideration that is summarized in the table is the melting and boiling temperatures of the coolant. Mercury has the clear disadvantage that its boiling temperature is fairly low, which would ultimately result in a low thermodynamic

doi:10.1088/978-0-7503-6069-2ch11

Table 11.1. Melting and boiling temperatures of some potential fast neutron reactor coolants at 1 atmosphere pressure.

Coolant	Melting temperature (°C)	Boiling temperature (°C)	Liquid range (°C)
Mercury	−38.8	+356.7	395.5
NaK	−11	+785	796
Sodium	+97.8	+882.9	786.1
Tin	+231.9	+2602	2370.1
Lead	+327.5	1749	1421.5

Table 11.2. Comparison of properties of sodium and water.

Property	Sodium (Na)	Water (H_2O)
Density (kg m^{-3})	856 (400 °C, 0.1 MPa)	660 (327 °C, 15 MPa)
Thermal neutron absorption cross section (b)	0.53	0.66
Thermal conductivity (W $(m{\cdot}K)^{-1}$)	72.2 (400 °C)	0.5 (327 °C, 15 MPa)
Molar heat capacity (J $(mol{\cdot}K)^{-1}$)	23.2	75.4

efficiency of the reactor for electricity production. The relatively small difference between the melting and boiling temperatures (liquid range in the table) means that the flexibility of the reactor's operating conditions is fairly limited.

As seen in table 11.1, sodium becomes liquid at a relatively low temperature and has a good range of potential operating temperatures. Sodium is also comprised of only a single naturally occurring isotope, ^{23}Na, which has an acceptably low neutron absorption cross section. The neutron absorption by sodium that does occur leads to the reaction,

$$n + {}^{23}\mathrm{Na} \rightarrow {}^{24}\mathrm{Na} + \gamma$$

which is followed by the β^- decay,

$$^{24}\mathrm{Na} \rightarrow {}^{24}\mathrm{Mg} + e^- + \bar{\nu}$$

with a half-life of 15 h. ^{24}Mg is stable and has a small neutron cross section, so that neutrons that do interact with sodium do not lead to by-products that are of any concern.

Sodium has a high thermal conductivity and acceptable density, viscosity, and heat capacity. A comparison of the physical properties of sodium and water as they relate to reactor coolants are given in table 11.2.

Beginning around the mid-1950s, a number of experimental and prototype (as well as some commercial) SFRs were constructed, some of the more significant of which are discussed in section 11.3. By the mid-1970s, however, the commercial success of pressurized light-water reactors and boiling water reactors resulted in a waning of interest in SFR. In more recent years, interest in liquid metal-cooled fast reactors, and in SFRs in particular, has grown. This is particularly the case since the Generation IV Forum identified SFRs as a possible technology. This chapter deals with such reactors, while chapter 12 deals with fast reactors that are cooled with liquid lead.

11.2 General design of sodium-cooled fast reactors

The basic design of a liquid metal-cooled fast breeder reactor is shown by the schematic in figure 11.1. Two designs are illustrated. The design on the left is known as a pool-type reactor and has the primary coolant (liquid sodium) contained in the reactor vessel, which also contains the reactor core and a heat exchanger to transfer heat to an intermediate loop. The design on the right is known as a loop-type reactor and has the heat exchanger outside of the reactor vessel. Both designs have been used in experimental and demonstration reactors that have been constructed and both have provided acceptable performance. The details of some of these designs are presented in the next section.

The design of the core of a typical SFR follows from the discussion of the design of gas-cooled fast reactors in the previous chapter. Pin-type fuel rods are separated by wire windings or other appropriate spacers to allow for coolant flow around the fuel. Fuel tubes have a fission gas plenum at the top to avoid over-pressuring the tube. Tubes are arranged in hexagonal fuel assemblies, which are combined to create the core. Control rods are interspersed with fuel rods in the core that follow along the lines of the gas-cooled core in figure 10.6. Breeder blankets can be configured axially or radially, or both, as discussed in chapter 10.

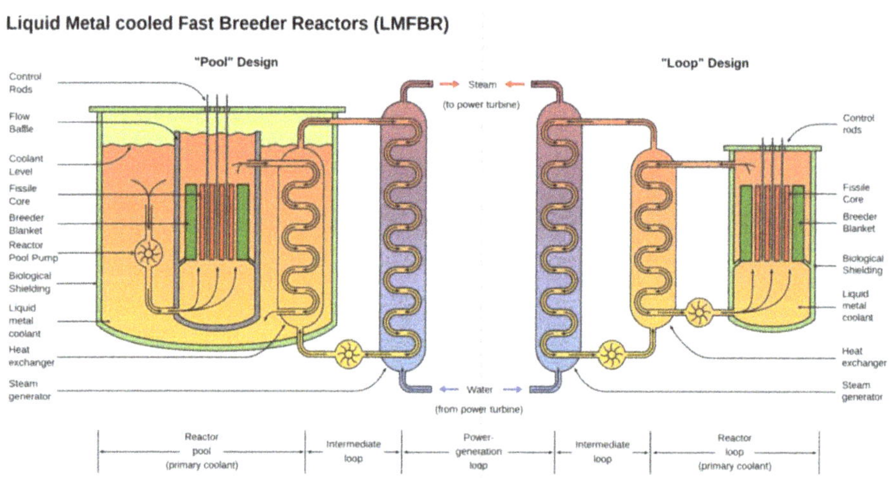

Figure 11.1. Two typical designs for a fast breeder reactor, in this case a liquid metal-cooled fast breeder reactor. The diagrams show two methods of transferring the heat from the coolant to water to produce steam to operate a turbine/generator. This schematic of the two types of liquid metal fast breeder reactor (LMFBR), a fast neutron reactor designed to breed fuel by producing more fissile material than it consumes image has been obtained by the author from the Wikimedia website where it was made available by Graevemoore (2008) under a CC BY-SA 3.0. licence. It is included within this chapter on that basis. It is attributed to Graevemoore.

SFRs have some advantages, and also some disadvantages, over conventional thermal neutron light-water reactors. Some of the significant advantages are,

- The coolant circulates at near atmospheric pressure, and this minimizes the possibilities of leaks.
- The high boiling temperature of sodium allows for flexibility in reactor design and provides a safety margin for operation.
- The secondary sodium coolant loop isolates the primary sodium coolant, which can be radioactive due to neutron capture or contamination from the fuel in the event of a core incident.
- The high operating temperature provides good thermodynamic efficiency.
- The high outlet temperature can provide process heat for industrial applications.
- The high thermal conductivity and reasonable heat capacity of sodium provides good decay heat removal in the event of loss of coolant circulation.
- The inclusion of a breeder blanket provides better fuel utilization and the opportunity to operate the reactor with a closed fuel cycle.
- Sodium does not corrode steel.

A major disadvantage of an SFR is the reactivity of sodium. When sodium comes in contact with water it reacts and produces hydrogen by the reaction,

$$2Na + 2H_2O \rightarrow 2NaOH + H_2.$$

The hydrogen that is liberated by this reaction can lead to explosions. In addition, sodium at high temperature ignites when it contacts oxygen. Both of these situations have occurred in some of the experimental and demonstration SFRs that have been constructed and some of these incidents are described in the next section.

Another concern about SFRs is their tendency to have a positive void coefficient. The void coefficient is a measure of the change in reactivity that results from a void in the coolant in the reactor core. This, for example, can result if a liquid coolant boils in the core. A positive coefficient means that the reaction rate increases as a result of a void in the coolant. While this situation represents an anomalous operating condition, it can be a safety concern if it results in an overheating of the reactor core. Generation IV SFRs must take this potential situation into account by implementing designs that minimize positive void coefficients and measures that mitigate their potential impact on reactor operation.

11.3 History of sodium-cooled fast reactors

Of the six reactor designs considered by the Generation IV International Forum, the SFR has the longest prior history of development. A thorough review of reactors in this category has been given by the International Atomic Energy Commission (2006) and Aoto *et al* (2014). A tabulation of the most significant SFRs constructed to date is available in Schulenberg (2022). A summary of previous SFRs is given in table 11.3 and an overview of some of the most important reactors from the table follows.

Table 11.3. Summary of previously constructed SFRs ordered by date of first criticality.

Reactor	Country[a]	Class[b]	Type[c]	Thermal power (MW_{th})	Electrical power (MW_e)	First criticality	Shutdown
BR-10	Russia	E	L	8	0	1958	2003
EBR-II	USA	E	P	62.5	20	1961	1994
Fermi	USA	E	L	200	61	1963	1975
Rapsodie	France	E	L	40	0	1967	1983
BOR-60	Russia	E	L	55	12	1969	2020
KNK-II	Germany	E	L	58	20	1972	1991
BN-350	Kazakhstan	D	L	750	130	1972	1999
Phénix	France	D	P	563	255	1973	2010
PFR	UK	D	P	650	250	1974	1994
Jōyō	Japan	E	L	140	0	1977	Operational
FFTF	USA	E	L	400	0	1980	1996
BN-600	Russia	C	P	1470	600	1980	Operational
FBTR	India	E	L	40	13	1985	Operational
Superphénix	France	C	P	2990	1242	1985	1998
Monju	Japan	D	L	714	280	1994	2017
CEFR	China	E	P	65	23.5	2010	Operational
BN-800	Russia	C	P	2100	880	2014	Operational

[a] At time of shutdown (or present).
[b] (E) experimental, (D) demonstration or prototype, and (C) commercial.
[c] (P) Pool and (L) loop. Data adapted from Schulenberg (2022) and International Atomic Energy Agency (2006).

11.3.1 EBR-II

The Experimental Breeder Reactor-2 (EBR-II) followed the development of the EBR-I which was located near Arco, Idaho and operated from 1951 to 1964. The EBR-I was the world's first breeder reactor and one of the first to generate electricity. It utilized liquid NaK as a coolant and uranium enriched to around 90% ^{235}U as fuel. A blanket of natural uranium was used to breed new fissile ^{239}Pu and successfully demonstrated the basic principles of a fast breeder reactor (International Atomic Energy Agency 2012a).

The EBR-II was an SFR and was constructed at the National Reactor Testing Station in Idaho. The reactor containment building is shown in figure 11.2. Construction began in 1956 and dry criticality (that is criticality without coolant) was achieved in 1961. Wet criticality (i.e., criticality with a coolant) was achieved in 1963. First electricity transmission to the grid occurred in 1964.

The fuel for EBR-II consisted of uranium enriched to 67% in ^{235}U. Fuel rods were enclosed in thin-walled stainless-steel tubes. The tubes also contained a small quantity of sodium metal to act as a heat transfer agent. Fuel tubes were welded shut with a void at the top to collect fission gases. Clusters of fuel pins were combined into hexagonal fuel assemblies. The core contained 637 hexagonal

Figure 11.2. Containment building of the EBR-2 sodium-cooled fast reactor (c. 1970). Reproduced from Idaho National Laboratory (2009). CC BY 2.0.

assemblies, of which 53 were fuel assemblies (referred to as driver fuel assemblies). The core also contained 570 breeder blanket assemblies and 14 assemblies containing control or safety rods (Bostelmann *et al* 2021).

EBR-II operated for more than 30 years and typically ran at an average annual capacity factor of greater than 70%. It was fully functional in 1994 when it was shutdown due to the termination of the United States Department of Energy's fast reactor development program. During its operation, EBR-II experienced only one significant incident (in the late-1960s) when the secondary sodium cooling loop developed a leak that resulted in a fire. Damage was minimal and the incident resulted in revisions to the procedures for dealing with sodium.

While EBR-II provided power to the grid, its main purpose was to test the operation of a sodium-cooled fast breeder reactor. One of its most important accomplishments resulted from the investigation of passive safety features. In 1986, the reactor was shutdown in order to simulate a loss of coolant circulation accident. EBR-II demonstrated its ability to successfully remove decay heat through the natural convection of the sodium.

11.3.2 BN-350

The BN-350 reactor was an SFR constructed at the Mangyshlak Nuclear Power Plant in Shevchenko, Soviet Union (now Aktau, Kazakhstan) on the Caspian Sea. It became operational in 1972 and ran until 1999. The reactor was originally designed to produce 1000 MW_{th} but was only operated at a lower capacity of 750 MW_{th}. While the reactor could, in principle, produce 350 MW_e, it was operated at a reduced electrical output of 130 MW_e and the remaining thermal energy, in the form of output steam, was used to supply a seawater desalination plant (see section 7.6.1). Since the city of Shevchenko/Aktau has no natural sources of fresh water, the desalination facility provided all of its necessary potable water. Thermal energy

Figure 11.3. Nuclear desalination plant using multiple effect distillation associated with the BN-350 sodium-cooled fast neutron reactor. Reprinted from Aoto *et al* (2014), Copyright (2014), with permission from Elsevier.

from the reactor supplied 80 000 m^3 per day of fresh water and this was supplemented with an additional 40 000 m^3 per day supplied using heat produced from oil and gas boilers (International Atomic Energy Agency 2012a). The desalination facility is illustrated in figure 11.3.

The BN-350 reactor was a loop-type fast breeder reactor with sodium primary and secondary loops. The primary loop provided sodium at an outlet temperature of 437 °C and the secondary loop produced steam by means of a steam generator. The steam supplied a turbine/generator for electricity production and provided thermal energy for the desalination plant.

The core of the BN-350 reactor was 1.06 m in diameter and 1.55 m in height. It was surrounded by axial breeder blankets of 0.6 m thickness above and below the core, and a radial breeder blanket of 0.45 m thickness surrounding the core. To ensure that the power distribution over the radius of the core was as constant as possible, the core was divided (radially) into zones with fuel of different fissile enrichments. The original design used two zones with ^{235}U enrichments of 17% (inner zone) and 26% (outer zone). This was later modified to include an intermediate zone with an enrichment of 21% ^{235}U between the inner and outer zones.

The BN-350 reactor was instrumental in establishing the feasibility of using a nuclear reactor for cogeneration to provide both electricity and to supply heat for a desalination plant.

11.3.3 BN-600

The two currently operating commercial SFRs, BN-600 and BN-800 (described below), are located at the Beloyarsk Nuclear Power Station near Zarechny, Sverdlovsk Oblast, Russia (shown in figure 11.4). The BN-600 reactor has been

Figure 11.4. Beloyarsk Nuclear Power Station as seen from the Beloyarskoye Reservoir near Zarechny, Sverdlovsk Oblast, Russia. This main building of Beloyarsk Nuclear Power Station as seen from the Beloyarskoye Reservoir near Zarechny, Sverdlovsk Oblast, Russia image has been obtained by the author from the Wikimedia website where it was made available by Hardscarf derivative of Memorino (2010). Image stated to be in the public domain. It is included within this chapter on that basis. It is attributed to Hardscarf.

Figure 11.5. Scale model of the core of a BN-600 liquid sodium-cooled fast breeder reactor on display at the Beloyarsk Nuclear Power Station. This model of a BN-600 reactor, displayed in the Beloyarsk Nuclear Power Station, Russia image has been obtained by the author from the Wikimedia website where it was made available by Nucl0id (2009a) under a CC BY-SA 3.0 licence. It is included within this chapter on that basis. It is attributed to Nucl0id.

operational since 1980 and is rated at 600 MW_e. A model of the BN-600 reactor core is shown in figure 11.5.

The core of the BN-600 reactor is 1.03 m in diameter and 2.058 m high. It contains 369 vertically mounted fuel assemblies. Figure 11.6 shows a model of the

Figure 11.6. Life-size model of a fuel assembly from the BN-600 reactor on display at the Beloyarsk Nuclear Power Station. This nuclear fuel of a BN-600 reactor image has been obtained by the author from the Wikimedia website where it was made available by Nucl0id (2009b) under a CC BY-SA 3.0 licence. It is included within this chapter on that basis. It is attributed to Nucl0id.

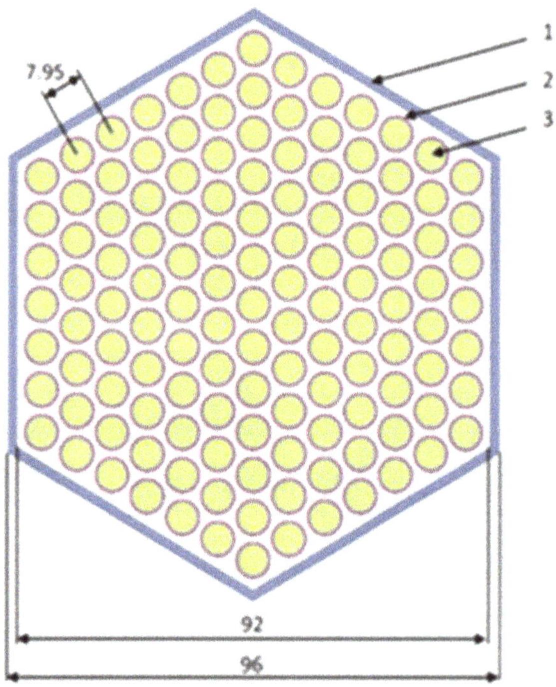

Figure 11.7. Design of the BN-600 fuel assembly: (1) jacket, (2) fuel cladding (0.4 mm thick), and (3) fuel rod (6.1 mm diameter). Dimensions are in mm. Reproduced from Korobeynikov *et al* (2023). CC BY 4.0.

fuel assembly and figure 11.7 shows the arrangement of the 127 enriched uranium dioxide fuel rods in each assembly. Like the BN-350 reactor core, the core of the BN-600 reactor has a radially varying fuel enrichment with a lower enrichment in the inner core, a higher enrichment in the outer core and an intermediate zone in between. The distribution of fuel assemblies is shown in figure 11.8 and the specifications of the different fuel zones are given in table 11.4. There are axial breeder blankets on the top (0.30 m thick) and bottom (0.35 m thick) of the core but no radial breeder blankets. The BN-600 is also a test pad for new fuel designs, see discussion in section 12.5 on nitride fuel for the BREST-300 lead-cooled reactor.

The BN-600 reactor has operated since 1980 and in 2020 had its license renewed until 2025. Although it has been a commercial success, it has not been without issues. Between 1980 and 1991, there were 12 intercircuit leaks (i.e., leaks between sodium and water/steam in the steam generator). Between 1980 and 1997, there were 27 external sodium leaks (i.e., leaks of sodium into air). These latter leaks resulted in 14 fires. Only one fire involved primary loop sodium, and hence resulted in a small release of radioactive material. All other fires involved non-radioactive secondary loop sodium (International Atomic Energy Agency 2012a, Pakhomov 2018). In more recent years, the reactor has had a good record of reliability.

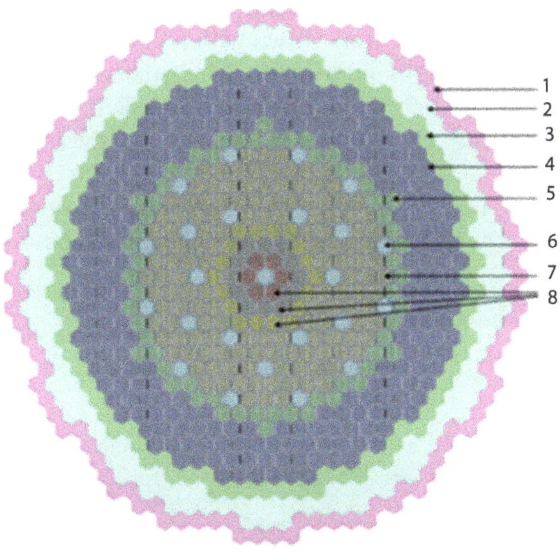

Figure 11.8. Layout of fuel assemblies in the BN-600 reactor core. (1) Radial reflector, (2,3) shielding, (4) 26% ^{235}U fuel assemblies, (5) 21% ^{235}U fuel assemblies, (6) control rods, and (7,8) 17% ^{235}U fuel assemblies. Reproduced from Korobeynikov *et al* (2023). CC BY 4.0.

Table 11.4. Specifications of the three fuel zones in the BN-600 core.

Zone	Number fuel assemblies	^{235}U enrichment (%)
Inner	136	17
Middle	94	21
Outer	139	26

11.3.4 Superphénix

Superphénix, shown in figure 11.9, was a pool-type SFR. With a thermal output of 2990 MW_{th} and an electrical output of 1242 MW_e, Superphénix was the world's largest SFR. A model of the reactor inside its containment structure is shown in figure 11.10. The stainless-steel reactor vessel contained the reactor core along with four primary cooling pumps and eight intermediate heat exchangers. The core of Superphénix consisted of 364 fissile fuel assemblies containing MOX fuel with 15% PuO_2. There were axial depleted uranium blankets above and below the fissile core, as well as a radial depleted uranium blanket and steel neutron reflectors.

Construction on the Superphénix reactor began in 1976 as part of France's growing dependence on nuclear generated electricity and concerns for the long-term availability of uranium supplies. It first achieved criticality in 1985 and was connected to the grid in 1986. Superphénix was shutdown on 24 December 1996 for maintenance and was not restarted prior to being closed permanently in September 1998 for political reasons.

Superphénix did not garner significant public support and was the subject of two violent incidents. In 1977, during the early stages of construction, anti-nuclear protesters at a public demonstration clashed with police resulting in the death of one protester and over 100 serious injuries. In 1982, the reactor was attacked by a militant group (see section 6.7). Five rocket propelled grenades were launched towards the reactor and caused minor damage to the concrete containment structure.

During the period from 1985 to 1996, Superphénix produced power for 53 months (mostly not near its maximum capacity), and was shutdown for 25 months

Figure 11.9. Superphénix reactor in Creys-Malville, France. Reproduced from International Atomic Energy Agency (2012b). CC BY-SA 2.0.

Figure 11.10. Model of the Superphénix reactor. This model of the Superphenix nuclear power station, a now closed fast breeder reactor image has been obtained by the author from the Wikimedia website where it was made available by Marshall Astor (2006) under a CC BY-SA 2.0 licence. It is included within this chapter on that basis. It is attributed to Marshall Astor.

for technical problems and for 66 months due to administrative or political problems. Superphénix experienced 80 operational events. Seventy-two of these were minor and were classified as deviations on the INES scale (see section 6.2). However, six events were classified as level 1 on the INES scale and two were classified as level 2. The two level 2 incidents were a fuel storage leak in 1987 and a primary sodium coolant leak in 1990.

As a result of its experience with Superphénix, France discontinued its fast reactor program and concentrated on light-water thermal reactors, with which it now produces the majority of its electricity. In recent years, France has been involved in cooperative development programs with other European countries, such as the ALLEGRO gas-cooled fast reactor (see section 10.4).

11.3.5 Monju

Monju was a loop-type sodium-cooled fast breeder reactor located near the Tsuruga Nuclear Power Plant, Fukui Prefecture, Japan (see figure 11.11). It was rated at 714 MW_{th} and 280 MW_e. The primary sodium coolant pump and intermediate heat exchanger were located outside of the reactor vessel but inside the reactor confinement structure, as illustrated in figure 11.12. The reactor used MOX (UO_2-PuO_2) fuel, and the core was divided into two zones with the inner zone containing 15% PuO_2 and the outer zone containing 20% PuO_2.

Construction on the Monju reactor began in 1986 and the reactor achieved first criticality in 1994. On 8 December 1995, a leak occurred in the secondary sodium loop on the return line just outside of the reactor confinement structure. Several

Figure 11.11. Monju fast breeder reactor. Reproduced from International Atomic Energy Agency (2013a). CC BY-SA 2.0.

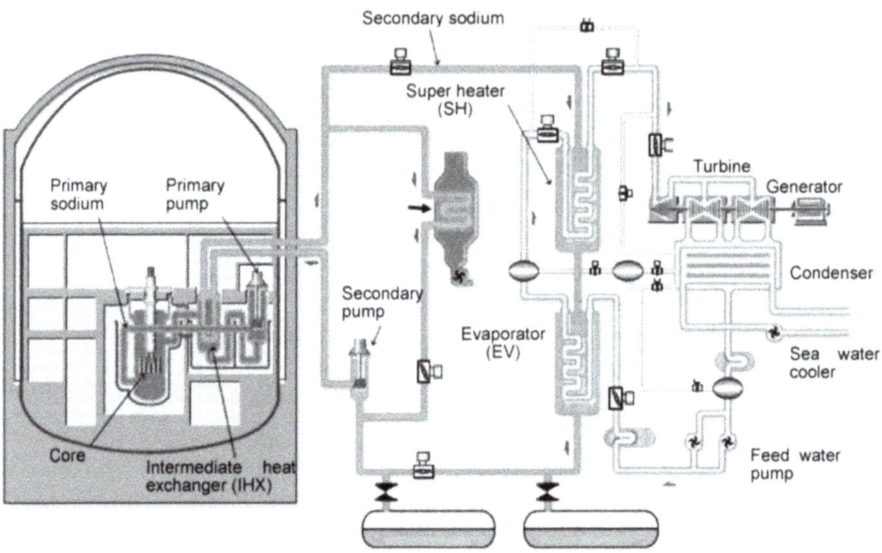

Figure 11.12. Layout of the Monju SFR. Reprinted from Mochizuki and Takano (2009), Copyright (2009), with permission from Elsevier.

hundred kilograms of sodium were released and ignited on contact with air. This incident was categorized as INES level 1. The reactor was shutdown for repairs and an extensive safety review. The reactor was finally restarted in 2010 and achieved criticality on 8 May 2010. On 26 August 2010, a 3.3-tonne fuel transfer machine fell

Figure 11.13. Construction of the BN-800 reactor in 2010. This Строительная площадка энергоблока БН-800 Белоярской АЭС image has been obtained by the author from the Wikimedia website where it was made available by Denisporubov (2010). Image stated to be in the public domain. It is included within this chapter on that basis. It is attributed to Denisporubov.

into the reactor vessel following a scheduled fuel replacement. The damaged equipment could not be retrieved from the reactor vessel until June 2011. However, the Fukushima accident in March of 2011 initiated a detailed reconsideration of Japan's nuclear energy policies. As a result of various technical and administrative issues, Monju was not restarted after the 2010 incident and was officially shutdown as of 2017. During its 23-year lifetime, Monju provided electricity to the grid for approximately one hour.

11.3.6 BN-800

The BN-800 SFR, along with the BN-600 reactor, are located at the Beloyarsk Nuclear Power Plant near Zarechny, Sverdlovsk Oblast, Russia (see figure 11.4). The BN-600BN-800 is a pool-type reactor rated at 2100 MW_{th} and 880 MW_{e}. Preliminary work on BN-800 began as early as 1983, but after the Chernobyl accident the project was put on hold. Construction restarted in 2006 (see figure 11.13) and first criticality was achieved in 2014.

The BN-800 reactor vessel is shown in figure 11.14. Unlike the BN-600 reactor, which utilized enriched uranium dioxide fuel, BN-800 was originally designed to operate on MOX (UO_2-PuO_2) fuel with 20.5% PuO_2. This fuel was chosen so that the reactor could make use of surplus weapons-grade plutonium. However, the design of the core is flexible and allows for other fuel configurations. To date, the reactor has operated with a hybrid core using a mixture of enriched UO_2 and UO_2-PuO_2 (Pakhomov 2018), and future plans include the use of mixed nitride fuels (MNUP = UN-PuN) (Khomyakov *et al* 2020). This approach is of interest because nitrides have better thermal conductivity than oxides and have 30%–40% higher fissile density (Ekberg *et al* 2018).

A major goal of the BN-800 reactor is to demonstrate the viability of a close fuel cycle. While the design of the BN-800 reactor follows from experience gained during the operation of the BN-600 reactor (and before that the BN-350 reactor), there are

Figure 11.14. BN-800 liquid sodium-cooled fast breeder reactor at the Beloyarsk Nuclear Power Station. This Reactor BN-800 image has been obtained by the author from the Wikimedia website where it was made available by Rosatom. Empresa Estatal de Energía Atómica Rusa (2016) under a CC0 licence. It is included within this chapter on that basis. It is attributed to Rosatom. Empresa Estatal de Energía Atómica Rusa.

several additional design features that improve the safety and reliability of the reactor. These include,

- A core catcher in the event of a core meltdown.
- A passive emergency shutdown rod system.
- A sodium cavity above the core to reduce the positive void coefficient.
- A single turbine/generator system.
- A decay heat removal system.

The BN-800 reactor has operated well since its start-up with a minimal number of difficulties. In 2020, for example, it operated with an 82% capacity factor (World Nuclear News 2022).

11.4 India's nuclear energy plan

Due to India's large population and developing economy, the implementation of a viable energy infrastructure is essential. While India is not a member of the Generation IV Forum, nuclear power plays an important role in its future energy vision and it has been active in pursuing the development of reactor designs that fit within the Generation IV guidelines. In particular, India's substantial thorium resources (see table 11.5) provide the opportunity to undertake the development of a closed cycle breeder program (see e.g., Tongia and Arunachalam 1998, Gopalakrishnan 2002). This approach involves a three-stage plan that was first proposed by the Indian physicist Homi Bhabha (1909–1966) in 1954 and officially adopted by the Indian government in

Table 11.5. Estimated thorium reserves by country. Data adapted from Nuclear Energy Agency (2016)

Country	Thorium reserves (10^3 tonnes)
India	846
Brazil	632
Australia	595
United States	595
Egypt	380
Turkey	374
Venezuela	300
Canada	172
Russia	155
South Africa	148
China	100
Rest of world	2058
Total	6355

1958. Work on the three-stage plan began in 1972 with the establishment of the Reactor Research Centre and the construction of the Fast Breeder Test Reactor (FBTR). The three stages of the plan are described below.

Stage 1 Thermal Uranium Reactors: This stage involves the production of ^{239}Pu in conventional thermal neutron reactors. Neutrons produced by induced fission in ^{235}U breed fissile ^{239}Pu through their capture by fertile ^{238}U,

$$n + {}^{238}\mathrm{U} \rightarrow {}^{239}\mathrm{U} + \gamma$$

followed by the β^- decays,

$${}^{239}\mathrm{U} \rightarrow {}^{239}\mathrm{Np} + e^- + \bar{\nu}$$

and

$${}^{239}\mathrm{Np} \rightarrow {}^{239}\mathrm{Pu} + e^- + \bar{\nu}.$$

India has operated thermal reactors since the 1960s. At present there are 19 operating thermal reactors in the country. Two are pressurized light-water reactors of Russian design and the remainder are heavy water reactors, mostly Indian reactors of a design derived from the CANDU reactor.

Stage 2 Fast Breeder Reactors: This stage uses fissile ^{239}Pu from stage 1 to breed more ^{239}Pu from fertile ^{238}U in a fast neutron reactor. When sufficient fissile material is available, neutrons from the fission of ^{239}Pu will be used to breed fissile ^{233}U from fertile ^{232}Th. Neutron capture by ^{232}Th produces fissile ^{233}U by the reaction,

$$n + {}^{232}\mathrm{Th} \rightarrow {}^{233}\mathrm{Th} + \gamma$$

followed by the β^- decays,

$$^{233}\mathrm{Th} \rightarrow {}^{233}\mathrm{Pa} + e^- + \overline{\nu}$$

and

$$^{233}\mathrm{Pa} \rightarrow {}^{233}\mathrm{U} + e^- + \overline{\nu}.$$

India has constructed an experimental fast breeder reactor (the FBTR as described below) and is in the process of constructing a larger fast breeder reactor (the PFBR, see below).

Stage 3 Thermal Thorium Breeder Reactors: The final stage uses induced fission of fissile ^{233}U produced in Stage 2 to generate energy,

$$n + {}^{233}\mathrm{U} \rightarrow {}^{234}\mathrm{U} \rightarrow \text{fission}.$$

Additional ^{232}Th input into the reactor will breed new fissile fuel from excess neutrons, as in Stage 2. The fuel cycle will be self-sustaining with only the input of new fertile thorium.

11.4.1 Fast Breeder Test Reactor

The FBTR is a 40 MW_{th} sodium-cooled fast breeder reactor whose design was derived from the French Rapsodie reactor (International Atomic Energy Agency 2012a, Srinivasan *et al* 2006). Construction began in 1972 and the reactor has been operational since 1985. An image of the inside of the reactor containment building is shown in figure 11.15.

The FBTR is a sodium-cooled loop-type reactor with a secondary sodium loop that transfers heat though a heat exchanger to produce steam. The flow diagram for the reactor is illustrated in figure 11.16, showing the two sodium loops and the steam loop.

Figure 11.15. Interior of the FBTR building at the Kalpakkam Nuclear Complex in India showing the top of the reactor vessel. Reproduced from International Atomic Energy Agency (2013b). CC BY-SA 2.0.

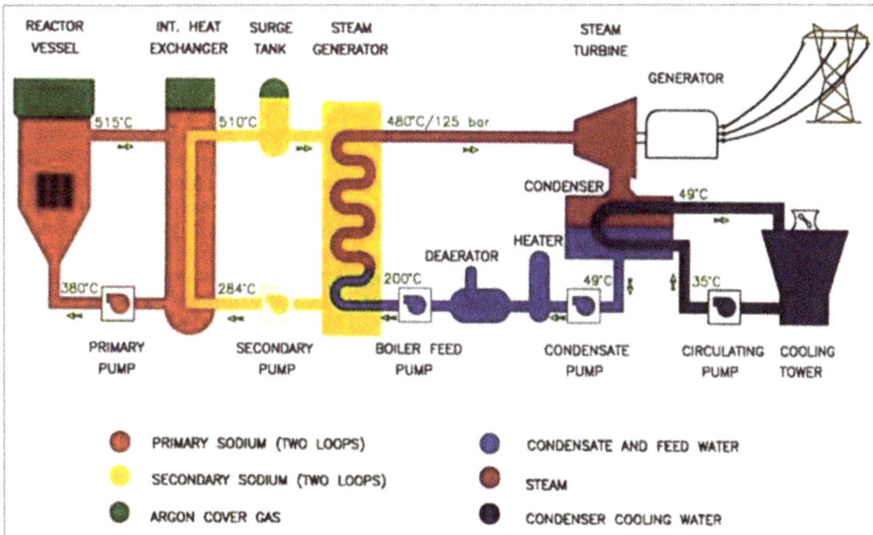

Figure 11.16. Flow diagram for the FBTR. Reproduced from Rajan (2021). CC BY 3.0.

The core of the FBTR, as shown in figure 11.17, consists of 745 hexagonal assemblies. These locations contain fuel (at the center of the core) surrounded by nickel reflectors, thorium (in the form of ThO_2) blankets, and steel reflectors. Various fuel configurations have been tested in the FBTR. These include mixed oxide (MOX) fuel containing 30% PuO_2 and 70% UO_2 (enriched to 85% ^{235}U), as well as carbide fuel, 30% PuC, and 70% UC. The core design includes a radial breeder blanket, as shown in figure 11.17, as well as axial blankets on both the top and bottom of the core.

A major consideration for Stage 2 of India's nuclear energy plan is the length of time that this stage will require before the final stage can be started. This depends on the rate at which ^{233}U can be bred from the fissile thorium blanket. The doubling time, as discussed in section 3.8, is a useful measure of the timescale for ^{233}U production. There are two approaches to estimating the doubling time. The simple doubling time (SDT) is the doubling time when all fissile material is bred in a single reactor. The compound system doubling time (CSDT) is the doubling time related to a system of reactors which work together where fissile material produced in one reactor is used to fuel other reactors. Table 11.6 gives estimates of doubling times (SDT and CSDT) for $^{238}U \rightarrow {}^{239}Pu$ cycles and $^{232}Th \rightarrow {}^{233}U$ cycles (Tongia and Arunachalam 1998). As the table shows, doubling times depend on the form of the reactor fuel. While the results shown in the table are estimates for the ideal case, the actual doubling time will also depend on factors such as,

- Reactor capacity factors
- Fuel reprocessing time
- Fuel reprocessing losses
- New reactor construction times

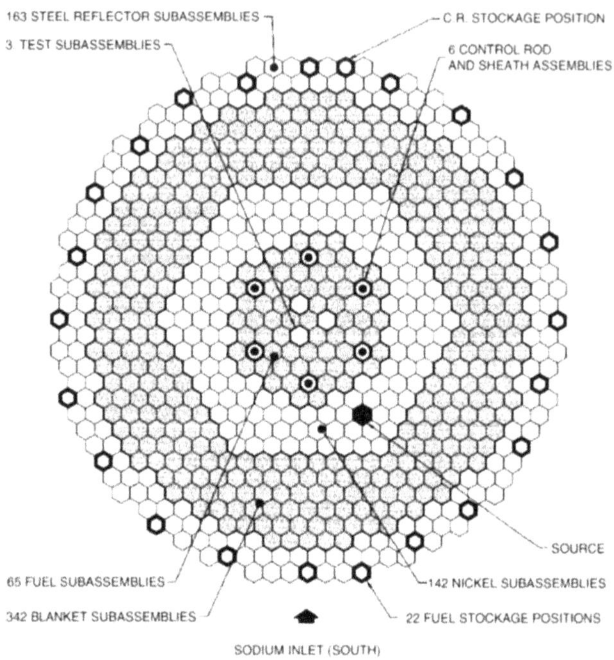

Figure 11.17. Core layout of the FBTR. Reprinted from Srinivasan *et al* (2006), Copyright (2006), with permission from Elsevier.

Table 11.6. Doubling times (see section 4.8) for different fuel process and different fuel forms. SDT = simple doubling time and CSDT = compound system doubling time, as described in the text. Data are adapted from Tongia and Arunachalam (1998).

Doubling time	Fuel form	$^{238}U \rightarrow {}^{239}Pu$ cycle	$^{232}Th \rightarrow {}^{233}U$ cycle
SDT (years)	Metal	12.3	108.3
	Carbide	14.7	101.0
	Oxide	25.7	155.8
CSDT (years)	Metal	8.5	75.1
	Carbide	10.2	70.0
	Oxide	17.8	108.0

The important information in table 11.6, however, is that thorium doubling times are quite long in comparison to uranium doubling times. In the best case, thorium doubling times are 70 years or more, meaning that the establishment of a self-sustaining closed cycle thorium thermal breeder reactor program, even after the construction of a suitable number of fast breeder reactors to produce ^{233}U, is a long-term endeavor.

11.4.2 Prototype Fast Breeder Reactor

The Prototype Fast Breeder Reactor (PFBR) is a Generation IV reactor that is being constructed at Kokkilamedu, near Kalpakkam, in Tamil Nadu state, India. It is a 1250 MW_{th} sodium-cooled fast breeder reactor that follows from the experience gained from the FBTR and is intended as a demonstration of a commercial-scale breeder reactor that is part of Stage 2 of India's nuclear energy plan.

The PFBR is a pool-type reactor, as shown in figure 11.18 (Chetal *et al* 2006, International Atomic Energy Agency 2012a). It has two primary sodium cooling loops with an outlet temperature of 547 °C. There are two secondary sodium cooling loops with four steam generators per loop. Total electrical output is 500 MW_e.

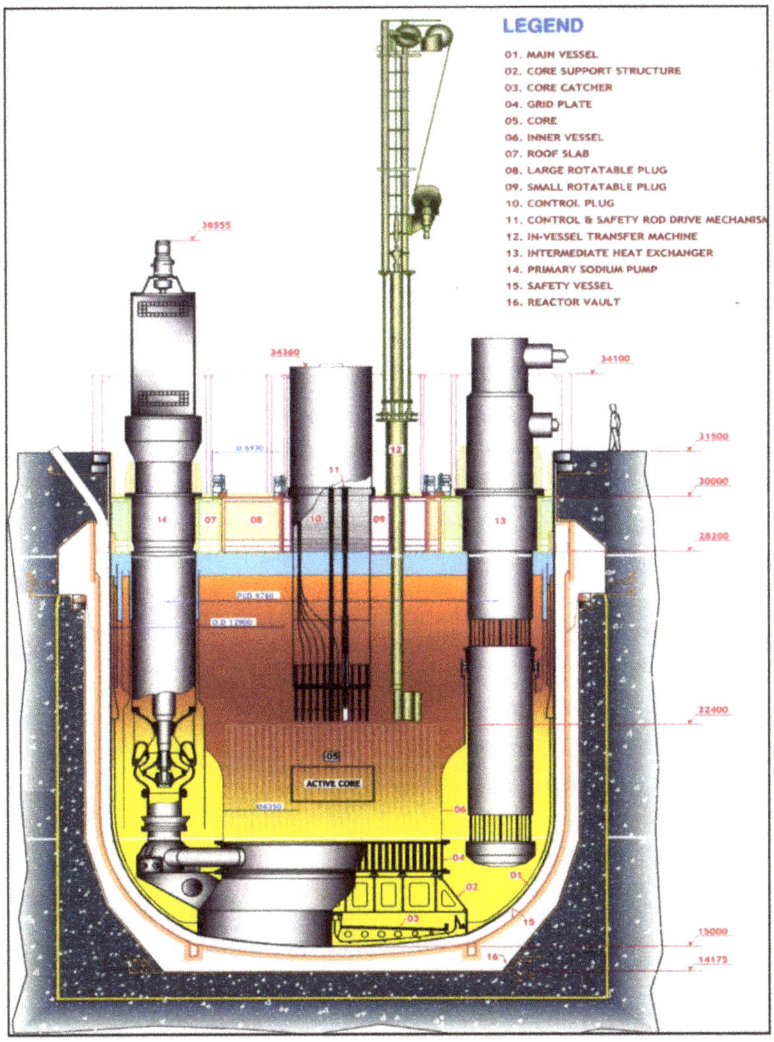

Figure 11.18. Diagram of the PFBR reactor. Reprinted from Sabih *et al* (2016), Copyright (2016), with permission from Elsevier.

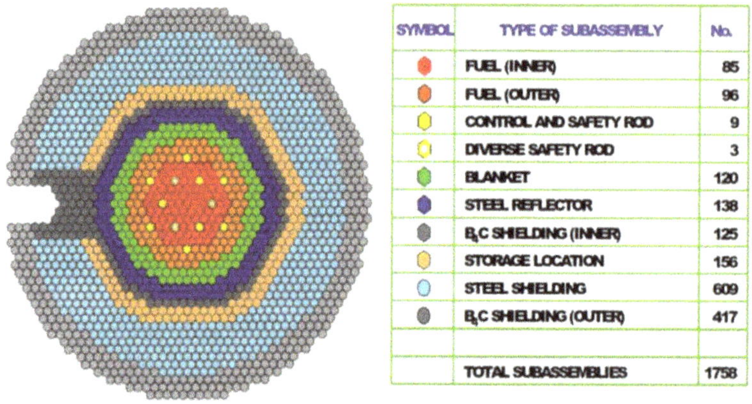

Figure 11.19. Core configuration for the PFBR. Reprinted from Sabih *et al* (2016), Copyright (2016), with permission from Elsevier.

The core of the PFBR is 1.9 m in diameter and 1.0 m height, and consists of 1758 assemblies, as shown in figure 11.19. There are 181 fuel assembles in the central region, along with 12 control/safety rods. The fuel assemblies are surrounded by 180 breeder blanket assemblies. The outer portion of the core consists of reflectors and shielding as shown in the figure.

The initial operation of the PFBR will utilize MOX fuel (PuO_2-UO_2) because substantial experience has already been gained with this form of fuel. Later, operation using metallic fuel is planned in order to take advantage of the improved doubling time (see table 11.6) for this form of fuel.

Construction on the PFBR began in 2004 and the reactor was originally planned to be operational in 2010. Over the years, construction problems, difficulty in obtaining reactor components, and bureaucracy have delayed the start-up of the PFBR. Speculations concerning the detailed reasons for the delays have appeared in the literature (e.g., Kale 2020, Spansen 2020, Ramana and Sharma 2021). Present information suggests a late-2024 date for the reactor to attain first criticality. Overall, it is interesting to note that the second stage of India's three-part nuclear energy plan began in 1972 with the beginning of the construction of the FBTR. Thus, a consideration of the timescale for Generation IV reactor development is important for long-term nuclear planning purposes.

11.5 CFR-600 (China)

There are several Generation IV SFR projects in progress around the world. In this section and the next two sections, we look at the details of three representative reactor development projects.

Following the operation of the CEFR reactor since 2010 (see table 11.3) China has undertaken the construction of a larger prototype SFR, the CFR-600. Two reactors are planned for construction on Changbiao Island, Xiapu County, Fujian

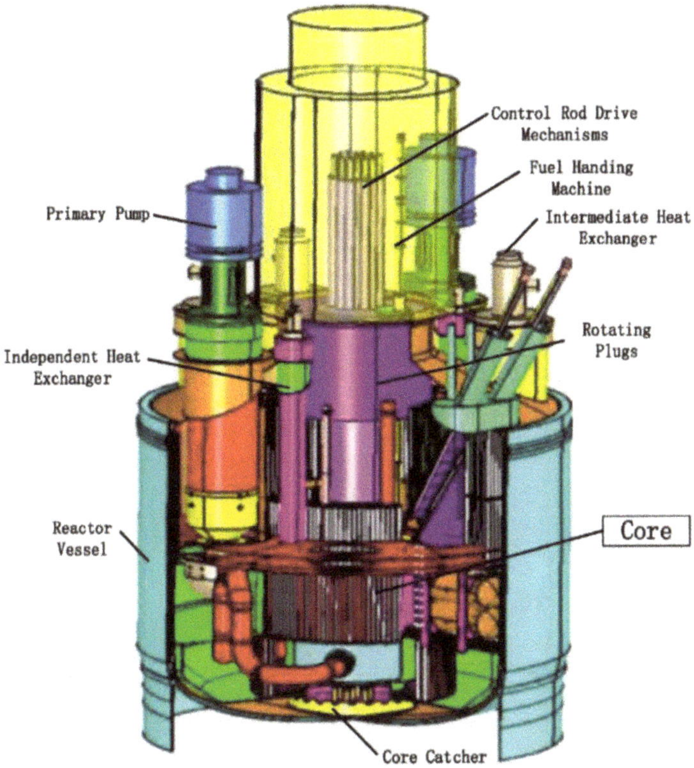

Figure 11.20. Diagram of the CFR-600 reactor. Reprinted from Wang *et al* (2021), Copyright (2021), with permission from Elsevier.

province, China. Construction on the first reactor (Xiapu-1) began in 2017 and construction on the second reactor (Xiapu-2) began in 2020.

The CFR-600 is a pool-type fast reactor (see figure 11.20) and is rated at 1500 MW_{th} and 600 MW_e (International Atomic Energy Agency n.d., World Nuclear Association 2021). Similar to other pool-type fast reactors, the CFR-600 has three cooling loops. The primary loop is inside the reactor vessel and heat is transferred to the secondary circuit through an intermediate heat exchanger (see figure 11.20). The core inlet temperature is 380 °C and the core outlet temperature is 550 °C.

In accordance with Generation IV Forum objectives, safety is a major consideration in the design of the CFR-600. These safety features will include,

- A core catcher
- A passive decay heat removal system
- A passive emergency shutdown system utilizing hydraulically suspended shutdown rods (Wang *et al* 2021)
- A consideration of the magnitude of the sodium void coefficient

The core of the CFR-600 is divided into three fuel zones, as illustrated in figure 11.21. Initially, the reactor will used enriched UO_2. After testing with UO_2, the core will be

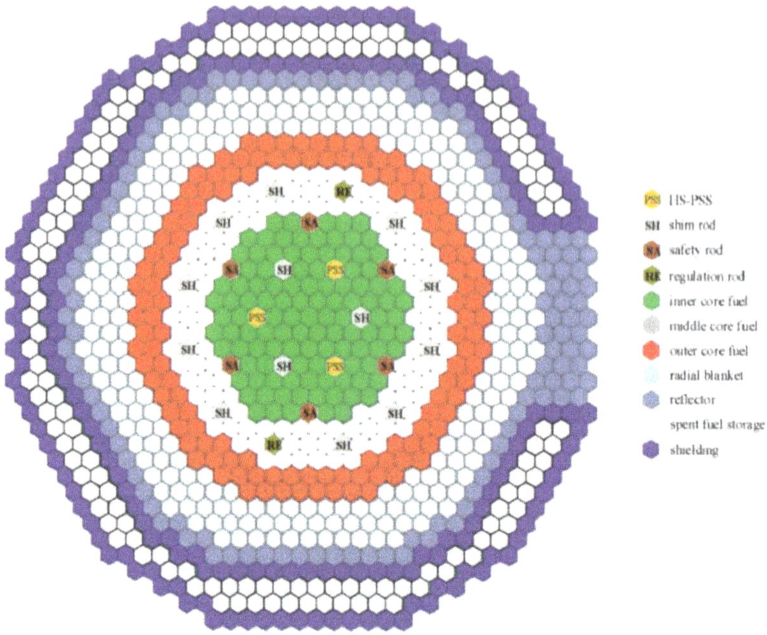

Figure 11.21. Core layout for the CFR-600 SFR. Reprinted from Wang *et al* (2021), Copyright (2021), with permission from Elsevier.

loaded with MOX fuel containing UO_2-PuO_2. The core layout in the figure shows that the fissile core is surrounded by a radial breeder blanket, and outside that is a neutron reflector and shielding. One of the main objectives of the CFR-600 reactor is to close the uranium fuel cycle. To this end, it is designed to have a breeding ratio of 1.2. With such a high breeding ratio, proliferation resistance is an important design consideration.

It has been reported that the first CFR-600 reactor has achieved criticality and since mid-2023 has been operating on low power. The timeline for grid connection is not known (Zhang 2023). This is in keeping with the original schedule for the reactor construction and operation. The second CFR-600 reactor is scheduled for completion in 2026. A future Chinese commercial-scale sodium-cooled breeder reactor is in the planning stages. It is expected to be completed around the mid-2030s and would have a capacity of 1000–1200 MW_e.

11.6 BN-1200 (Russia)

The BN-1200 SFR is a Generation IV reactor that is planned for construction at the Beloyarsk Nuclear Power Plant near Zarechny, Sverdlovsk Oblast, Russia. The BN-1200 is a pool-type reactor with an output of 2900 MW_{th} and 1220 MW_e. A cross section of the reactor vessel showing the location of the core, primary circulating pump, and intermediate heat exchanger is shown in figure 11.22. The primary

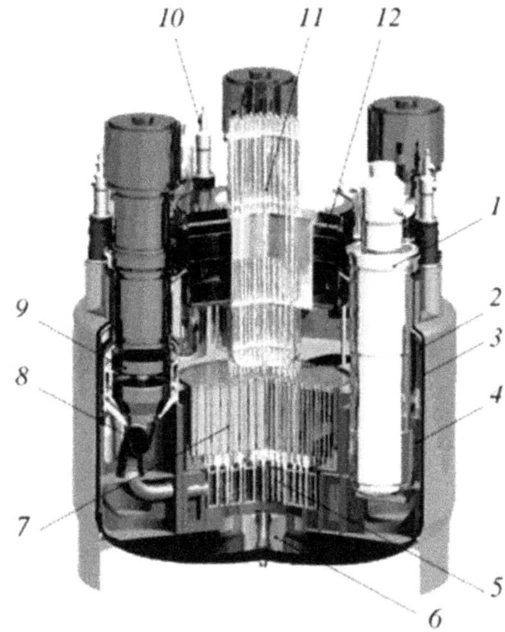

Figure 11.22. Interior design of the BN-1200 reactor vessel: (1) intermediate heat exchanger, (2) main reactor vessel, (3) back-up reactor vessel, (4) supporting girdle, (5) pressure tank, (6) core catcher, (7) core, (8) pressure pipeline, (9) main circulating pump, (10) fuel assembly loading machine, (11) control and safety rod actuators, and (12) rotating plugs. Reproduced from Rachkov *et al* (2010). CC BY 3.0.

cooling loop is entirely within the reactor vessel, and this ensures that any cooling leaks will not spread radioactive contamination outside the vessel. The secondary circuit contains four loops, each with its own circulating pump and steam generator (Vasiliev *et al* 2013).

The BN-1200 reactor is based on experience obtained from earlier BN-family reactors, particularly the BN-600 and BN-800 reactors. It incorporates safety features that have been tested on these smaller reactors. These features include,

- Four emergency heat removal loops
- A core catcher
- Hydraulically suspended shutdown rods that will release in the event of core overheating
- Design elements to minimize the positive sodium void coefficient

The BN-1200 will initially utilize MOX fuel but will ultimately use nitride fuel that has been tested on the BN-600 and BN-800 reactors (Vasiliev *et al* 2021). The use of nitride fuels and a closed fuel cycle will minimize use of uranium resources, minimize radioactive waste production, and ensure economic viability.

In early-2012, the Rosatom State Nuclear Energy Corporation announced its plans to construct the BN-1200 reactor at the Beloyarsk Nuclear Power Plant and in

June 2012 the Sverdlovsk regional government approved the plans for the construction. Construction was planned to start in 2015 with commercial operation around 2020. In 2015, experimental results from the BN-800 reactor required a redesign of the BN-1200 reactor fuel. Construction was put on hold and in January 2022, Rosatom announced a new estimated completion date for the BN-1200 reactor of 2035.

11.7 PGSFR (South Korea)

A long-term plan for the future of the nuclear power industry in South Korea was issued by the Korean Atomic Energy Commission in 2008. The plan includes the development of an SFR. The project to construct the Prototype Generation IV Sodium-cooled Fast Reactor (PGSFR) was initiated in 2012. The design phase of the project was completed in 2020 and an operational reactor is anticipated by 2028.

The PGSFR is a pool-type SFR rated at 600 MW_{th} and 150 MW_e. It differs from many other contemporary SFR designs in two ways: first, the PGSFR is a burner (not breeder) reactor; and, second, it uses metal (not oxide or nitride) fuel. While the PGSFR design includes typical Generation IV safety features, it also emphasizes nuclear non-proliferation (as evidenced by its lack of a ^{239}Pu-breeder blanket) and transuranic burn-up (as described below). For a detailed description of some of the design elements see Kim *et al* (2022), Yeom *et al* (2020), and Yoo *et al* (2016).

The general design of the PGSFR is shown in figure 11.23. The reactor core, primary circulating pumps, intermediate heat exchangers, and decay heat exchangers are contained within the 10.5 m diameter by 15.5 m high reactor vessel. There are two primary circulating pumps and four intermediate heat exchangers. The intermediate heat transfer system consists of two loops, each of which has two intermediate heat exchangers and one intermediate circulating pump, and is connected to one steam generator.

The decay heat removal system consists of two active decay heat removal loops and two passive decay heat removal loops. The active decay heat removal systems consist of a decay heat exchanger within the reactor vessel that transfers primary coolant heat to a secondary sodium coolant loop. The secondary sodium loop is cooled through a forced-draft sodium-to-air heat exchanger that utilizes a blower. The passive decay heat removal system is similar but uses a natural-draft sodium-to-air heat exchanger that is cooled by natural convection.

A diagram of the reactor core is shown in figure 11.24. The core contains of two fuel zones with nine control rods. The inner core consists of 52 fuel assemblies and the outer core consists of 60 fuel assemblies. Each fuel assembly contains 217 fuel rods. The active core region has a diameter of 1.58 m and a height of 0.90 m. The fuel region is surrounded by a neutron reflector and radiation shield.

There are three planned fuel configurations for the PGSFR. The initial core will use a U-Zr alloy (with 10% Zr) fuel with a uniform enrichment of 19.2% ^{235}U in both inner and outer fuel zones. This fuel will be used initially in the reactor in order to

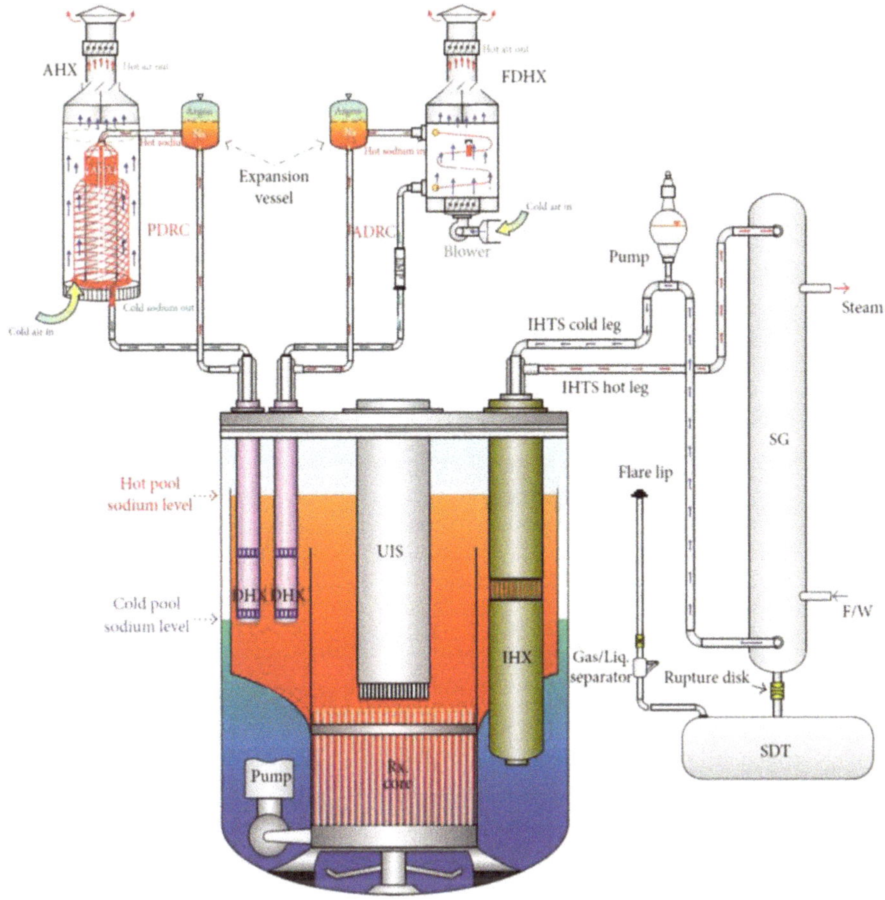

Figure 11.23. Configuration of the PGSFR core and heat transport system. ADRC = active decay heat removal circuit, AHX = natural-draft sodium-to-air heat exchanger, DHX = decay heat exchanger, FDHX = forced-draft sodium-to-air heat exchanger, IHTS = intermediate heat transfer system, IHX = intermediate heat exchanger, PDRC = passive decay heat removal circuit, SDT = steam dump tank, SG = steam generator, and UIS = upper internal structure. Reproduced from Kim *et al* (2013). CC BY 3.0.

test its operation with a known fuel composition. This fuel configuration will be followed by a core utilizing fuel that contains spent fuel from a light-water thermal reactor. The composition will be U-TRU-Zr, where TRU represents transuranics in the spent reactor fuel. Fuel enrichments of the inner and outer core zones are to be determined. This approach will test transuranic burn-up in a fast neutron reactor. The final fuel will contain spent light-water reactor fuel along with self-recycled fuel from the PGSFR reactor.

The results of the PGSFR operation intend to demonstrate the utilization of spent light-water reactor fuel in a fast neutron reactor. This provides a means of both ensuring proliferation resistance and minimizing radioactive waste.

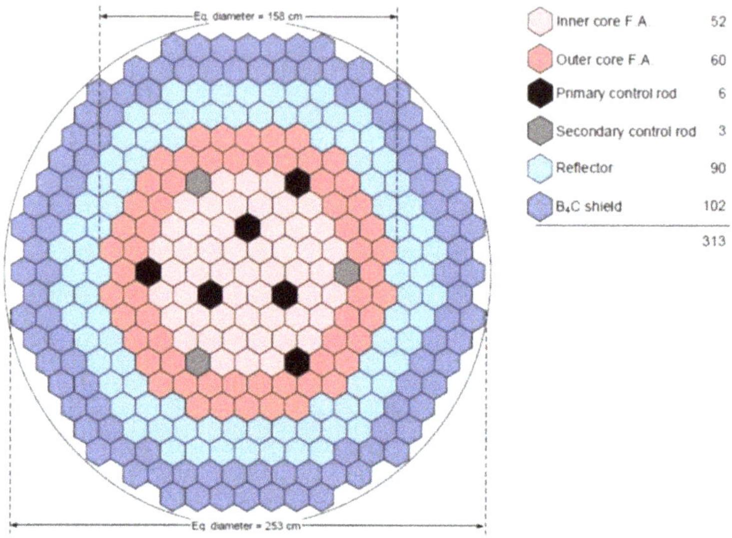

Figure 11.24. Layout of the core of the PGSFR reactor. Reprinted from Yoo *et al* (2016), Copyright (2016), with permission from Elsevier.

References

Aoto K, Dufour P, Hongyi Y, Glatz J P, Kim Y, Ashurko Y, Hill R and Uto N 2014 A summary of sodium-cooled fast reactor development *Prog. Nucl. Energy* **77** 247–65

Astor M 2006 A model of the Superphenix nuclear power station, https://commons.wikimedia.org/wiki/File:Superphenix.jpg

Bostelmann F, Ilas G and Wieselquist W A 2021 Nuclear data sensitivity study for the EBR-II fast reactor benchmark using SCALE with ENDF/B-VII.1 and ENDF/B-VIII.0 *J. Nucl. Eng.* **2021** 345–67

Chetal S C, Balasubramaniyan V, Chellapandi P, Mohanakrishnan P, Puthiyavinayagam P, Pillai C P, Raghupathy S, Shanmugham T K and Sivathanu Pillai C 2006 The design of the prototype fast breeder reactor *Nucl. Eng. Des.* **236** 852–60

Denisporubov 2010 Construction site of the BN-800 power unit of the Beloyarsk NPP https://commons.wikimedia.org/wiki/File:BN-800_construction.jpg

Ekberg C, Ribeiro Costa D, Hedberg M and Jolkkonen M 2018 Nitride fuel for Gen IV nuclear power systems *J. Radioanal. Nucl. Chem.* **318** 1713–25

Gopalakrishnan A 2002 Evolution of the Indian nuclear power program *Ann. Rev. Energy Environ.* **27** 369–95

Graevemoore 2008 A schematic of the two types of liquid metal fast breeder reactor (LMFBR) https://en.wikipedia.org/wiki/File:LMFBR_schematics2.svg

Hardscarf 2010 Main building of Beloyarsk Nuclear Power Station https://commons.wikimedia.org/wiki/File:Beloyarsk_NNP-3.jpg

Idaho National Laboratory 2009 EBR-II www.flickr.com/photos/inl/3466167200/

International Atomic Energy Commission 2006 *Fast Reactor Database—2006 Update* IAEA-TECDOC-1531 (Vienna: IAEA) https://www-pub.iaea.org/MTCD/Publications/PDF/te_1531_web.pdf

International Atomic Energy Agency 2012a *Status of Fast Reactor Research and Technology Development* IAEA-TECDOC-1691 (Vienna: IAEA) https://www-pub.iaea.org/MTCD/Publications/PDF/TE_1691_CD/PDF/IAEA-TECDOC-1691.pdf

International Atomic Energy Agency 2012b Superphénix nuclear power plant. Creys-Malville, France www.flickr.com/photos/iaea_imagebank/8134138979/

International Atomic Energy Agency 2013a Monju, fast breeder reactor www.flickr.com/photos/iaea_imagebank/8389261826/

International Atomic Energy Agency 2013b A fast-breeder test reactor at the Kalpakkam Nuclear Complex, India www.flickr.com/photos/iaea_imagebank/8386517544/

International Atomic Energy Agency n.d. CFR-600 (China Institute of Atomic Energy, China) https://aris.iaea.org/PDF/CFR-600.pdf

Jurney E T 1954 *The Los Alamos Fast Plutonium Reactor* (Los Alamos, NM: Los Alamos Scientific Laboratory of the University of California) https://catalog.hathitrust.org/Record/100167298

Kale R D 2020 India's fast reactor programme—a review and critical assessment *Prog. Nucl. Energy* **122** 103265

Khomyakov Y S, Zherebtsov A A, Shadrin A Y, Khaperskaya A V, Vasiliev B A and Farashkin M R 2020 Program for CNFC R&D using BN-800 *J. Phys.: Conf. Ser.* **1475**

Kim K-S, Kim J-B and Park C-G 2022 Conceptual designs and characteristic of the fuel handling and transfer system for 150 MWe PGSFR and 1400 MWe SFR burner reactor *Nucl. Eng. Technol.* **54** 4125–33

Kim Y-I, Lee Y B, Lee C B, Chang J and Choi C 2013 Review design concept of advanced sodium-cooled fast reactor and related R&D in Korea *Sci. Technol. Nucl. Install.* **2013** 290362

Korobeynikov V V, Kolesov V V and Ignatiev I A 2023 Computational simulation of minor actinide burning in a BN-600 reactor with fuel without uranium and plutonium *Nucl. Eng. Technol.* **9** 59–64

Mochizuki H and Takano M 2009 Heat transfer in heat exchangers of sodium cooled fast reactor systems *Nucl. Eng. Des.* **239** 295–307

Nucl0id 2009a Model of a BN-600 reactor https://commons.wikimedia.org/wiki/File:BN-600_nuclear_reactor.jpg

Nucl0id 2009b Nuclear fuel of a BN-600 reactor https://commons.wikimedia.org/wiki/File:Nuclear_fuel_of_a_BN-600_reactor.jpg

Nuclear Energy Agency 2016 *Uranium 2016: Resources, Production and Demand* NEA No. 7301 (Paris: OECD Publishing) www.oecd-nea.org/jcms/pl_15004

Pakhomov I 2018 BN-600 and BN-800 operating experience www.gen-4.org/gif/upload/docs/application/pdf/2019-01/gifiv_webinar_pakhomov_19_dec_2018_final.pdf

Rachkov V I *et al* 2010 Concept of an advanced power-generating unit with a BN-1200 sodium-cooled fast reactor *At. Energy* **108** 254–9

Rajan K K 2021 A study on sodium—the fast breeder reactor coolant *IOP Conf. Ser.: Mater. Sci. Eng.* **1045** 012013

Ramana R V and Sharma N 2021 Problems with the prototype fast breeder reactor *The India Forum* (24 February 2021) www.theindiaforum.in/article/problems-prototype-fast-breeder-reactor

Rosatom 2016 Reactor BN-800 https://commons.wikimedia.org/wiki/File:BN-800_reactor.jpg

Schulenberg T 2022 *The Fourth Generation of Nuclear Reactors—Fundamentals, Types and Benefits Explained* (Berlin: Springer) ch 6 pp 93–113

Sabih M *et al* 2016 Design and performance evaluation of core flow monitoring mechanisms for PFBR *Ann. Nucl. Energy* **94** 732–41

Spansen 2020 Now, December 2021—India's Prototype Fast Breeder Reactor [PFBR] to go critical www.spansen.com/2020/08/december-2021-india-prototype-fast-breeder-reactor-pfbr.html

Srinivasan G, Suresh Kumar K V, Rajendran B and Ramalingam P V 2006 The Fast Breeder Test Reactor—design and operating experiences *Nucl. Eng. Des.* **236** 796–811

Tongia R and Arunachalam V S 1998 India's nuclear breeders: technology, viability and options *Curr. Sci.* **75** 549–58

Vasiliev B A, Shepelev S F, Ashirmetov M R and Poplavsky V M 2013 BN-1200 reactor power unit design development (Paris, 4–7 March 2013) https://inis.iaea.org/collection/NCLCollectionStore/_Public/45/089/45089480.pdf

Vasiliev B A, Belov S B, Kiselev A V, Eliseev V A, Khomiakov Iu S and Rodina E A 2021 Unification of the BN-1200 reactor core designs with MOX and MNUP fuel *Nucl. Eng. Des.* **382** 111387

Wang X, Kuang B, Liu P, Hou J, Hu W and Li Y 2021 Performance test of conceptually designed hydraulically suspended passive shut down subassembly for CFR600 *Prog. Nucl. Energy* **140** 103906

World Nuclear Association 2021 Fast neutron reactors https://world-nuclear.org/information-library/current-and-future-generation/fast-neutron-reactors.aspx

World Nuclear News 2022 Beloyarsk BN-800 fast reactor running on MOX (13 September 2022) www.world-nuclear-news.org/Articles/Beloyarsk-BN-800-fast-reactor-running-on-MOX

Yeom S, Han J W, Ryu S, Kim D, Eoh J H and Choi S R 2020 Design and evaluation of reactor vault cooling system in PGSFR *Nucl. Eng. Des.* **365** 110717

Yoo J, Chang J, Lim J-Y, Cheon J-S, Lee T-H, Kim S K, Lee K L and Joo H-K 2016 Overall system description and safety characteristics of prototype Gen IV sodium cooled fast reactor in Korea *Nucl. Eng. Technol.* **48** 1059–70

Zhang H 2023 China started operation of its first CFR-600 breeder reactor *IPFM Blog* (15 December 2023) https://fissilematerials.org/blog/2023/12/china_started_operation_o.html

IOP Publishing

Generation IV Nuclear Reactors
Design, operation and prospects for future energy production
Richard A Dunlap

Chapter 12

Lead-cooled fast reactors (LFRs)

Chapter 12 discusses the development of Generation IV lead-cooled fast reactors. The advantages and disadvantages of lead and lead-bismuth eutectic (LBE) as reactor coolants is discussed. Because of the high boiling point of these coolants, the negative void coefficient is not as much of a concern as it is in sodium-cooled reactors. The early history of the use of LBE-cooled reactors for marine use is discussed. The experience gained from these reactors has provided insight into the design of safe Generation IV reactors. This chapter discusses several lead-cooled reactor development projects in Russia, the European Union, and China. Most current developments in this area utilize lead rather than LBE as a coolant in order to avoid potential problems with the build-up of toxic ^{210}Po in the coolant.

12.1 Introduction

Sodium-cooled fast reactors, as discussed in the previous chapter, have received considerable attention, dating back to the 1950s. While some experimental sodium-cooled reactors have achieved a good level of success, others have not performed very well. Although there are a number of sodium-cooled reactor projects that are currently underway, there remain two significant concerns about the use of sodium as a fast reactor coolant: the first is the reactivity of sodium on contact with water or air, which has led to numerous reactor fires in the past; and the second is the positive void coefficient that is typical of sodium-cooled reactors. This latter characteristic can lead to reactor instability in the event of the coolant boiling as a result of core overheating. The use of a different liquid metal coolant can help to mitigate these concerns. Lead and lead alloys are attractive alternatives to sodium (Schulenberg 2022, Smith and Cinotti 2023) and are discussed in this chapter.

12.2 Advantages and disadvantages of lead as a coolant

Table 12.1 presents a comparison of the relevant properties of lead and sodium as related to their use as a fast reactor coolant. Based on the information in the table, as well as the

doi:10.1088/978-0-7503-6069-2ch12

Table 12.1. Comparison of the properties of lead and sodium.

Property	Lead (Pb)	Sodium (Na)
Atomic weight	207.20	22.99
Density (kg m^{-3})	10 660 (328 °C, 0.1 MPa)	856 (400 °C, 0.1 MPa)
Thermal neutron absorption cross section (b)	0.17	0.53
Thermal conductivity (W $(m\cdot K)^{-1}$)	17.7 (500 °C)	72.2 (400 °C)
Molar heat capacity (J $(mol\cdot K)^{-1}$)	26.65	23.2
Melting temperature (°C)	327.5 (0.1 MPa)	97.8 (0.1 MPa)
Boiling temperature (°C)	1749 (0.1 MPa)	882.9 (0.1 MPa)

Table 12.2. Abundances of naturally occurring lead isotopes and their thermal neutron absorption (n, γ) cross sections.

Isotope	Natural abundance (%)	Neutron cross section (b)
^{204}Pb	1.4	0.7
^{206}Pb	25.2	0.03
^{207}Pb	21.7	0.72
^{208}Pb	51.7	0.0005

general chemical properties of these two elements, we can summarize the advantages and disadvantages of lead (particularly in comparison to sodium) as a reactor coolant.

12.2.1 Advantages

Reactivity: Lead does not undergo any rapid chemical reactions on exposure to air or water at high temperatures. Therefore, any leaks into the steam circulation system or outside of the reactor do not pose a fire hazard.

Neutron moderation: Neutrons interact with lead nuclei, primarily by elastic scattering (see sections 2.7 and 3.4). Since the atomic weight of the lead atom is much greater than that of the sodium atom, the neutron energy loss per collision is much less for lead. Overall, this means that the neutron spectrum in the lead-cooled reactor is harder (i.e., more energetic) than in the sodium-cooled reactor. The low level of moderation also allows for increased distance between the fuel elements and allows for more effective cooling.

Neutron reflection: The large lead nuclear mass also improves lead's ability to reflect neutrons and provides better neutron utilization in the core.

Neutron absorption: As noted in the table, lead has a very low thermal neutron absorption cross section. Those neutrons that are absorbed via the (n, γ) reaction do not produce any undesirable by-products. We can see this by looking at the reactions involving all of the naturally occurring isotopes of lead, which are summarized in table 12.2.

Neutron absorption by ^{204}Pb produces ^{205}Pb, which is a long-lived (3×10^7 y half-life) nuclide that decays to the ground state of ^{205}Tl by electron capture. Neutron capture by ^{206}Pb and ^{207}Pb produces other stable Pb isotopes. Neutron absorption by ^{208}Pb (which has a very low cross section, see table 12.2) produces ^{209}Pb, which decays to stable ^{209}Bi by β^- decay with a half-life of 3.23 h. However, see the discussion concerning bismuth impurities in the next section.

Gamma-ray shielding: Because of its high atomic number, lead atoms interact strongly with photons (including gamma-rays). This reduces the gamma flux from fission by-products outside of the core.

Operating temperature range: Because of the wide range of temperatures over which lead remains in the liquid phase, there is considerable flexibility in the reactor operating conditions.

High boiling temperature: The high boiling temperature of lead means that the formation of voids due to coolant boiling is much less likely than it is for sodium. Thus, concerns over the effects of a positive void coefficient are not as relevant.

Thermal efficiency: The ability to operate the reactor at high temperature improves the thermodynamic efficiency for electricity Generation.

Natural circulation: The ability to operate the cooling system near atmospheric pressure and the high density, thermal conductivity, and heat capacity of lead provide a high degree of natural circulation and a good level of passive decay heat removal.

12.2.2 Disadvantages

Density: The high density of lead means that the reactor core will be heavier than that of a fast reactor cooled by molten salt, gas, or sodium. This means that additional structural support must be incorporated into the design. The high density of lead (in the liquid state) compared to the density of many reactor components means that possible buoyancy effects must be considered in the reactor design.

Nickel solubility: Nickel is soluble in liquid lead (and lead alloys). This means that reactor components that come in contact with the lead coolant cannot contain nickel. This applies particularly to fuel cladding that has traditionally been made of stainless steel (which contains nickel). This must be considered in the reactor design by choosing appropriate materials or utilizing appropriate coatings to protect materials.

Erosion: The circulation of lead, particularly if it contains impurities such as oxides, can erode metallic components in the cooling system.

Opacity: Liquid lead is opaque to visible light (as is liquid sodium). This is in contrast to other reactor coolants such as helium, water, CO_2, etc. This opacity means that observing potential problems within the reactor and servicing these components is more complex.

Solidification: Since lead has a much higher melting temperature than (for example) sodium, it is essential during reactor operation to ensure that the temperature of the lead coolant does not drop below this temperature. If this occurs, then plugs in the cooling system can form leading to a stoppage of the coolant circulation.

Oxygen contamination: Although lead does not undergo a violent reaction when exposed to air, as does sodium, it does chemically react with oxygen at elevated temperatures. This reaction results in the formation of PbO. If oxygen is present in the lead coolant, then PbO can form, and this will precipitate out of the coolant and can clog passages in the cooling system.

12.3 Comparison of lead-bismuth eutectic with lead

While the use of pure liquid lead as a fast reactor coolant is specified as a possible design feature of a Generation IV reactor, a number of projects in the past and at present utilize a lead-bismuth alloy of the eutectic composition 44.5% Pb + 55.5% Bi. Table 12.3 compares the basic thermal properties of lead and LBE. While LBE shares many of the same advantages and disadvantages as pure lead for a fast reactor coolant, there are some differences based on the thermodynamic properties shown in table 12.3, as well as some differences related to other properties of the bismuth-containing alloy. These differences are discussed below.

12.3.1 Advantages of LBE over lead

Melting temperature: The substantially lower melting temperature of LBE compared to pure lead is a considerable advantage to reactor operation. The LBE is easier to melt and also allows a greater degree of safety because accidental freezing and blockage of the coolant circulation is of much less concern. The similar boiling temperatures of lead and LBE ensures that accidental coolant boiling and void formation are not a concern.

12.3.2 Disadvantages of LBE compared to lead

Cost and availability: Lead is inexpensive and readily available. Bismuth currently cost about ten times as much as lead. In the event of widespread use of LBE-cooled reactors, there may be concern about the availability of bismuth.

Thermal expansion: LBE experiences a crystallographic phase transition just below its melting temperature. This phase transition causes LBE to expand as it cools below the freezing point. Thus, it is important to ensure that the coolant does not solidify in locations (e.g., the core) where this expansion could be detrimental to the equipment.

Table 12.3. Transition temperatures for lead and LBE compared to sodium. LBE = 44.5% Pb + 55.5% Bi. Properties of sodium are shown for comparison.

Coolant	Melting temperature	Boiling temperature	Liquid range
Lead	327.5 °C	1749	1421.5 °C
LBE	123.5 °C	1670 °C	1546.5 °C
Sodium	+97.8 °C	+882.9 °C	786.1 °C

Thermal conductivity: LBE has a slightly smaller thermal conductivity (at 500 °C), 14.3 W $(m \cdot K)^{-1}$, compared to 17.7 W $(m \cdot K)^{-1}$ for pure lead.

Neutron absorption: Neutron capture by 100% naturally abundant ^{209}Bi is probably the most serious concern related to the use of LBE as a fast reactor coolant. The total thermal neutron absorption cross section for ^{209}Bi is 39.6 mb (Borella *et al* 2011). This corresponds to the reaction,

$$n + {}^{209}\text{Bi} \rightarrow {}^{210}\text{Bi} + \gamma.$$

This is followed by the β^- decay of ^{210}Bi,

$${}^{210}\text{Bi} \rightarrow {}^{210}\text{Po} + e^- + \overline{\nu}$$

with a half-life of 5.0 d. The ^{210}Po decays by α-decay with a half-life of 138.4 d,

$${}^{210}\text{Po} \rightarrow {}^{206}\text{Pb} + \alpha. \tag{12.1}$$

The presence and decay of ^{210}Po are major concerns related to the use of Bi-containing coolant. Because of its half-life, ^{210}Po will reach an equilibrium concentration in the reactor core. For a small (80 MW_{th}) reactor, Cinotti *et al* (2011) have estimated that the equilibrium concentration will be 2 kg of ^{210}Po. ^{210}Po is a radiation hazard and is highly toxic, and would represent a serious concern in the event of a primary coolant leak. A second concern deals with the decay of ^{210}Po. The α-particles emitted in equation (12.1) have an energy of 5.3 MeV and contribute to the heat generated in the core. For the 80 MW_{th} reactor, it has been estimated that, in the event of a core shutdown, the heat produced by the ^{210}Po decay would be equal to the decay heat after a period of 5 days (Cinotti *et al* 2011). This must be factored into the analysis of decay heat removal.

The use of pure lead does not completely eliminate the ^{210}Po problem for two reasons. First, bismuth is a common impurity in lead and even high purity lead can contain around 10^{-3}% bismuth (all ^{209}Bi), which contributes to ^{210}Po production, as described above. Second, as noted above, neutron capture by 51.7% naturally abundant ^{208}Pb produces ^{209}Pb which β^- decays to ^{209}Bi, which again contributes to ^{210}Po production by neutron capture. Compared to LBE, however, equilibrium ^{210}Po concentrations in the primary coolant of a pure Pb cooled reactor core are several orders of magnitude smaller.

12.4 History of early lead-cooled fast reactors

To date, no reactor cooled with pure lead has been constructed. However, several LBE-cooled reactors have been in use, dating back to the late-1950s. These reactors were constructed in Russia (or the previous Soviet Union) and relatively little detailed information about their design and operation has been released to the general public. However, some general accounts of their operation have appeared in the literature (e.g., GlobalSecurity.org n.d., Schulenberg 2022, Smith and Cinotti 2023, Troyanov *et al* 2022, World Nuclear Association 2021, Zrodnikov *et al* 2000) and the present section provides a brief overview of this previous work.

The world's first LBE-cooled fast reactor (designated 27/VT) became operational on 25 December 1958 at the Institute of Physics and Power Engineering in Obninsk, Kaluga Oblast, Soviet Union. This is the location where the Obninsk Nuclear Power Plant became operational in 1954 (see section 4.6.1). The 27/VT reactor produced 73 MW_{th} and was a prototype for the VT-1 reactor that was used in the Soviet nuclear powered submarine development program (designated as Project 645). In 1962, the Soviet nuclear submarine K-27 equipped with two VT-1 73 MW_{th} lead-bismuth cooled reactors, was launched. (Note that the Soviet Union's first nuclear submarine, the K-3 Leninsky Komsomol, used a water-cooled reactor and was launched in 1957.) In 1968, oxygen contamination in the LBE-coolant of the reactor in K-27 led to the precipitation of lead oxide, which became stuck in the reactor core. This caused the core to overheat and some of the fuel rods melted. Excess radiation exposure resulted in nine deaths. K-27 was subsequently decommissioned and in 1982, after removal of the reactor, was sunk in the Kara Sea.

Subsequent to the development of the VT-1 reactor, two more advanced LBE-cooled marine reactors were developed in the Soviet Union. These reactors were designated OK-550 (Project 705) and BM-40A (Project 705K), and both had an output of 155 MW_{th}. These were used for the construction of seven additional nuclear submarines, labeled 'Alfa-class submarines', which went into service beginning in 1968. The final Alfa-Class submarine was decommissioned in 1995, ending the use of LBE-cooled nuclear reactor for marine propulsion. Two incidents occurred during the lifetime of the Alfa-class submarines. In 1971, one of the OK-550 reactors experienced corrosion of the primary circuit cooling pipes due to moisture build-up from leaks in the steam circuit. In 1982, one of the BM-40A reactors experienced a leak of steam into the primary cooling loop. This resulted in the release of about 150 l of ^{210}Po contaminated coolant into the submarine. There is no record of any deaths that resulted from incidents in the Alfa-class submarines. As a result of these experiences with LBE-cooled submarine reactors, all subsequent nuclear submarines worldwide use water-cooled reactors.

The experience (both good and bad) gained from the use of LBE-coolant for submarine reactors, along with two prototype land-based reactors that were constructed as part of these projects, led the way for future Russian development of Generation IV LBE-cooled power reactors, as described below.

12.5 BREST-300 and BREST-1200 (Russia)

There are several heavy liquid metal cooled Generation IV reactor development projects around the world. These include projects in Europe, the United States, Russia, and China. While a few use LBE-coolant, the majority use pure lead. The present section and the following two sections describe some of the more notable design projects that are currently being pursued.

Based on the discussion in the previous section, it is clear that Russia has the most experience in the development and construction of lead or LBE-cooled fast reactors. This expertise has been put to use in the development of future Generation IV reactors. Beginning in the early-2000s, Russia undertook two reactor design projects that utilized lead-based coolants. The first was the development of the SVBR-100 reactor (Svintsovo-Vismutovyi Bystryi Reaktor—lead-bismuth fast reactor), a small modular LBE-cooled reactor rated at 280 MW_{th} and 100 MW_e. The second project was the development of a lead-cooled reactor, BREST (Bystry REaktor so Svintsovym Teplonositelem—fast reactor with lead coolant). The SVBR project was discontinued in 2018. The BREST development program is working on two reactors, the BREST-300 (sometimes called BREST-OD-300), a 300 MW_e demonstration reactor and a 1200 MW_e commercial-scale version, BREST-1200 (sometimes called BREST-OD-1200), which is intended to follow the successful testing of BREST-300 (Adamov *et al* 2021).

The basic design of the BREST-300 reactor is illustrated in figure 12.1. A major design feature of the BREST-300 reactor is the inclusion of the reactor vessel containing the core in a concrete containment structure or vault. Each of the four steam generators and main circulating pump loops is similarly contained in a separate concrete vault. The layout of the reactor, circulating pump, and steam generator for each loop is illustrated in figure 12.2. It can be seen that the steam generator is placed above the level of the reactor core (i.e., by 6 m). This means that natural convection will aid the main pumps in circulating the lead coolant through the primary loop. Feed water that is input into the steam generator is maintained at 337 °C (above the melting temperature of lead) in order to ensure that the primary coolant does not freeze.

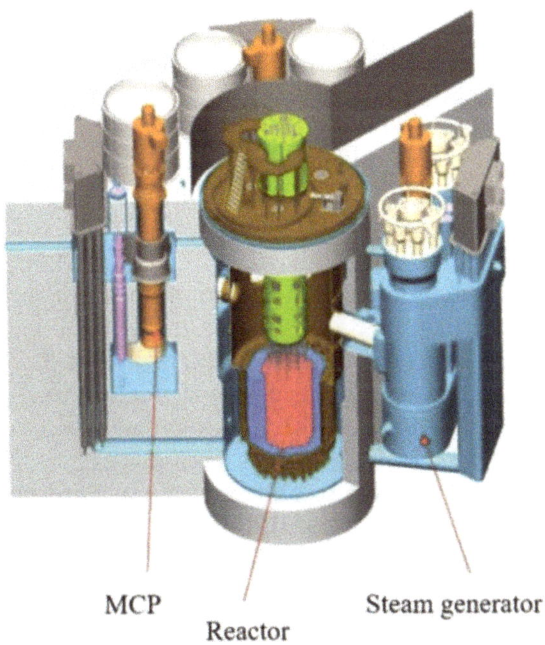

Figure 12.1. Cutaway of the BREST-300 reactor showing the reactor (core), main circulating pump (MCP), and steam generator. Reproduced from Chudinova and Nikonov (2020). CC BY 3.0.

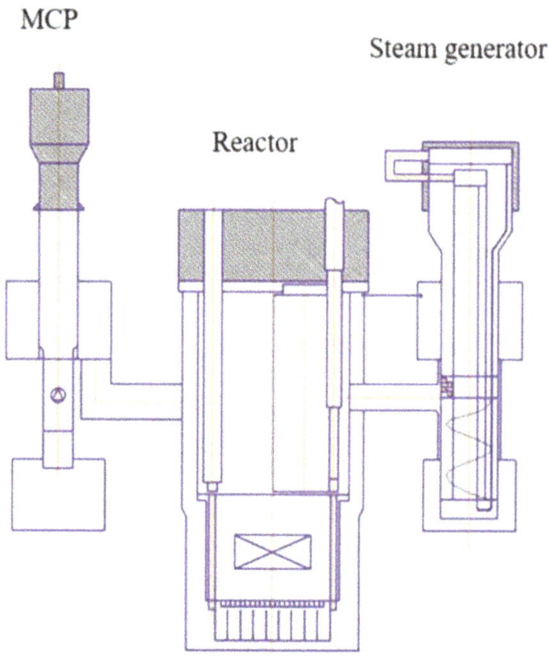

Figure 12.2. Layout of the components of the BREST-300 reactor. Reproduced from Chudinova and Nikonov (2020). CC BY 3.0.

The BREST-300 reactor is designed to use mixed nitride fuel (SNUP), i.e., UN-PuN, produced by mixing recycled plutonium and depleted uranium, with a fissile enrichment of about 13.5%. This fuel has been extensively tested in the BN-600, as described in the previous chapter. The nitrogen in the nitride fuel is enriched in ^{15}N (from the natural abundances of 99.63% ^{14}N and 0.37% ^{15}N). This avoids radioactive ^{14}C build-up in the core that is produced from ^{14}N by the (n, p) reaction,

$$n + {}^{14}N \rightarrow {}^{14}C + p.$$

Fuel pins are contained in stainless-steel cladding with a lead gap to ensure effective heat transfer. The fuel tubes have a 0.9 m void at the top to collect fission gases.

A sample core layout is shown in figures 12.3 and 12.4. The core is divided into several fuel zones. Each zone has a similar enrichment but uses a different fuel rod diameter in order to flatten the radial power distribution. There is no breeder blanket. All breeding occurs in the core and ^{239}Pu that has undergone fission is replaced with new ^{239}Pu bred from fertile ^{238}U in depleted uranium. The core breeding ratio is estimated to be 1.05 (Orlov and Gabaraev 2023). The active core is surrounded by axial and radial lead neutron reflectors.

The BREST-300 (and the future BREST-1200) design is consistent with the objectives of the Generation IV Forum. The relevant criteria are described below.

Sustainability: The BREST-300 reactor makes effective use of natural uranium by breeding new fissile material in the core with a breeding ratio greater than unity.

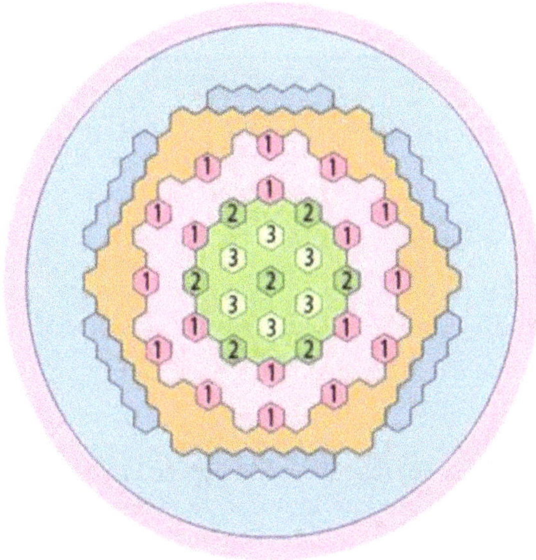

Figure 12.3. Radial cross section of a sample core design of the BREST-300 reactor. Green and purple—central part, orange and darker blue—peripheral part. (1) Control rods, (2) emergency shutdown rods, and (3) compensation reactivity rods. Reproduced from Karazhelevskaya *et al* (2020). CC BY 3.0.

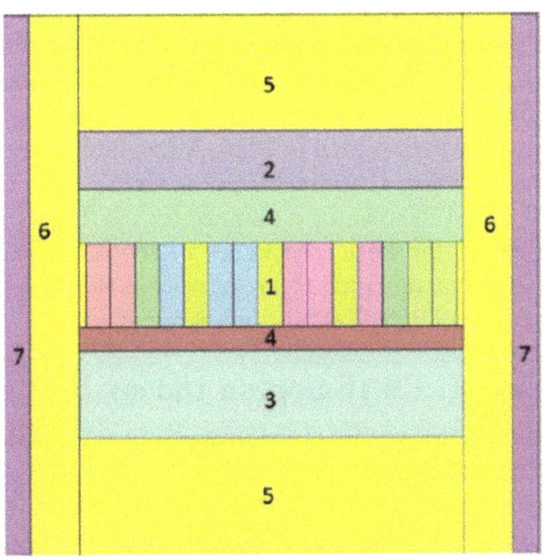

Figure 12.4. Axial cross section of the core of the BREST-300 reactor: (1) reactor core, (2) fuel assembly top fitting, (3) fuel assembly bottom fitting, (4) gas cavity, (5) end reflector, (6) side reflector, and (7) steel core barrel. Reproduced from Karazhelevskaya *et al* (2020). CC BY 3.0.

Table 12.4. Comparison of basic properties of the BREST-300 and BREST-1200 reactors. Data adapted from Bokova *et al* (2021) and references therein.

Property	BREST-300	BREST-1200
Thermal power (MW_{th})	700	2800
Electrical power (MW_e)	300	1200
Active core diameter (m)	2.3	4.755
Active core height (m)	1.1	1.1
Number fuel assemblies	185	332
Primary coolant inlet (°C)	420	420
Primary coolant outlet (°C)	540	540
Net efficiency (%)	43	43

Safety: The lead coolant is not reactive when exposed to water or air (as is sodium) and the coolant circulates at near atmospheric pressure so that leaks are not augmented by a pressure differential. The high boiling temperature of lead minimizes the risk of core instability due to void formation and maintaining the temperature in the steam loop above the melting temperature of lead eliminates the risk of primary coolant blocks caused by freezing. The vaulted concrete containment structure serves as a back-up reactor vessel in the event of a core accident.

Non-proliferation: Since there is no breeder blanket, the quantity of fissile ^{239}Pu in the reactor core over the lifetime of the reactor is much smaller than that in a traditional fast breeder reactor and much less accessible. In addition, the ^{239}Pu content in the core is less than 20%, meaning that the fuel is of minimal use for weapons purposes.

Construction on the BREST-300 reactor began in 2021 in Seversk, Tomsk Oblast, Russia. It is expected to be operational in 2026. A fuel fabrication facility is planned for construction at the Siberian Chemical Combine in Seversk. Construction of the commercial-scale BREST-1200 reactor is planned for the future. Table 12.4 provides a comparison of some of the basic properties of the demonstration version and commercial version of the lead-cooled fast reactor. The BREST-1200 will follow the same basic design as the BREST-300 but on a larger scale.

12.6 ALFRED and ELFR (European Union)

The Advanced Lead-cooled Fast Reactor European Demonstrator (ALFRED) is a Generation IV lead-cooled demonstrator reactor that is being developed as a step towards the European Lead-cooled Fast Reactor (ELFR) commercial lead-cooled reactor. ALFRED is being designed by the international consortium Fostering ALFRED construction (abbreviated FALCON), which includes contributors from Italy's National Agency for New Technologies, Energy and the Environment, Ansaldo Nucleare, and Romania's Reinvent Energy (World Nuclear News 2021). The reactor will be constructed in Mioveni, Romania near Pitesti, where a nuclear fuel manufacturing plant produces fuel for Romania's CANDU reactors.

In addition to serving as a testing ground for design concepts and materials for the ELFR reactor, ALFRED will, in particular, develop passive safety features that will not require a connection to an external grid in the event of a reactor incident (Frignani *et al* 2019). Table 12.5 summarizes the basic characteristics of the ALFRED reactor.

A diagram illustrating a cross section of the ALFRED reactor is shown in figure 12.5. The reactor vessel contains the core along with the associated eight

Table 12.5. Characteristics of the ALFRED reactor. Data adapted from Alemberti *et al* (2020), Ibrahim *et al* (2021), Luzzi *et al* (2014), and Orlov and Gabaraev (2023).

Property	Value
Thermal power (MW_{th})	300
Electrical power (MW_e)	125
^{239}Pu enrichment (inner zone)	20.5%
^{239}Pu enrichment (outer zone)	26.2%
Pins per fuel assembly	127
Fuel assemblies (inner zone)	57
Fuel assemblies (outer zone)	78

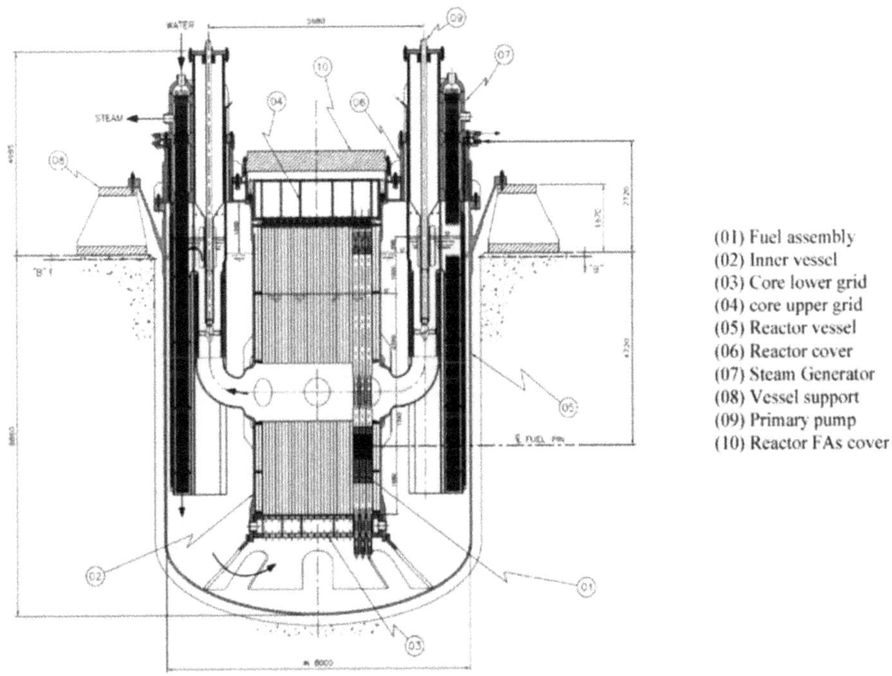

Figure 12.5. Axial cross section of the ALFRED reactor. Reprinted from Frignani *et al* (2019), Copyright (2019), with permission from Elsevier.

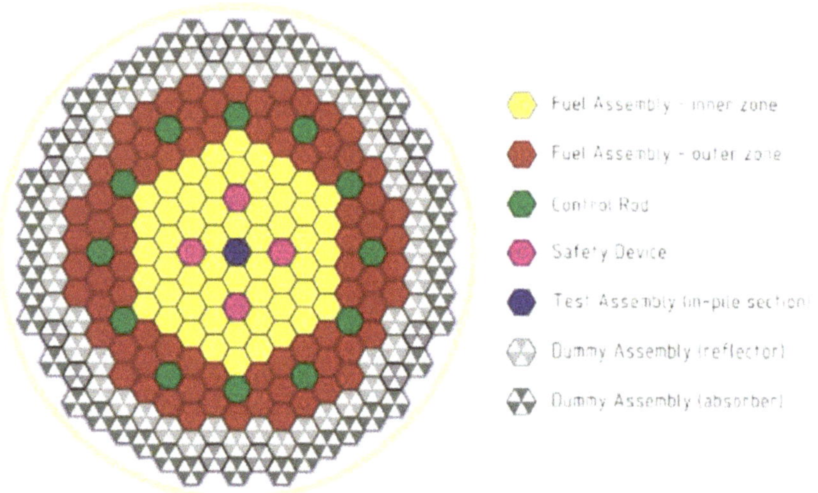

Figure 12.6. ALFRED core configuration for Stage 3 (see below) showing 57 inner fuel assemblies, 78 outer fuel assemblies, 12 control rods, 4 safety rods, and 108 reflector/absorber assemblies. Reprinted from Alemberti *et al* (2020), Copyright (2020), with permission from Elsevier.

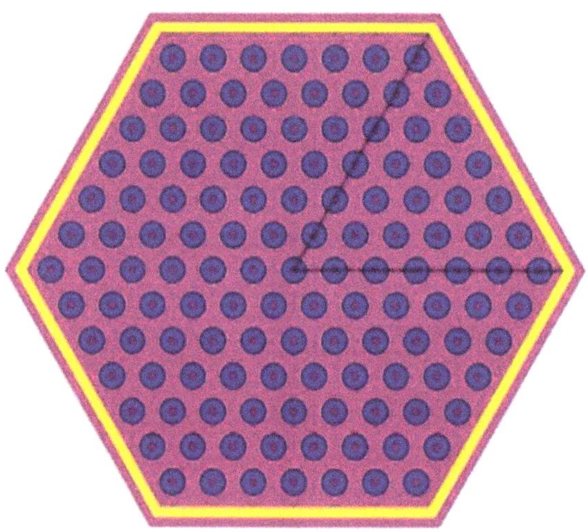

Figure 12.7. Radial cross section of a fuel element from ALFRED. Reproduced from Ibrahim *et al* (2021). CC BY 3.0.

primary circulating pumps and eight steam generators. The core is divided into two zones, as indicated in table 12.5 and as shown in figure 12.6. There is no breeder blanket and fissile fuel is bred in the core from fertile ^{238}U in mixed oxide fuel with ^{239}Pu enrichment in the two zones as given in table 12.5. The fuel assemblies contain 127 fuel pins, as illustrated in figure 12.7.

Table 12.6. Stages of operation of ALFRED. Data adapted from Alemberti *et al* (2020) and Orlov and Gabaraev (2023).

Stage	Condition	Core inlet (°C)	Core outlet (°C)	Power (MW_{th})
Stage 0	Start-up	390	390	0
Stage 1	Low power	390	430	100
Stage 2	Medium power	400	480	200
Stage 3	High power	400	520	300

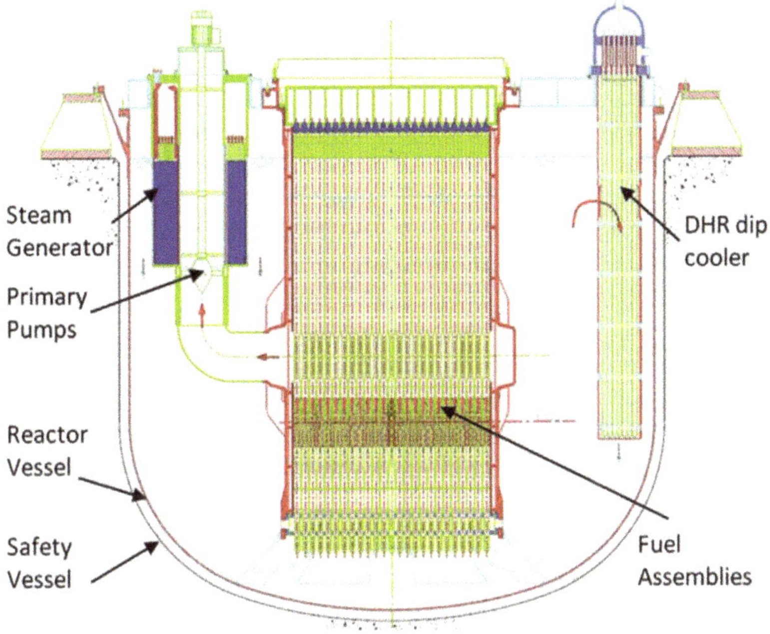

Figure 12.8. Diagram of the ELFR reactor. Reprinted from Alemberti *et al* (2014), Copyright (2014), with permission from Elsevier.

ALFRED is designed for staged operation, as summarized in table 12.6, where the reactor thermal output will be increased over time in stages. Modifications to the core configuration and operating conditions will be made, as appropriate, in order to increase thermal power output. At present, the timeline for the development of ALFRED puts Stage 3 operation somewhere between 2035 and 2040 (Orlov and Gabaraev 2023).

Operational experience from ALFRED will provide input into the final design of the commercial-scale ELFR reactor. ELFR, as shown in figure 12.8, will follow the same general reactor vessel layout as ALFRED. The reactor vessel will contain the

Table 12.7. Characteristics of the ELFR reactor. Data adapted from Stanisz *et al* (2016).

Property	Value
Thermal power (MW_{th})	1500
Electrical power (MW_e)	600
^{239}Pu enrichment	~18%
Pins per fuel assembly	169
Fuel assemblies (inner zone)	157
Fuel assemblies (middle zone)	96
Fuel assemblies (outer zone)	174

reactor core along with the eight primary coolant pump/steam generator units. As the diagram also shows, the decay heat removal (DHR) system is included in the vessel and consists of four passive units as shown. Four additional DHR units are connected to four of the eight steam generators. The basic characteristics of the ELFR reactor are given in table 12.7 and can be compared to those for the ALFRED reactor from table 12.5. Like the ALFRED reactor, ELFR will use MOX fuel and has no breeder blanket. Breeding is accomplished in the core with a breeding ratio of 1.07 (International Atomic Energy Agency n.d.). The details of the ELFR reactor project will depend on the success of ALFRED and an operational reactor is planned for some time between 2040 and 2050.

12.7 SNCLFR-100 (China)

SNCLFR-100 is a small modular Generation IV lead-cooled fast reactor that is being developed by the University of Science and Technology of China (USTC) in Hefei, Anhui, China (see e.g., Chen *et al* 2016, Duan *et al* 2021). An important aspect of the design of the SNCLFR-100, as shown in figure 12.9, is its use of natural circulation for cooling. The reactor core and four heat exchangers (steam generators) are contained in the reactor pool inside the reactor vessel. The hot plenum (pool) at the top of the reactor vessel is split into four radial sections that supply the four steam generators. The hot lead falls through the once-through heat exchanger (blue arrows in the figure), thereby creating steam in the secondary (water) loop to supply the turbine/generators. The lead, as it cools, falls to the cold plenum (pool) at the bottom of the reactor vessel, where it is heated again by the core and rises (red arrows in the figure) through the center of the vessel. The natural circulation is adequate to cool the reactor core in both normal and anomalous operating conditions, and the reactor does not require the use of cooling pumps. This approach means that loss of coolant flow as a result of circulating pump failure is not a safety concern.

The layout of the SNCLFR-100 core is shown in figure 12.10. This figure shows that the core consists of three fuel zones surrounded by a reflector and an outside radiation shield. Control rods are interspersed with fuel assemblies, as shown in the

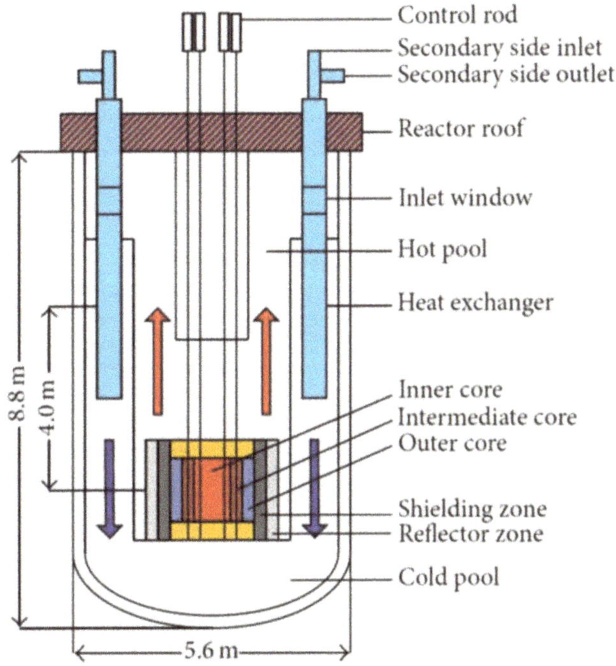

Figure 12.9. Cross section of the SNCLFR-100 reactor showing the flow of the primary coolant by natural circulation. Reproduced from Guo *et al* (2021). CC BY 4.0.

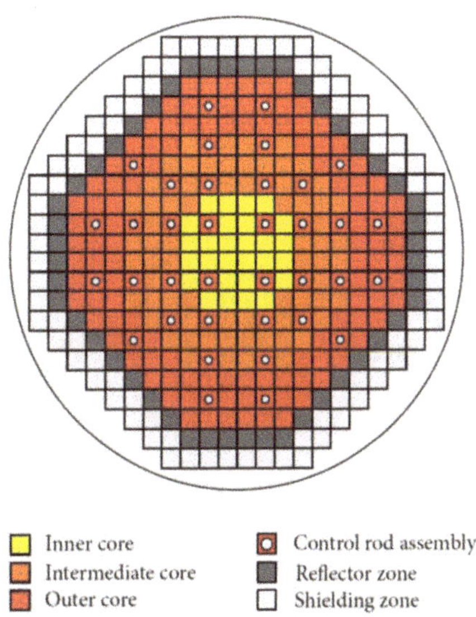

Figure 12.10. Layout of the SNCLFR-100 core showing the three zones and locations of the control rods. The reflector zone and shielding zone are shown surrounding the active core. Reproduced from Zhao *et al* (2016). CC BY 3.0.

figure. The number of different assemblies in the core is given in table 12.8. Details of the fuel assemblies without and with control rods are shown in figures 12.11 and 12.12, respectively. Those assemblies without control rods contain 81 pins, while those with control rods contain 72 pins and the control rod which replaces nine pins at the center of the assembly. The fuel assemblies of the SNCLFR-100 are not the typical hexagonal cross section seen in the designs discussed above. These fuel

Table 12.8. Properties of the SNCLFR-100 reactor. Data adapted from Guo *et al* (2021) and Zhao *et al* (2016).

Property	Value
Thermal power (MW_{th})	100
Electrical power (MW_e)	40
Number fuel assemblies	204
Number control assemblies	36
Number reflector assemblies	48
Number shielding assemblies	84
Pins per fuel assembly	81
Pins per control assembly	72
^{239}Pu enrichment (inner zone)	16%
^{239}Pu enrichment (middle zone)	19%
^{239}Pu enrichment (outer zone)	24%
Inlet temperature (°C)	400
Outlet temperature (°C)	480

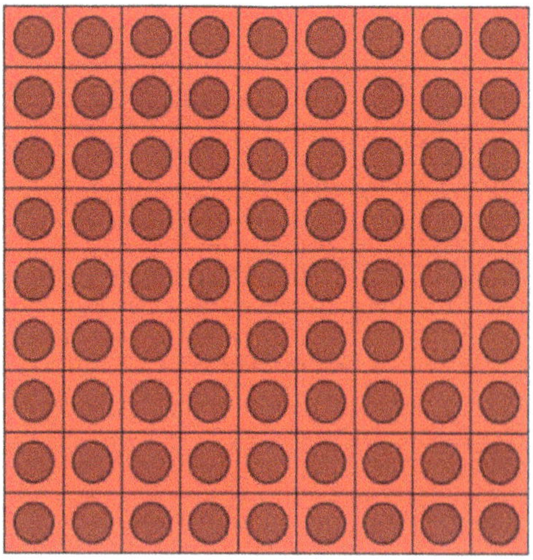

Figure 12.11. SNCLFR-100 fuel assembly containing 81 fuel pins. Reproduced from Zhao *et al* (2016). CC BY 3.0.

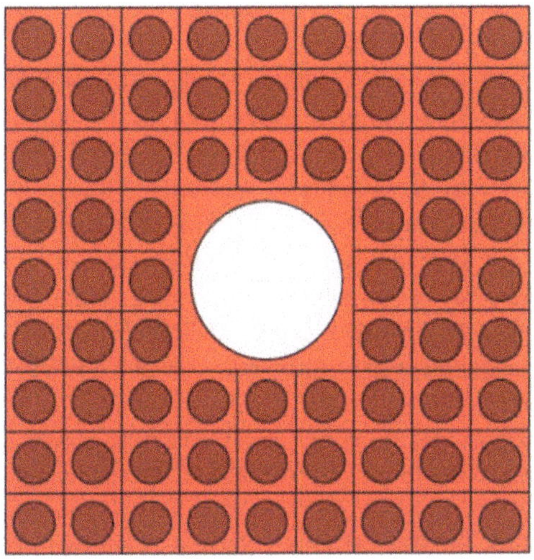

Figure 12.12. SNCLFR-100 fuel assembly containing control rod and 72 fuel pins. Reproduced from Zhao *et al* (2016). CC BY 3.0.

assemblies are square as seen in figures 12.10–12.12, and the pins are arranged on a square lattice within the fuel assembly. This arrangement, rather than a hexagonal arrangement, allows for additional space around the fuel pins for coolant as an aid for the natural circulation. In addition, the SNCLFR-100 reactor is inherently safe because it is designed with a negative coefficient of reactivity, while the void coefficient is not a safety concern because of the high boiling temperature of lead.

The fuel for the SNCLFR-100 reactor is mixed oxide (UO_2-PuO_2). The enrichment, in ^{239}Pu, for the three core zones is given in table 12.8 and, as is the case for several reactors discussed above, the enrichment is greater for the outer zone(s) in order to even out the power distribution in the core. As illustrated in figure 12.10 there is no breeder blanket and the SNCLFR-100 reactor will breed new fissile ^{239}Pu from ^{238}U within the core. This approach, as with other lead-cooled fast reactors, will improve the reactor's proliferation resistance.

Table 12.8 shows that the SNCLFR-100 reactor has a small 100 MW_{th} and 40 MW_e output. This small modular design is appropriate for remote use for isolated communities. This application is consistent with its inherently safe design and its low maintenance 10-year service life. The timeline for the construction of the SNCLFR-100 has not yet been announced.

References

Adamov E O, Kaplienko A V, Orlov V V, Smirnov V S, Lopatkin A V, Lemekhov V V and Moiseev A V 2021 BREST lead-cooled fast reactor: from concept to technological implementation *At. Energy* **129** 179–87

Alemberti A, Caramello M, Frignani M, Grasso G, Merli F, Morresi G and Tarantino M 2020 ALFRED reactor coolant system design *Nucl. Eng. Des.* **370** 110884

Alemberti A, Smirnov V, Smith C F and Takahashi M 2014 Overview of lead-cooled fast reactor activities *Prog. Nucl. Energy* **77** 300–7

Bokova T A, Meluzov A G, Bokov P A, Volkov N S and Marov A R 2021 Variants of nuclear power plants of small and medium power with heavy liquid-metal coolants *Open J. Microphys* **11** 53–71

Borella A, Belgya T, Kopecky S, Gunsing F, Moxon M, Rejmund M, Schillebeeckx P and Szentmiklósi L 2011 Determination of the ^{209}Bi(n, γ)^{210}Bi and ^{209}Bi(n, γ)^{210m,g}Bi reaction cross sections in a cold neutron beam *Nucl. Phys.* **850** 21

Chen H *et al* 2016 Conceptual design of a small modular natural circulation lead cooled fast reactor SNCLFR-100 *Int. J. Hydrog. Energy* **41** 7158–68

Chudinova V A and Nikonov S P 2020 Modeling the core of a lead-cooled reactor *J. Phys.: Conf. Ser.* **1689** 012052

Cinotti L, Smith C F, Sekimoto H, Mansani L, Reale M and Sienicki J J 2011 Lead-cooled system design and challenges in the frame of generation iv international forum *J. Nucl. Mater.* **415** 245–53

Duan W-S, Zou Z-R, Luo X and Chen H-L 2021 Startup scheme optimization and flow instability of natural circulation lead-cooled fast reactor SNCLFR-100 *Nucl. Sci. Tech.* **32** 133

Frignani M, Alemberti A and Tarantino M 2019 ALFRED: A revised concept to improve pool related thermal-hydraulics *Nucl. Eng. Des.* **355** 110359

GlobalSecurity.org n.d. OKB Gidropress – Naval Reactors www.globalsecurity.org/military/world/russia/gidropress.htm

Guo C, Zhao P, Deng J and Yu H 2021 Safety analysis of small modular natural circulation lead-cooled fast reactor SNCLFR-100 under unprotected transient *Front. Energy Res.* **9** 678939

Ibrahim M, Ibrahim A, Aziz M, Saudi H A and Hassaan M Y 2021 Comparative neutronic study for heterogeneous and homogeneous fuel assembly in a lead-cooled fast reactor *IOP Conf. Ser.: Mater. Sci. Eng.* **1171** 012009

International Atomic Energy Agency n.d. ELFR (Ansaldo Nucleare, Europe/Italy) https://aris.iaea.org/PDF/ELFR.pdf

Karazhelevskaya Y E, Levon M A, Terekhova A M and Zlobin A S 2020 Irregularity of plutonium isotopic composition of the BREST-OD-300 initial load *J. Phys.: Conf. Ser.* **1689** 012049

Luzzi L, Cammi A, Di Marcello V, Lorenzi S, Pizzocri D and Van Uffelen P 2014 Application of the TRANSURANUS code for the fuel pin design process of the ALFRED reactor *Nucl. Eng. Des.* **277** 173–87

Orlov A I and Gabaraev B A 2023 Heavy liquid metal cooled fast reactors: peculiarities and development status of the major projects *Nucl. Energy Tech.* **9** 1–18

Schulenberg T 2022 *The Fourth Generation of Nuclear Reactors—Fundamentals, Types, and Benefits Explained* (Berlin: Springer)

Smith C F and Cinotti L 2023 Lead-cooled fast reactors (LFRs) *Handbook of Generation IV Nuclear Reactors* 2nd edn ed I L Pioro (Cambridge, MA: Woodhead Publishing) ch 6 pp 195–230

Stanisz P, Oettingen M and Cetnar J 2016 Monte carlo modeling of lead-cooled fast reactor in adiabatic equilibrium state *Nucl. Eng. Des.* **301** 341–52

Troyanov V M, Toshinsky G I, Stepanov V S and Petrochenko V V 2022 Lead-bismuth cooled reactors: history and the potential of development. Part 1. History of development *Nucl. Energy Technol.* **8** 187–95

World Nuclear Association 2021 Nuclear Power in Russia https://world-nuclear.org/information-library/country-profiles/countries-o-s/russia-nuclear-power.aspx

World Nuclear News 2021 Contract for Romanian lead-cooled reactor research facility (23 November 2021) https://world-nuclear-news.org/Articles/Contract-for-Romanian-lead-cooled-reactor-research

Zhao P, Shi K, Li S, Feng J and Chen H 2016 CFD analysis of the primary cooling system for the small modular natural circulation lead cooled fast reactor SNRLFR-100 *Sci. Tech. Nucl. Install.* **2016** 9612120

Zrodnikov A V, Chitaykin V I, Gromov B F, Grigoryv O G, Dedoul A V, Toshinsky G I, Dragunov Yu G and Stepanov V S 2000 Use of Russian technology of ship reactors with lead-bismuth coolant in nuclear power *Advisory Group Meeting on Small Power and Heat Generation Systems on the Basis of Propulsion and Innovative Reactor Technologies; Obninsk (Russian Federation, 20–24 Jul 1998)* (Vienna: Int. Atomic Energy Agency) pp 127–55 IAEA-TECDOC-1172

IOP Publishing

Generation IV Nuclear Reactors
Design, operation and prospects for future energy production
Richard A Dunlap

Chapter 13

The future of nuclear energy

Chapter 13 is a summary of the current development of six generation IV nuclear reactors: the very high-temperature reactor (VHTR), the molten salt reactor (MSR), the supercritical water-cooled reactor (SCWR), the gas-cooled fast reactor (GFR), the sodium-cooled fast reactor (SFR), and the lead-cooled fast reactor (LFR). This summary includes a review of each of the Generation IV designs with an analysis of the advantages and disadvantages of each. The status of research on these reactor technologies is described, along with an analysis of the major challenges facing their commercial implementation. In the final section, each of the six Generation IV designs is assessed in terms of the ways in which they satisfy the Generation IV objectives, sustainability, safety and reliability, economics, and proliferation resistance. Finally, the ways in which Generation IV reactors can be integrated into our existing nuclear power infrastructure and can complement the development of other low-carbon energy sources is considered.

13.1 Introduction

It is clear from chapter 1 that the accelerated development of low-carbon and carbon-free sources of energy is essential for the future sustainability of our environment and that nuclear fission energy can play a significant role in this development. The Generation IV International Forum has identified six nuclear reactor technologies for development as commercially viable Generation IV fission reactors. The choice of reactor technologies was based on the evaluation of four basic criteria: sustainability, economics, safety and reliability, and proliferation resistance and physical protection. The significance of each of these criteria for the design of Generation IV reactors is considered below.

Sustainability: As discussed in section 5.10, it is clear that uranium and thorium resources are not renewable. Current estimates of uranium resources suggest that at the current rate of use nuclear energy can supply electricity for 85–110 years.

doi:10.1088/978-0-7503-6069-2ch13

Certainly, new resources can be discovered and that would extend this time. However, as fossil fuel use is phased out, the need for nuclear power could grow. It is clear that more efficient utilization of our uranium resources and also the implementation of technologies that can make use of fertile thorium are necessary in order to make nuclear energy a long-term component of our energy needs. The first aspect of the sustainability of many Generation IV reactors includes the use of closed fuel cycles, where a much greater proportion of the energy content of natural uranium is utilized. In some reactors, thorium is also included in the fuel. Generation IV reactors can utilize fuel recycling technology (see section 5.11) or may be designed to breed and burn. Some Generation IV reactors are capable of burning spent fuel from traditional light water reactors. The use of reactors that produce higher temperatures increases the utilization of nuclear fuel resources by increasing the thermodynamic efficiency of electricity generation and also by providing opportunities for cogeneration. This may be in the form of residential heating or in the form of heat for industrial processes. This latter category can include the production of hydrogen, which can provide industrial chemicals or be used as a source of additional energy.

The sustainability issue also deals with the management of nuclear waste. Generation IV reactors deal with this by a combination of reducing total radioactive waste by incorporating a higher degree of burn-up and by burning residual fission actinides to substantially reduce the lifetime of radioactive waste.

Economics: The cost of the construction of a nuclear power plant is a major factor in the successful implementation of Generation IV reactor technologies. The commercialization of new nuclear reactor technologies requires both funding for reactor development and testing as well as the construction of commercial generating facilities. Investment in nuclear power requires that the financial risk should not exceed that for the development of other alternative energy technologies. While larger reactors are more economical in terms of the cost per unit capacity, smaller reactors are less expensive overall and have applications where larger reactors may not be suitable. These applications include cases where power is required in remote locations, where cogeneration requires a nuclear facility in close proximity to an industrial facility or community, and for countries with developing economies where cost is an important consideration. Finally, the bottom line for the consumer will be the levelized cost of electricity (see chapter 1) and nuclear energy must be competitive with other sources of low-carbon energy.

Safety and reliability: An important factor in the development of Generation IV nuclear reactors is their safety. Future reactors need to be safe, but they also need to be perceived as safe. Confidence in the safety of future nuclear power plants needs to outweigh the negative public perception of nuclear accidents, particularly Chernobyl and Fukushima. As stressed in the previous chapters, the inclusion of passive safety features and redundant safety systems are an essential aspect of Generation IV reactors. It is essential that reactors are not only passively safe but that they are resistant against operator error, such as that which played a major role in the Chernobyl accident. While peripheral systems are important from a safety standpoint, it is the core where accidents typically cause the greatest concern. Generation IV reactors are required to minimize both the risk for core damage and the severity of such damage should it occur. Passive decay heat removal systems are an important approach to minimizing the risk of core damage and core catchers (as are employed in many Generation IV designs) are a last defense for preventing the spread of radioactive material outside of the reactor vessel. It is also expected that Generation IV reactor accidents can all be dealt with internally without the need for any off-site assistance.

While reliability and safety are linked, it is not always true that safe reactors are reliable, and vice versa. The overall capacity factor is a good measure of reliability. As indicated in table 1.1, current nuclear power stations (containing Generation II and III reactors) are excellent examples of reliability with typical capacity factors greater than 90%. While many factors, particularly for renewable energy sources, affect capacity factors but are not related to reliability, the high capacity factor for nuclear power clearly indicates that current reactor technology is reliable. A major goal of Generation IV reactors is to reach the level of reliability of Generation II and III reactors or exceed it. It is common for experimental and prototype nuclear reactors to experience reliability issues and this has been the case (for example) for Superphénix and Monju. However, other prototype or demonstrator reactors utilizing Generation IV concepts have performed reasonably well, such as Peach Bottom, EBR-II, and BN-800.

Proliferation resistance and physical protection: Many fission reactors constructed in the 1970s and 1980s were designed as dual-purpose reactors, i.e., for both the production of electricity for commercial use and for the production of weapons-grade plutonium for military use. The philosophy of Generation IV reactors is just the opposite and these reactors are designed so that weapons-grade material cannot be removed from the core by any reasonable means. This is seen in the design of many of the fast neutron reactors that have been reviewed in the previous chapters, which do not include a breeder blanket where fissile plutonium is accumulated but rather breed and utilize fissile material within the core itself in a breed and burn fuel cycle.

Generation IV reactor designs also provide increased physical protection of the reactor in the event of an act of terrorism. This physical protection includes a suitable containment structure that offers adequate resistance against foreseeable methods of attack.

The following six sections review the progress that has been made in the development of each of the six Generation IV reactor technologies. The final section provides an overall picture of how these technologies can contribute to our future energy needs.

13.2 Very high-temperature reactors (VHTRs)

The development of high-temperature reactors dates back to the 1960s when the United Kingdom, the United States, and Germany all developed experimental reactors that used TRISO-type fuel consisting of uranium-based or uranium/thorium-based fuel coated in a high-temperature ceramic. These fuel particles were encapsulated in graphite, which acted as a moderator and reactor cores were either of a prismatic design or a pebble bed design. The identification of

Table 13.1. Typical properties of Generation IV VHTRs. Data adapted from Generation IV International Forum (2002, 2014), and Pioro and Rodriguez (2023). LEU = low enriched uranium.

Property	Value
Abbreviation	VHTR
Coolant	Helium
Moderator	Graphite
Neutron spectrum	Thermal
Fuel form	TRISO (LEU UO_2)
Fuel cycle	Open
Outlet temperature (°C)	900–1000
Typical electrical capacity (MW_e)	250–300
Typical efficiency (%)	>55
Gen IV reactors[a]	HTR-PM (China)[b] HTR-PM600 (China) GTHTR300C (Japan) Xe-100 (USA)

[a] Examples of Generation IV VHTRs under development or under construction.
[b] In commercial operation as of 6 December 2023.

suitable materials to withstand the high temperatures produced by these reactors has been a major challenge for their successful development and operation.

A summary of the typical properties of Generation IV VHTRs is given in table 13.1. These are currently under development as small modular reactors in several countries, as indicated in the table, and this technology is generally considered to be the most advanced of the six Generation IV reactors. In fact, China put two HTR-PM VHTRs at the Shidao Bay Nuclear Power Plant in Shandong Province into commercial service on 6 December 2023 and these are the first Generation IV reactors to be fully operational commercially (China Nuclear News Corporation 2023, World Nuclear News 2023).

The VHTR is the only Generation IV design where there is no active development of reactors with closed fuel cycles. However, depending on the experienced gained with once-through fuel cycles for these reactors, future development may consider other fuel cycles (e.g., those utilizing thorium) that can be closed. At present, VHTRs concentrate on the benefits of the temperature of the reactor core output. This high temperature is beneficial for two reasons: it allows for improved thermodynamic efficiency for electricity generation and it optimizes the potential for cogeneration. Both benefits result in a more efficient overall use of uranium fuel resources than in a traditional light water reactor. As noted in chapter 7, cogeneration can involve the use of surplus heat for industrial processes, in addition to the generation of electricity. The most likely applications for excess heat from a VHTR are desalination or hydrogen production, and the small modular design of the VHTR is compatible with locating the reactor near where there is a need for the heat resources. The high temperature of the steam available from VHTRs provides the opportunity for various approaches to hydrogen production. Hydrogen produced using excess heat from a nuclear power plant can be used for various industrial process, such as production of ammonia for fertilizer manufacturing. It can also be used as a direct source of energy by means of a fuel cell or it can be used for the production of carbon-free synthetic fuels, such as methane or methanol (see e.g., Dunlap 2023). In all cases, the use of excess nuclear reactor heat can reduce the energy requirements for industrial processes from other energy sources.

13.3 Molten salt reactors (MSRs)

Molten salt-cooled reactors have been of interest since the 1950s, but the only previous experimental reactors of this type were constructed in the 1960s at Oak Ridge National Laboratory (see section 8.5). Generation IV MSRs include a wide variety of significantly different designs. All utilize molten salt as a coolant, but these reactors may be either thermal neutron reactors or fast neutron reactors and may utilize a uranium or a thorium fuel cycle. Many designs utilize fuel that is dissolved in the molten salt coolant, although some use fuel dissolved in salt which does not act as the primary coolant but transfers heat to an additional molten salt cooling loop. Still other designs use a graphite-moderated TRISO fuel that is cooled by non-radioactive molten salt. A summary of the typical properties of Generation IV MSRs is given in table 13.2. As discussed in chapter 8, the designs that are currently

Table 13.2. Typical properties of Generation IV MSRs. Data adapted from Generation IV International Forum (2002, 2014), and Pioro and Rodriguez (2023).

Property	Value	
Abbreviation	MSR	
Coolant	Fluoride or Chloride Salt (in some designs with dissolved uranium)	
Moderator	Graphite	None
Neutron spectrum	Thermal	Fast
Fuel form	UF_4, PuF_4, ThF_4, UCl_3, actinide chloride, TRISO	
Fuel cycle	Open	Closed
Outlet temperature (°C)	700–800	
Typical electrical capacity (MW_e)	1000	
Typical efficiency (%)	50	
Gen IV reactors[a]	IMSR (Canada)	
	SSR-W (Canada)	
	TMSR-LF (China)	
	TMSR-SF (China)	
	MOSART (Russia)	
	FLEX (UK)	
	MCFR (USA)	

[a] Examples of Generation IV MSRs under development or under construction.

under development worldwide cover a variety of different approaches to using molten salt as a reactor coolant. These include circulating fuel dissolved in molten salt, static molten salt fuel with separate non-radioactive molten salt coolant, and molten salt-cooled pebble bed reactors.

MSRs can have a traditional closed fuel cycle by using a breeder blanket and fuel reprocessing. They may also have a once-through fuel cycle that breeds and burns fissile fuel in the core. A major benefit of MSRs is their ability to burn actinides. This reduces the quantity and longevity of the radioactive waste that is produced. It also allows MSRs to utilize spent fuel from traditional light water reactors. Reactor designs that utilize fuel that is dissolved in molten salt have the benefit that there is no requirement for fuel fabrication. In fact, reactors do not need to be shut down for refueling. Rather additional fissile material can be added to the molten fuel and unwanted fission products can be removed while the reactor is in operation.

Although outlet temperatures are not as high as for VHTRs, the use of molten salt as a coolant allows for the transfer of heat to a molten salt storage tank. This stored thermal energy can be utilized for further electricity generation or industrial processes as needed.

In keeping with Generation IV guidelines, MSRs have inherent safety features. They are the only fast spectrum reactor that has a negative void coefficient. In fact, this characteristic can be utilized for reactor control by introducing helium bubbles into the fuel to lower the reactivity. In addition, reactors that utilize circulating fuel dissolved in molten salt have the option to divert the fuel to a dump tank in the event of a core operational anomaly.

Perhaps the major drawback concerning MSRs is the corrosivity of the salt. This presents materials issues in the reactor design and considerable development work is still required to ensure the reliability of all components. A number of MSR development projects are underway worldwide. The TMSR-SF that is being developed in China is a graphite-moderated pebble bed reactor that uses molten salt as a coolant and benefits from the thermal properties of this material. Reactors that utilize fuel dissolved in molten salt are still in the relatively early stages of development and much work is still needed for the implementation of effective online salt processing systems.

13.4 Supercritical water-cooled reactors (SCWRs)

SCWRs are either pressure vessel or pressure tube design, and some of their important characteristics are summarized in table 13.3. These two designs follow closely along the lines of conventional pressurized light water reactors and heavy water moderated CANDU reactors, respectively. The design of these reactors relies substantially on knowledge concerning the nuclear reactor core design that has been gained from pressurized light water reactors and CANDU reactors, and knowledge concerning the supercritical water cooling system that has been gained from supercritical water-cooled coal fired generating stations.

Advantages of SCWRs include good thermal efficiency compared to normal pressurized water reactors because of the higher steam output temperature. The high

Table 13.3. Typical properties of Generation IV SCWRs. Data adapted from Generation IV International Forum (2002, 2014), and Pioro and Rodriguez (2023).

Property	Value
Abbreviation	SCWR
Coolant	Supercritical water
Moderator	Water or heavy water
Neutron spectrum	Thermal or fast
Fuel form	UO_2, MOX, $(Pu, Th)O_2$
Fuel cycle	Open or closed
Outlet temperature (°C)	510–625
Typical electrical capacity (MW_e)	300–1500
Typical efficiency (%)	45–50
Gen IV reactors[a]	Canadian SCWR (Canada)
	CSR1000 (China)
	HPLWR (European Union)
	Super FR (Japan)

[a]Examples of Generation IV SCWRs under development or under construction.

temperature also increases the possibility of cogeneration applications. However, the major advantage of SCWRs is the simplicity of their design, which decreases the capital cost of construction. These simplifications result from the use of supercritical steam because the greater enthalpy gained by the coolant in the core means that a smaller turbine can produce the same output as a larger turbine for a traditional pressurized water reactor and the primary cooling steam can be used to drive the turbine directly. In addition, the lack of a phase transition in the steam means that typical pressurized water reactor components, such as a steam separator and dryer, are not necessary. Estimates of capital cost place the SCWR about 20% to 30% lower than a comparably sized pressurized water reactor.

The similarity of particular components to those of a pressurized water reactor and a supercritical steam coal plant means that significant technology from those systems can be incorporated into the design of a SCWR. Thus, the design is more evolutionary than revolutionary and can be done incrementally. However, there are still some challenges to the implementation of a viable design. These include materials issues, particularly those related to fuel cladding, and the proper evaluation of concerns related to corrosion at elevated temperatures. This latter issue also involves the more detailed study of water chemistry at anticipated operating temperatures. While other Generation IV reactor designs incorporate significant passive safety features, these still need to be refined for the SCWR. The basic light water moderated version of the SCWR utilizes a standard once-through uranium fuel cycle. Although the reactor benefits from a higher efficiency than Generation III and earlier reactors, as well as the possibility of cogeneration, improvements in sustainability should be considered for future designs. This is the case for the Canadian heavy water moderated SCWR, where an advanced Pu/Th fuel cycle is being implemented.

13.5 Gas-cooled fast reactors (GFRs)

The GFR is the least developed of all of the Generation IV designs. No operational GFR has been constructed. Therefore, this technology does not have knowledge based on past operational experience to provide input into new Generation IV designs. A summary of the typical properties of GFRs is given in table 13.4.

Several possible coolants have been considered for these reactors, including helium, supercritical carbon dioxide, steam, and N_2O_4. N_2O_4 undergoes a dissociation/recombination transition in the cooling loop. This has the advantage that additional heat is transported out of the core by the coolant. However, N_2O_4 is very corrosive, and this must be considered in any possible design. Steam has the disadvantage that H_2O moderates the neutron spectrum because of neutron scattering from the hydrogen nuclei. In the event of a partial core meltdown, the volume of steam coolant could increase, leading to additional neutron moderation and an increase in reactivity. Thus, there is no negative feedback mechanism in place in the event of a core accident. Carbon dioxide has the advantage of a large negative thermal coefficient of reactivity and is a possible coolant for these reactors. However, at present, virtually all designs for a Generation IV GFR utilize helium as a coolant.

The GFR has some clear advantages. A positive void coefficient, as must be considered in the design of (for example) the SFR, is not a concern because the coolant is already in the gaseous state and cannot form voids by boiling. Since the design of the GFR using helium as a coolant is very similar to the VHTR described above and in chapter 7, except without the incorporation of a graphite moderator, much of the research that has been conducted on this other type of Generation IV reactor is also applicable to the GFR. This includes (for example) the development of materials for fuel cladding and reactor vessel construction that are tolerant of the temperature and neutron flux in the reactor core. The fuel cycle is similar to that of

Table 13.4. Typical properties of Generation IV GFRs. Data adapted from Generation IV International Forum (2002, 2014), and Pioro and Rodriguez (2023).

Property	Value
Abbreviation	GFR
Coolant	Helium
Moderator	None
Neutron spectrum	Fast
Fuel form	MOX, (U, Pu)C
Fuel cycle	Closed
Outlet temperature (°C)	850
Typical electrical capacity (MW_e)	1200
Typical efficiency (%)	>50
Gen IV reactors[a]	ALLEGRO (European Union) EM^2 (USA)

[a] Examples of Generation IV GFRs under development or under construction.

other fast reactors, such as the SFR and LFR, and the development of these reactors will also benefit fuel design for the GFR. The high output temperature provides opportunities for cogeneration and the closed fuel cycle provides good fuel burn-up with minimal radioactive waste, which benefits overall sustainability.

The main disadvantage of the GFR concept is the difficulty in designing an adequate passive decay heat removal system. This difficulty results from the low heat capacity of helium as a primary coolant. Heat transfer from the core is ensured under normal operating conditions by a large flow rate of the helium gas. In typical Generation IV GFR designs, there are several redundant active and passive decay heat removal loops to cool the reactor core in the event of a loss of coolant pressure event.

13.6 Sodium-cooled fast reactors (SFRs)

SFRs are one of the most mature Generation IV technologies. Since the late-1950s, at least 17 operational sodium-cooled reactors have been constructed and operated in the United States, Europe, Russia, and Asia. These reactors, some of which are still operational, have provided a wealth of knowledge about their functioning and also about their potential problems. Because of this considerable experience, SFRs are considered to be one of the Generation IV designs that can be deployed in the shortest time (Locatelli *et al* 2013). A summary of the typical properties of Generation IV SFRs is given in table 13.5.

SFRs offer a number of characteristics that provide motivation for their further development. These reactors can be implemented on a variety of different sizes, including small modular reactors with a capacity of 50 MW_e for remote site applications and large power plant reactors with capacities up to 1500 MW_e. This

Table 13.5. Typical properties of Generation IV SFRs. Data adapted from Generation IV International Forum (2002, 2014), and Pioro and Rodriguez (2023).

Property	Value
Abbreviation	SFR
Coolant	Sodium
Moderator	None
Neutron spectrum	Fast
Fuel form	MOX, UN-PuN, UC-PuC, U-Zr, U-TRU-Zr, UC-ThC
Fuel cycle	Closed
Outlet temperature (°C)	500–550
Typical electrical capacity (MW_e)	50–1500
Typical efficiency (%)	40
Gen IV reactors[a]	CFR-600 (China)
	PFBR (India)
	BN-1200 (Russia)
	PGSFR (South Korea)

[a] Examples of Generation IV SFRs under development or under construction. TRU = transuranium.

allows for the design of reactor facilities that match particular applications. The use of a closed fuel cycle is advantageous from a resource utilization standpoint. The SFR design is, perhaps, most suitable of the Generation IV technologies to implement as an integral fast reactor. In this design, fuel from a fast reactor with a breeding ratio of greater than unity is reprocessed on-site to produce new fissile reactor fuel. Typically, electrorefining is the most appropriate on-site fuel reprocessing technique and separates different elements by means of large-scale electroplating techniques. Spent fuel can be reprocessed multiple times and this provides for efficient resource utilization, as well as minimizing the quantity and longevity of radioactive waste. However, the use of on-site reprocessing facilities causes concerns related to nuclear proliferation. It may be necessary to compromise fuel efficiency to some extent by minimizing plutonium inventory in order to improve proliferation resistance.

The outlet temperature of 500 °C–550 °C is lower than that of several of the other Generation IV reactors. While this limits the application of excess heat for cogeneration purposes, it places less strict requirements on the properties of materials for reactor components and allows for the utilization of materials that have been previously developed for sodium-cooled reactor applications.

The properties of liquid sodium as a coolant have both benefits and disadvantages. The thermal properties of liquid sodium allow for the design of effective passive decay heat removal systems to improve the safety characteristics of the reactor in the event of cooling system issues. However, the reactive nature of sodium and the presence of a positive void coefficient must be carefully considered in the reactor design and operation.

Much work has been done in recent years on the development of a Generation IV SFRs. The experimental FBTR SFR has been operational in India since 1985. A larger version, the prototype PFBR reactor is presently under construction. These reactors form an integral component of India's three-stage nuclear energy plan, as described in chapter 11. The Indian reactor design uses a loop type geometry, compared to a pool-type geometry that is more commonly in use for other Generation IV SFR projects. A demonstrator reactor, CFR-600, has been under construction in China since 2017. The BN-1200 reactor is planned for construction in Russia but has been put on hold. Several other Generation IV SFR development projects are in progress, e.g., in Canada, South Korea, and the United States.

13.7 Lead-cooled fast reactors (LFRs)

The LFR is an alternative to the more mature SFR that eliminates some potential difficulties with this older technology, such as the reactivity of sodium in the event of a leak and the significance of the positive void coefficient. A summary of the typical properties of Generation IV LFRs is given in table 13.6.

Early reactors of this type used lead-bismuth eutectic (LBE) as a coolant and were designed for use in Soviet submarines, as described in detail in section 12.4. Most Generation IV reactors currently under development use pure lead as a coolant in order to avoid having to deal with the potential hazards of ^{210}Po.

Table 13.6. Typical properties of Generation IV LFRs. Data adapted from Generation IV International Forum (2002, 2014), Pioro and Rodriguez (2023).

Property	Value
Abbreviation	LFR
Coolant	Lead, Lead-Bismuth Eutectic (LBE)
Moderator	None
Neutron spectrum	Fast
Fuel form	UN-PuN, MOX
Fuel cycle	Closed
Outlet temperature (°C)	480–570
Typical electrical capacity (MW_e)	20–1000
Typical efficiency (%)	41–43
Gen IV reactors[a]	SNCLFR-100 (China)
	ALFRED (European Union)
	ELFR (European Union)
	BREST-1200 (Russia)

[a] Examples of Generation IV LFRs under development or under construction.

The LFR serves much the same purpose as gas-cooled and SFRs and is designed for efficient fuel utilization and actinide burn-up. Overall, lead-cooled reactors have several advantages (compared to GFRs or SFRs), as follows:

- The high boiling temperature eliminates concerns about the positive void coefficient.
- The large nuclear mass minimizes neutron moderation.
- Lead acts as a neutron reflector and helps to maintain the neutron density in the core.
- (n, γ) reactions are relatively unimportant.
- Lead acts as an effective γ-ray shield.
- Molten lead does not significantly react with air or water.
- High coolant temperatures provide good thermodynamic efficiency.
- Passive decay heat removal systems are effective.
- Coolant circulates at near atmospheric pressure.

However, LFRs have a number of potential drawbacks, as follows:

- Potential coolant clogging from oxygen contamination.
- High density of the coolant leads to excessive reactor core weight.
- Coolant blockage due to freezing.
- Corrosion or erosion of some metals at high temperature.
- Optical opacity of molten lead.
- Buoyancy of reactor core components in high density molten lead.

While LBE-cooled fast reactors share many of the same benefits and drawbacks as LFRs, there are a few differences. The lower melting temperature of LBE means that there is less likelihood of the coolant freezing. However, the production of ^{210}Po

as a by-product of neutron absorption by ^{209}Bi poses a significant safety hazard in the event of a coolant leak and can complicate the approach to waste fuel management. Finally, the cost of bismuth may be prohibitive from an economic standpoint, particularly for larger reactors.

An attractive feature of LFR technology is the ability to construct reactors over a wide range of sizes. There is particular interest in small modular LFRs. These follow from the Soviet era LBE-cooled marine reactors. Small modular reactors may be suitable for remote locations or for cogeneration applications of hydrogen production or desalination. These small LFRs are also suitable for sale to countries that do not have an active nuclear industry. Reactors can have a sealed core with a closed fuel cycle of up to 20 years or more. There is no need for refueling, rather the core module can be changed and the spent module can be reprocessed by the supplier.

13.8 The future of nuclear energy

The six Generation IV reactor technologies that have been discussed in this book illustrate approaches that can resolve concerns related to present reactor designs. All of the six Generation IV technologies deal with the four criteria specified by the Generation IV Forum, i.e., sustainability, safety and reliability, economy, and proliferation resistance. Some designs excel in one area while others excel in different areas. Approaches to satisfying these design criteria also differ. The ways in which Generation IV reactor designs can satisfy the different criteria are summarized below, along with typical examples of reactor technologies that excel in each of these areas.

Sustainability

- Increased thermodynamic efficiency from high output temperature (e.g., VHTR)
- Closed fuel cycles (e.g., LFR)
- Waste burning (e.g., MSR)
- Use of thorium resources (e.g., MSR)
- Thermal storage (e.g., MSR)

Safety and reliability

- Core dump (e.g., MSR)
- Passive decay heat removal (e.g., SFR)
- Modular design (e.g., LFR)

Economics

- Direct cycle (e.g., SCWR)
- Simplified cooling loops (e.g., SCWR)
- Minimal operator intervention needed (e.g., LFR)
- Options for cogeneration (e.g., VHTR)

Proliferation resistance

- Minimal fissile material in core (e.g., SCWR)
- Difficult to separate fissile material (e.g., MSR)

In order to optimize benefits from Generation IV reactors, the six specified designs can complement one another. For example, VHTRs provide good efficiency and cogeneration opportunities. Spent fuel from VHTRs can be further utilized in MSRs in order to extract additional energy and burn-up waste actinides. ^{233}U produced from ^{232}Th in fast breeder reactors can further be used as fuel in thermal neutron reactors. At present, the operating commercial nuclear reactors are virtually all Generation II or Generation III, and these will continue to represent the major component of nuclear power for the next few decades. The future implementation of Generation IV reactors with waste burning capabilities can provide a means of dealing with spent fuel from these current operational reactors.

We have also seen that the development of the different Generation IV reactor technologies is at different stages. There are no full-scale commercial Generation IV reactors of any design that are currently operational. However, as noted above and in previous chapters, several experimental, prototype, or demonstrator Generation IV reactors have been constructed. The timelines for operational demonstrator reactors as predicted by the Generation IV Forum Roadmap in 2002 and by the revised roadmap (2014) are given in table 13.7. This table also summarizes the progress in each of these technologies as of 2024.

The first Generation IV demonstrator VHTR (HTR-PM) became operational in 2021 (World Nuclear News 2021) and has been providing electricity to the grid as of December 2023. A demonstrator LFR (BREST-300) has been under construction since 2020. Although somewhat slower in development, there is a prototype Generation IV SFR (CFR-600) under construction and an experimental Generation IV MSR (TMSF-LF1) that has recently become operational.

Full-scale commercial Generation IV reactors are still many years or, in most cases, several decades down the road. Their successful implementation depends on continued research and the operation of demonstrator reactors that provide evidence of their safety and reliability. In addition, demonstrator reactors must prove the economic viability of the Generation IV technology. In the short term at least, Generation IV technologies must compete against proven reliable and economical (if not entirely sustainable) Generation III reactor technologies, as well as other low-

Table 13.7. Predicted timeline for operational demonstrator Generation IV reactors from 2002 (Generation IV International Forum 2002) and 2014 (Generation IV International Forum 2014), along with actual progress as of 2024.

Reactor	2002 prediction	2014 prediction	2024 progress
VHTR	2015	2022	Demonstrator operational 2021
MSR	2020	>2030	Experimental reactor operational 2023
SCWR	2020	2025	Under development
GFR	2020	>2030	Under development
SFR	2015	2022	Prototype under construction 2017
LFR	2020	2021	Demonstrator under construction 2020

carbon sources such as wind and solar, although it is important for overall future energy infrastructure that these technologies can be used to complement each other.

The implementation of commercial Generation IV reactors depends not only on the development of suitable technology that satisfies the Generation IV objectives but also on the desire for governments to establish policies that facilitate the use of nuclear energy. The establishment of suitable government initiatives to promote nuclear power depends to a great extent on the public perception of nuclear energy. Here, we look briefly at the changing opinions of nuclear energy and the ways in which it has influenced the development of nuclear power.

The growth of the nuclear power industry has been discussed in section 5.9. The major nuclear accidents—Three Mile Island in 1979, Chernobyl in 1986, and Fukushima in 2011, as discussed in detail in chapter 6—have certainly had an impact on this growth. In fact, a close examination of figure 5.26, which shows the annual newly installed capacity, illustrates these effects. There is a brief but clear decrease in installed nuclear capacity following the Three Mile Island accident corresponding to shutdowns and a reassessment of safety regulations. There is a significant decrease in newly installed capacity following the Chernobyl accident. This reflects both changes in the nuclear power industry that resulted from the accident, as well as environmental concerns that had been growing prior to 1986 (Dunlap 2021). The effects of the Fukushima accident are also quite obvious. The shutdown of all nuclear reactors in Japan following the accident (see section 6.6 and in particular figure 6.16) is responsible for the large negative contribution to newly installed capacity for 2011.

The trends in the total number of reactors, as well as the total generating capacity, also provides some insight into the use of nuclear power over the years. Figure 13.1 shows these trends since the beginning of the nuclear power industry. In the bottom portion of the figure, the effects of Chernobyl and Fukushima can be seen. It is also clear that the total number of operational reactors has not changed significantly since around the mid-1990s. This is because newly operational reactors approximately compensate for those that were constructed (mostly) in the 1960s that have been decommissioned at the end of their lives. It is interesting to note a slight increase in capacity shown in the top portion of the figure over the same period of time. This is due to smaller reactors being shut down and larger reactors coming online.

Despite the concerns over the Chernobyl and Fukushima accidents, there has been renewed interest in nuclear energy (Nuttall 2022) since the low point in the early-2000s (see figure 5.26). This also seen in figure 13.2, which shows a slightly increasing overall trend in the number of new reactor starts worldwide since 2000. At present there are 62 nuclear reactors that are under construction worldwide and 118 that are in the planning stages (World Nuclear Association 2024). Of the reactors that are under construction, all are Generation II reactors except for three or four prototype or demonstrator Generation IV reactors, as have been previously discussed in the text.

It is interesting to note the different approaches that are taken to nuclear energy development in different countries. Twenty six of the 42 reactors that are under

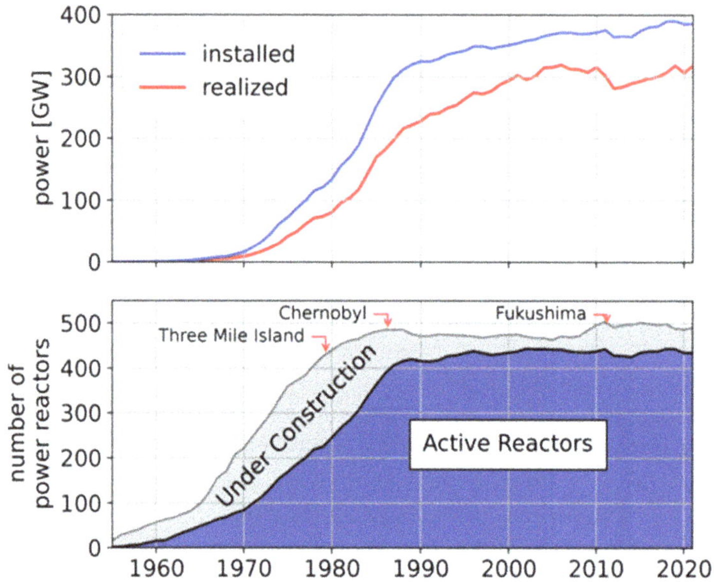

Figure 13.1. Timeline of nuclear power in the world. Top panel: installed capacity and energy generated (the ratio gives the capacity factor). Bottom panel: number of operational reactors and reactors under construction. This worldwide nuclear power timeline graph image has been obtained by the author from the Wikimedia website where it was made available by Geek3 (2022) under a CC BY-SA 4.0 licence. It is included within this chapter on that basis. It is attributed to Geek3.

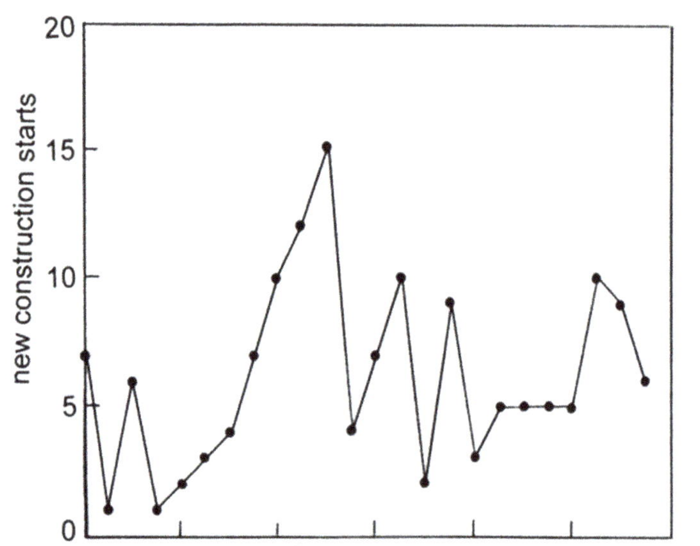

Figure 13.2. Number of new reactor construction starts per year since 2000.

construction and 42 of the 118 that are planned are in China. India, although they are not part of the Generation IV International Forum, are also very active in the development of nuclear energy, including reactors with Generation IV designs (see section 11.4). Some European countries have taken the opposite approach. Three countries, in particular, that previously had nuclear reactor programs have phased out nuclear power completely, although as members of the European Union they are participants (through Euratom) in the Generation IV Forum. Italy shut down its last reactor in 1990. However, two Italian organizations, National Agency for New Technologies, Energy and the Environment (ENEA) and Ansaldo Nucleare, have taken an active role in the development of the ALFRED LFR (see section 12.6). Lithuania shut down its last nuclear reactor in 2009. Finally, Germany shut down its last nuclear reactor in 2023.

In the United States, the situation falls between these two extremes. Although the United States (particularly the Department of Energy) has been a major contributor to the establishment of the Generation IV International Forum and the development of advanced reactors, the revival of the nuclear power industry has been slow. In the United States, no new nuclear reactors were commissioned between 1997 and 2015, and since then only two new reactors have started operation, i.e., Watts Bar Unit 2 (Spring City, Tennessee) in 2016 and Vogtle Unit three (Waynesboro, Georgia) in 2023. At present, only one reactor (Vogtle Unit four) is under construction and scheduled to start operation in 2024. It is interesting to consider the changes in the public perception of nuclear power in the United States in recent years. Table 13.8 shows the results of recent polls in the United States concerning nuclear power. In all cases, there is a clear indication that over the past eight years there has been a significant increase in the support for the development of nuclear power. All polls show that at present the majority of adults in the United States favor such development.

The present book has emphasized the need for a low-carbon source of electricity that can take the place of fossil fuel burning power plants. Generation IV nuclear power plants can provide a reliable and long-term solution by building on the success of previous reactors and learning from some unfortunate accidents. As table 13.7 indicates, the early expectations (from 2002) of Generation IV reactor

Table 13.8. Comparison of results from 2016 and 2023 for several polls concerning the percentage of adults in the United States who favor the further development of nuclear energy. Data adapted from Bisconti (2023), Brenan (2023), and Leppert and Kennedy (2023).

		Favor development	
Poll	Reference	2016	2023
Bisconti	Bisconti (2023)	66%	76%
Gallup	Brenan (2023)	44%	55%
Pew	Leppert and Kennedy (2023)	43%	57%

development were overly optimistic. However, predictions from the 2014 Generation IV Roadmap (Generation IV International Forum 2014) have been somewhat more accurate. The recent commissioning of the two 250 MW_{th} HTR-PM VHTRs is consistent with Generation IV Forum predictions and shows the success of the high-temperature pebble bed design, as well as the functionality of small modular reactors, in general. Progress has also been made, although somewhat more slowly, on the implementation of MSRs, SFRs, and LFRs. The development of SCWRs and GFRs is behind that of other Generation IV reactors and in the case of SCWRs will likely require much longer than anticipated in 2014. Major issues to be resolved in these two cases are the development of high-temperature corrosion resistant materials and the development of effective passive decay heat removal systems, respectively.

Overall, substantial progress in the development of advanced nuclear reactors has been made since the organization of the Generation IV Forum in the early-2000s, both by members of the Forum as well as other countries (e.g., India). Although the complete implementation of Generation IV reactor concepts will, no doubt, take several decades, the recent success of the first operational commercial Generation IV reactor (HTR-PM) marks an important milestone in the development of advanced nuclear reactors and their integration into our existing and developing energy systems, particularly previous generation nuclear reactors and renewable technologies. This achievement, along with the various experimental and prototype designs that have been constructed, has provided clear evidence for the safety and reliability of the Generation IV concepts. The design of advanced fuel cycles will ensure sustainable operation and proliferation resistance. The near-term deployment of demonstrator reactors should provide the economic incentive for the construction of full-scale commercial reactors.

References

Bisconti A S 2023 National nuclear energy public opinion survey: public support for nuclear energy stays at record level for third year in a row www.bisconti.com/blog/public-opinion-2023

Brenan M 2023 Americans' support for nuclear energy highest in a decade https://news.gallup.com/poll/474650/americans-support-nuclear-energy-highest-decade.aspx

China Nuclear News Corporation 2023 The world's first HTR-PM starts commercial operation (07 December 2023) https://en.cnnc.com.cn/2023-12/07/c_945661.htm

Dunlap R A 2021 *Energy from Nuclear Fusion* (Bristol: IOP Publishing)

Dunlap R A 2023 *Transportation Technologies for a Sustainable Future—Renewable Energy Options for Road, Rail, Marine and Air Transportation* (Bristol: IOP Publishing)

Geek3 2022 Timeline of nuclear power in the world https://commons.wikimedia.org/wiki/File:Nuclear-energy-timeline.svg

Generation IV International Forum 2002 *A Technology Roadmap for Generation IV Nuclear Energy Systems* www.gen-4.org/gif/jcms/c_40481/technology-roadmap

Generation IV International Forum 2014 *Technology Roadmap Update for Generation IV Nuclear Energy Systems* www.gen-4.org/gif/jcms/c_60729/technology-roadmap-update-2013

Leppert R and Kennedy B 2023 Growing share of Americans favor more nuclear power (Pew Research Center 18 August 2023) www.pewresearch.org/short-reads/2023/08/18/growing-share-of-americans-favor-more-nuclear-power/

Locatelli G, Mancini M and Todeschini N 2013 Generation IV nuclear reactors: current status and future prospects *Energy Policy* **61** 1503–20

Nuttall W J 2022 *Nuclear Renaissance—Technologies and Policies for the Future of Nuclear Power* 2nd edn (Boca Raton, FL: CRC Press)

Pioro I L and Rodriguez G H 2023 Generation IV international forum *Handbook of Generation IV Nuclear Reactors* 2nd edn ed I L Pioro (Cambridge, MA: Woodhead Publishing) ch 2 pp 111–32

World Nuclear Association 2024 Plans for New Reactors Worldwide https://world-nuclear.org/information-library/current-and-future-generation/plans-for-new-reactors-worldwide.aspx

World Nuclear News 2021 China's HTR-PM reactor achieves first criticality (13 September 2021) www.world-nuclear-news.org/Articles/Chinas-HTR-PM-reactor-achieves-first-criticality

World Nuclear News 2023 China's demonstration HTR-PM enters commercial operation (6 December 2023) www.world-nuclear-news.org/Articles/Chinese-HTR-PM-Demo-begins-commercial-operation

www.ingramcontent.com/pod-product-compliance
Ingram Content Group UK Ltd.
Pitfield, Milton Keynes, MK11 3LW, UK
UKHW051903100926
8556IPUK00005B/742

9 780750 360678